T0237787

Lecture Notes in Mathematics

Edited by A. Dold, F. Takens and B. Teissier

Editorial Policy
for the publication of monographs

1. Lecture Notes aim to report new developments in all areas of mathematics – quickly, informally and at a high level. Monograph manuscripts should be reasonably self-contained and rounded off. Thus they may, and often will, present not only results of the author but also related work by other people. They may be based on specialized lecture courses. Furthermore, the manuscripts should provide sufficient motivation, examples and applications. This clearly distinguishes Lecture Notes from journal articles or technical reports which normally are very concise. Articles intended for a journal but too long to be accepted by most journals, usually do not have this "lecture notes" character. For similar reasons it is unusual for doctoral theses to be accepted for the Lecture Notes series.

2. Manuscripts should be submitted (preferably in duplicate) either to one of the series editors or to Springer-Verlag, Heidelberg. In general, manuscripts will be sent out to 2 external referees for evaluation. If a decision cannot yet be reached on the basis of the first 2 reports, further referees may be contacted: the author will be informed of this. A final decision to publish can be made only on the basis of the complete manuscript, however a refereeing process leading to a preliminary decision can be based on a pre-final or incomplete manuscript. The strict minimum amount of material that will be considered should include a detailed outline describing the planned contents of each chapter, a bibliography and several sample chapters.

Authors should be aware that incomplete or insufficiently close to final manuscripts almost always result in longer refereeing times and nevertheless unclear referees' recommendations, making further refereeing of a final draft necessary.

Authors should also be aware that parallel submission of their manuscript to another publisher while under consideration for LNM will in general lead to immediate rejection.

3. Manuscripts should in general be submitted in English.
Final manuscripts should contain at least 100 pages of mathematical text and should include
- a table of contents;
- an informative introduction, with adequate motivation and perhaps some historical remarks: it should be accessible to a reader not intimately familiar with the topic treated;
- a subject index: as a rule this is genuinely helpful for the reader.

Continued on back inside cover

Lecture Notes in Mathematics 1739

Editors:
A. Dold, Heidelberg
F. Takens, Groningen
B. Teissier, Paris

Subseries: Fondazione C. I. M. E., Firenze
Adviser: Arrigo Cellina

Springer
Berlin
Heidelberg
New York
Barcelona
Hong Kong
London
Milan
Paris
Singapore
Tokyo

R. Burkard   P. Deuflhard   A. Jameson
J.-L. Lions   G. Strang

Computational Mathematics Driven by Industrial Problems

Lectures given at the 1st Session of the Centro Internazionale Matematico Estivo (C.I.M.E.) held in Martina Franca, Italy, June 21–27, 1999

Editor: V. Capasso   H. Engl   J. Periaux

Fondazione
C.I.M.E.

Springer

Authors

Rainer E. Burkard
Technische Universität Graz
Institut für Mathematik
Steyrergasse 30
8010 Graz, Austria
E-mail: burkard@opt.math.tu-graz.ac.at

Antony Jameson
Dept. of Aeronautics and Astronautics
Stanford University
Durand 279
Stanford, CA, 94305-4035, USA
E-mail: jameson@baboon.stanford.edu

Gilbert Strang
Dept. of Mathematics
Massachusetts Institute of Technology
Room 2-240
77 Massachusetts Avenue
Cambridge, MA 02139-4307, USA
E-mail: gs@math.mit.edu

Peter Deuflhard
Konrad-Zuse-Zentrum
Takustrasse 7
14195 Berlin-Dahlem, Germany
E-mail: deuflhard@zib.de

Jacques-Louis Lions
Collège de France
3 rue d'Ulm
75231 Paris cedex 05, France

Editors

Vincenzo Capasso
MIRIAM - Milan Research Centre for
Industrial and Applied Mathematics
Department of Mathematics
University of Milan
Via C. Saldini, 50
20133 Milan, Italy
E-mail: vincenzo.capasso@mat.unimi.it

Jacques Periaux
Dassault Aviation
78 quai Marcel Dassault
92214 Saint Cloud, France
E-mail: periaux@rascasse.inria.fr

Heinz W. Engl
Industrial Mathematics Institute
Johannes Kepler University
Altenbergerstrasse 69
4040 Linz, Austria
E-mail: engl@indmath.uni-linz.ac.at

Library of Congress Cataloging-in-Publication Data

Computational mathematics driven by industrial problems : lectures given at the 1st
session of the Centro internazionale matematico estivo (C.I.M.E.) held in Martina
Franca, Italy, June 21-27, 1999 / R. Burkard ... [et al.] ; editors V. Capasso, H. Engel, J. Periaux.
    p. cm. -- (Lecture notes in mathematics, ISSN 0075-8434 ; 1739)
    Includes bibliographical references.
    ISBN 3540677828 (softcover : alk. paper)
    1. Mathematical models--Congresses. 2. Mathematics--Industrial
applications--Congresses. I. Burkard, Rainer E. II. Capasso, V. (Vincenzo), 1945- III.
Engl, Heinz W. IV. Periaux, Jacques. V. Lecture notes in mathematics (Springer-Verlag)

QA3 .L28 no. 1739
[QA401]
510 s--dc21
[511'.8]

                                                                00-063777

Mathematics Subject Classification (2000): 65-XX, 49-XX, 90CXX, 41-XX, 76-XX, 60D05, 60GXX, 62M30
ISSN 0075-8434
ISBN 3-540-67782-8 Springer-Verlag Berlin Heidelberg New York

Springer-Verlag Berlin Heidelberg New York
a member of BertelsmannSpringer Science+Business Media GmbH

© Springer-Verlag Berlin Heidelberg 2000
Printed in Germany

Typesetting: Camera-ready T_EX output by the authors
Printed on acid-free paper   SPIN: 10724313   41/3142-543210

Preface

The Centro Internazionale Matematico Estivo (CIME) organized a summer course on Computational Mathematics Driven by Industrial Problems from June 21-27, 1999 at the Ducal Palace in Martina Franca (a nice baroque village in Apulia, Italy).

The relevance of methods of advanced mathematics for innovative technology has become well-recognized ever since the 1984 publication of the "David Report" by an ad hoc committee chaired by the Vice-President of EXXON, Dr. Edward E. David, jr.

As a direct consequence of the "revolution" in information technologies, mathematics has become more and more visible. The truth is that mathematics is not just applied but rather continuously created to respond to the challenges of technological development and competitiveness.

Today traditional machines are no longer the core of industrial development; computer simulation of highly complex mathematical models substitute the traditional mechanical models of real systems. This poses challenging problems in the development of new mathematics and new computational methods.

This course was designed to offer young European scientists an opportunity of acquiring knowledge in selected areas of mathematics and advanced scientific computing which have benefited from the needs of competitive industry. The application of mathematics for the solution of industrial problems includes mathematical modelling of the real system; mathematical analysis of well posedness; computational methods; identification of models and of parameters (inverse problems); optimization and control. The extensive courses and seminars included all of these aspects. Furthermore, some case studies were presented in two-hour seminars on areas of industrial excellence in Europe, namely polymers and glass, for which there are two Special Interest Groups within the European Consortium for Mathematics in Industry (ECMI).

In this volume you will find the written account of all the contributions. We are grateful to all their authors. It may be of interest to the reader that the President (G.S) of SIAM (Society for Industrial and Applied Mathematics), three of the recent Presidents (V.C., H.E., and R.M.) of ECMI (European Consortium for Mathematics in Industry), and the former President (J.P.) of ECCOMAS (European Council on Computational Methods in Applied Sciences and Engineering) were among the lecturers in Martina Franca.

It is a pleasure to also thank Professor Jacques Louis Lions, who was impeded from participating at the last moment due to relevant commitments with the President of France. He kindly accepted our invitation to include the lecture notes he had prepared for the course in this monograph.

The Directors are most grateful to CIME for giving us the opportunity of bringing together excellent lecturers and a surprisingly large number of brilliant and enthusiastic mathematicians, physicists and engineers from 18 countries. It is a great pleasure to acknowledge the assistance of Dr. Sabrina Gaito and Dr. Alessandra Micheletti without whom the course could impossibily have run so smoothly. Dr. Daniela Morale's assistance in editing this volume is also kindly acknowledge. We extend our warmest thanks to them, to the lecturers and to all participants.

We gratefully acknowledge the financial support from CIME and the European Union. Finally, we would like to thank the Mayor, the General Secretariat and the complete staff of the Town Council of Martina Franca, which offered the XVII century Ducal Palace as site of the course, for the warm hospitality and continuous assistance.

Milano, Linz, Paris
March, 2000

Vincenzo Capasso
Heinz Engl
Jacques Periaux

CONTENTS

Trees and paths: graph optimisation problems with industrial applications *

Rainer E. Burkard

Technische Universität Graz, Institut für Mathematik, Steyrergasse 30, A-8010 Graz, Austria. Email: burkard@opt.math.tu-graz.ac.at

Contents

---

*This research has been supported by Spezialforschungsbereich F 003 "Optimierung und Kontrolle", Projektbereich Diskrete Optimierung.

1 Introduction

1.1 Graph optimization problems

Graphs offer a simple model for "connecting" objects. Thus they frequently occur in models from such different fields like telecommunication, traffic, location and relational analysis. And in many of these problems we want to connect objects in an "optimal" way. The underlying mathematical theory deals with optimization strategies for trees and paths in graphs. This will be the object of these lectures.

After introductory remarks on undirected and directed graphs we outline the basic notions of computational complexity. Complexity issues play a crucial role in the evaluation of the efficiency of algorithms. The second chapter is devoted to tree problems. We distinguish between spanning trees and Steiner trees which may use additional vertices, so-called Steiner points. We shall see that minimum spanning trees can be computed in polynomial time, whereas no polynomial time algorithm is known for the Steiner minimum tree problem. One of the basic algorithms for finding a minimum spanning tree in a connected graph will turn out as a basic tool in combinatorial optimization: Kruskal's algorithm is the prototype of a greedy algorithm.

In the third chapter we investigate different shortest path problems. We shall see how the positiveness of single links will influence the computational complexity of solution routines. And we shall see how an analysis of the algorithm leads to solution procedures for path problems stated in ordered algebraic structures like ordered semigroups. As an application of path problems we mention not only classical cases like shortest routes in traffic networks, but also the critical path problem for project networks and the 1-median problem in location theory.

Summarizing we see that even so simple structures like trees and paths lead to interesting mathematical questions in different fields like computational geometry, matroids, systolic arrays and computational complexity, just to mention a few ones.

1.2 Basic properties of graphs

1.2.1 Undirected graphs

An undirected graph $G = (V, E)$ is defined by its vertex set V and by its edge set E, where every edge $e \in E$ connects two vertices of set V. Throughout this lecture we shall always assume that

- the sets V and E are finite,

- there is no *loop*, i.e., every edge connects two different vertices u and v,

- there are no *parallel edges*, i.e., every edge is uniquely defined by the pair of vertices it connects.

A graph G fulfilling the above properties is called a *finite, simple graph.* If edge e connects the vertices u and v, we say, edge e is incident with u and v: $e = [u, v]$. The number of edges incident with a vertex v is called the *degree* $d(v)$ of v. The *complete graph* K_n on n vertices is a graph where all vertices are pairwisely connected. Fig. 1.1 depicts the complete graphs K_1 to K_5.

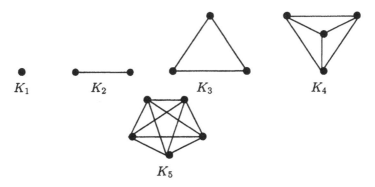

K_1 K_2 K_3 K_4

K_5

Figure 1.1: The complete graphs K_1–K_5

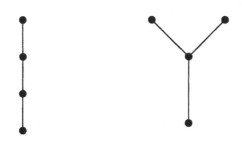

Figure 1.2: Two different trees on four vertices

A *path* from vertex u to vertex v can be described by a sequence of different vertices $u = v_0, v_1, ..., v_k = v$, where $[v_{i-1}, v_i]$ are edges for all $i = 1, 2, ..., k$. If we add the edge $[v_k, v_0]$ to this path, we get a *cycle*. Every cycle in G can be described in this way. A cycle is called *Hamiltonian cycle*, if it passes through all vertices of the graph just once. The question whether a given graph admits a Hamiltionian cycle or not, is one of the main problems in graph theory.

A graph is *connected*, if there exists a path between any two vertices of G. A graph $G = (V, E)$ is a *tree*, if G is connected and does not contain a cycle. Fig. 1.2 depicts two different trees on 4 vertices. Trees play a crucial role in graph optimisation. The following fundamental theorem exhibits several equivalent descriptions for trees.

Theorem 1.1 (*Characterization of trees)*
    Let $G = (V, E)$ be a finite graph. Then the following statements are
equivalent:

1. *G is a tree.*

2. *G is connected and* $|E| = |V| - 1$.

3. *G has no cycle and* $|E| = |V| - 1$.

4. *Between any two different vertices* $u, v \in V$ *there exists a unique path
   from u to v.*

5. *G is connected. If an edge is deleted, the graph is not any longer connected.*

6. *G has no cycle. If any edge is added, a cycle is created.*

This theorem can be proved by induction in a straightforward way. Note,
moreover, that $|E| = |V| - 1$ immediately implies that any (finite) tree has a
leaf, i.e., a vertex with degree 1.
    A graph without cycles which is not necessarily connected is called a
forest, i.e., a collection of trees.

1.2.2   Directed graphs and networks

A *directed graph* or *digraph* $G = (N, A)$ is defined by its set N of nodes and
by its arc set A. Every arc $e = (t, h) \in A$ can be represented as an ordered
pair (as an arrow in figures) with the *head node* $h \in N$ and the *tail node*
$t \in N$. As in the case of undirected graphs we shall assume in the following
that the sets N and A are finite, that there is no directed loop $e = (u, u)$
and that there are no parallel arcs. This means that every arc is uniquely
defined by its head and tail. The set of all nodes v for which there exists an
arc (u, v) is called the set $N(u)$ of *neighbors* of node u.
    A directed *path* in the digraph G is described by a sequence $(e_1, e_2, ..., e_k)$
of arcs with the property that the head of e_i equals the tail of e_{i+1} for all
$i = 1, 2, ..., k - 1$. A path is called *simple*, if all its nodes are different. If in
a simple path the head of e_k equals the tail of e_1, then the path is called a
(directed) *cycle*. A digraph is called *acyclic* if it does not contain a cycle.
    A *network* is a directed graph where two nodes are specified, the so-called
source s and the *sink t*. (In general, several sources and sinks can be given.
We restrict ourselves in this lecture to single-source single-sink problems).

1.3 Complexity

Numerical evidence has shown in the past that some optimization problems in graphs as finding a shortest path in a network are easy to solve whereas some other problems like the travelling salesman problem are rather difficult. This observation has been substantiated in the 70's with the development of the field of *complexity theory*, see e.g. the monograph of Garey and Johnson [11]. In this subsection we outline the main ideas of the complexity theory without becoming too technical.

In complexity theory we distinguish between problems and instances. A *problem* is a general task depending on formal parameters whose values are left unspecified. An *instance* of the problem is obtained by specifying values for all problem parameters. For example, finding a shortest path from some source s to some sink t in a directed network is a problem. If we specify the underlying network with all its nodes, arcs and arc lengths, we obtain an instance of this problem. The data of an instance constitute the input on a computer. When on a computer all data are binary encoded, the *length* of an instance is measured by the number of bits necessary to represent it, i.e., the length of an instance equals to the sum of the logarithms of all input data.

In complexity theory we distinguish between *decision* and *optimization* problems. A decision problem asks only for the answer "Yes" or "No", for example: does there exist a path from s to t in the given network? If the answer is "Yes", then there exists a feasible solution for the problem. If the answer is "No", the problem does not have a feasible solution. In an optimization problem we want to find the minimum value of an objective function over the set of feasible solutions, for example: find the shortest path from s to t.

An *algorithm* is a sequence of arithmetical and logical operations. We say, an algorithm solves a problem, if it gives for any problem instance the correct answer. An algorithm is called *polynomial*, if there exists a polynomial $p(x)$ such that for any instance of length L the number of operations used by the algorithm is bounded by $p(L)$. Usually an algorithm is called *exponential* if the number of operations of the algorithm is not bounded from above by a polynomial function in L.

Now we introduce the complexity classes $\mathcal{P}$ and $\mathcal{NP}$. A problem belongs to the complexity class $\mathcal{P}$, if there exists a polynomial algorithm for solving this problem. A decision problem is said to be in the class $\mathcal{NP}$, if given a "yes"-instance, the following two assertions hold:

1. There is a polynomially sized certificate for the problem. (A certificate is a binary encoding of the object that provides the answer.)

2. There is an algorithm which, using as input the instance of the problem and the certificate, can check in polynomial time whether or not the given "yes"-answer is indeed valid.

It is not difficult to see that $\mathcal{P} \subseteq \mathcal{NP}$. The question, whether $\mathcal{P} = \mathcal{NP}$, is one of the most important open problems in theoretical computer science. It is, however, generally conjectured that $\mathcal{P} \neq \mathcal{NP}$.

The problem, whether a graph G admits a Hamiltonian cycle or not, lies in the complexity class $\mathcal{NP}$. The input of this problem is defined by the vertex set V and the edge set E of graph G, see Figure 1.3. As certificate we consider a subset H of E which provides a Hamiltonian cycle.

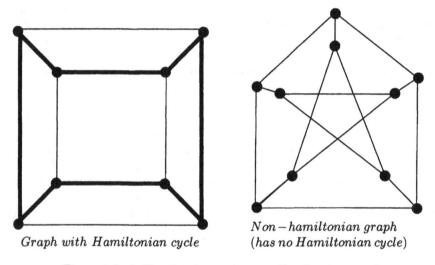

Graph with Hamiltonian cycle   *Non − hamiltonian graph*
                                          (has no Hamiltonian cycle)

Figure 1.3: A Hamiltonian and a non-Hamiltonian graph

Another fundamental notion in complexity theory is the notion of polynomial reducibility. A decision problem P can be *polynomially reduced* to a decision problem Q if and only if for every instance I of P an instance J of Q can be constructed in polynomial time (depending on the size of I) such that J is a "yes" instance if and only if I is a "yes" instance. A decision problem P is said to be $\mathcal{NP}$*-complete* if P is a member of the class $\mathcal{NP}$ and all other problems in the class $\mathcal{NP}$ can be polynomially reduced to P. An *optimization problem* is said to be $\mathcal{NP}$*-hard* if and only if the existence of a polynomial algorithm for solving it implies the existence of a polynomial algorithm for some $\mathcal{NP}$-complete decision problem.

In practice, optimization problems in the complexity class $\mathcal{P}$ are in the most cases easy to solve, whereas for $\mathcal{NP} - hard$ optimization problems of larger size often no optimal solution can be found due to the immense computational expense needed.

2 Shortest trees

2.1 Shortest connection of points

In practice we are often faced with the problem to connect n given points in the shortest possible way. This problem occurs, for example, if we want to connect the houses of a new estate by telephone lines or circuit lines. Such *local area networks* (see Fig. 2.1) can be modelled by an undirected graph $G = (V, E)$ with vertex set V and edge set E. The vertices of G correspond to the houses and the edges correspond to the lines. Since it should be possible to make a call from any house to any other house, the graph G must be *connected*.

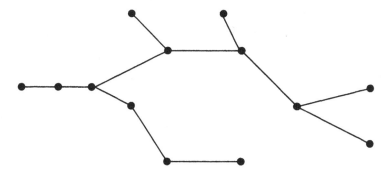

Figure 2.1: Local area networks

Moreover, the edges of our graph have some (positive) *length*. In our example we could choose as length $c(e)$ of an edge e between two sites x and y in the plane one of the following distances $d(x, y)$. Let (x_1, x_2) be the coordinates of x in the plane and let (y_1, y_2) be the coordinates of y.

- $d(x, y) := |x_1 - y_1| + |x_2 - y_2|$ ("Manhattan distance")

- $d(x, y) := \sqrt{(x_1 - y_1)^2 + (x_2 - y_2)^2}$ ("Euclidean distance")

- $d(x, y) := \max(|x_1 - y_1|, |x_2 - y_2|)$ ("Maximum distance")

- explicitly given distances due to geographical, geological or other circumstances.

Thus the total length of all edges in graph G becomes

$$c(G) := \sum_{e \in E} c(e).$$

Due to economic reasons, we would like to make the total length of G as small as possible. When every edge has a positive length, an optimal solution cannot contain a cycle: if we had a cycle, we could delete an arbitrary edge of this cycle and could still provide full service to all customers, but decrease the length of graph G. Thus an optimal solution of the problem to connect n given points in the shortest possible way is always a connected graph without cycles, i.e., a *tree*.

Let us consider the three vertices A, B and C of an equilateral triangle with side length 1 (see Fig. 2.2). There are two ways to connect the given three points: The left picture shows a *spanning tree* of length 2. Here the three vertices are connected by two edges of the triangle. The right picture, however, shows a so-called *Steiner tree*, where a new point, the so-called *Steiner point*, is introduced and then connected with the given three vertices. The length of this Steiner tree is $\sqrt{3}$ and hence it is shorter than the spanning tree. A spanning tree uses only the given points, whereas in a Steiner tree new points are selected where the tree can branch. A Steiner tree of minimum length connecting the four vertices of a square is shown in Fig. 2.3. Note that two Steiner points are introduced in this case, each incident with three edges.

Figure 2.2: Shortest connection of three points. Left: minimum spanning tree. Right: Steiner minimum tree.

Different applications require different solutions: whereas spanning trees are used for the design of local area networks, Steiner trees are used for example in VLSI design, see Section 2.5. If $\mathcal{P} \neq \mathcal{NP}$ (which is assumed by most of the researchers), then the problems of finding a minimum spanning tree and of finding a minimum length Steiner tree for a given point set have different complexity. Whereas there exist fast and simple algorithms for the construction of minimum spanning trees, the problem of finding an optimal Steiner tree is NP-hard. In the next section we shall describe minimum spanning tree algorithms. Steiner tree problems will be addressed in Section 2.5.

Local area networks are only one application in the broad context of network design problems in telecommunication. Important other questions are related to *network security* which means that precautions have to be made

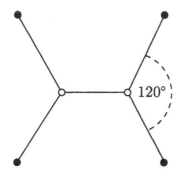

Figure 2.3: Steiner minimum tree connecting the four vertices of a square. Two new Steiner points are introduced.

that not the complete network fails if one edge or one vertex breaks down as it would happen in the case of spanning trees. Therefore one considers ring-topologies or even 3-connected graphs for a backbone network connecting the most important centers. See Figure 2.4 for a hierarchical telecommunication network with a backbone network on 4 main centers, 2-connected lower level networks and finally local area networks in form of spanning trees.

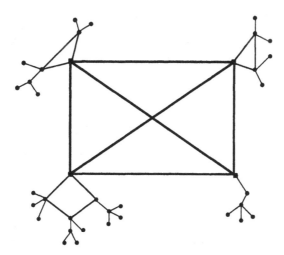

Figure 2.4: Network structure of a backbone network and local area networks

Further interesting research questions concern network expansion problems: how can an existing network be expanded in order to meet new or rising traffic needs?

2.2   Minimum spanning tree algorithms

In the last section we have seen how the problem of finding a local area
network with minimum length leads to a minimum spanning tree problem.
Here we state this problem more formally and we derive algorithmic principles
which are the basis for solution algorithms like Prim's algorithm and the
algorithm of Kruskal. Let a simple, connected graph $G = (V, E)$ with vertex
set V and edge set E be given. We assume that every edge e has a (positive)
length $c(e)$. A *spanning tree* $T = (V, E(T))$ of G is a tree on the vertex set
V whose edges are taken from the set E. The *length* $c(T)$ of the tree T is
defined by

$$c(T) := \sum_{e \in E(T)} c(e).$$

See Fig. 2.5 for a graph G and a spanning tree T of G.   The *minimum*

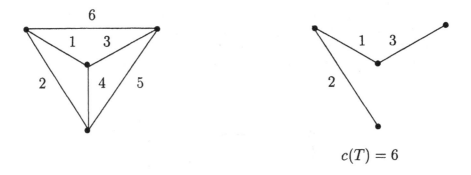

$$c(T) = 6$$

Figure 2.5: Graph G and corresponding minimum spanning tree.

spanning tree problem (MST) asks for a spanning tree T of G whose length is
minimum, i.e.,

$$\min_{T} c(T). \tag{1}$$

Minimum spanning tree algorithms rely on the notions of cycles and cuts
in the graph G. A *cut* (V_1, V_2) in G is a partitioning of the vertex set V into
two nonempty sets V_1 and V_2. We say an edge $e = [v_1, v_2]$ lies in the cut, if
$v_1 \in V_1$ and $v_2 \in V_2$. See Fig. 2.6 for an example.

The following two rules are basic for minimum spanning tree algorithms:

Inclusion rule (blue rule)
*Choose a cut without blue edge. Color the shortest uncolored edge in the
cut blue.*

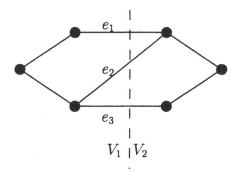

Figure 2.6: A cut (V_1, V_2). The three edges e_1, e_2, e_3 lie in this cut.

Exclusion rule (red rule)
Choose a cycle without red edge. Color the longest uncolored edge in the cycle red.

Theorem 2.1 *(Tarjan,1983) Applying the blue and the red rule in an arbitrary order (whenever possible), all edges of the graph can be colored and the blue edges form a minimum spanning tree.*

Sketch of the proof. First we show by induction that during the coloring process there always exists a minimum spanning tree which contains all blue edges. This is certainly true at the beginning, when no edges are colored at all. Now consider the case that an edge e is colored blue. Let T be the minimum spanning tree which contains all previously colored blue edges. If edge e belongs to T, everything is fine. Otherwise consider the cut C which led to the blue coloring of e. There is a path in T, joining the two endpoints of edge e and consisting only of blue or uncolored edges. Since e lies in C, there exists another edge e' in this path, which also lies in the cut. According to the assumption of the blue rule, e' is uncolored and since T has minimum length, $c(e') \leq c(e)$. But this implies $c(e') = c(e)$ and we can exchange e' with e thus getting a new minimum spanning tree.

Next we consider the case that e is colored red. If e does not lie in T, everything is fine. If e belongs to T, we get two subtrees T_1 and T_2 when we delete edge e. The vertex sets $V(T_1)$ and $V(T_2)$ define a cut in G. In this cut there is at least one uncolored edge e' of the cycle which led to the red-coloring of e. Due to the red rule, $c(e') \leq c(e)$ must hold. But then a new minimum spanning tree is obtained by exchanging e of T against e'. This new tree does not contain the red colored edge e.

An immediate consequence of this first part is that throughout the coloring process the blue edges always form a forest, i.e., there is no cycle consisting of only blue edges.

Now we show that all edges of G can be colored. Let $[u, v]$ be an uncolored edge. If the vertices u and v belong to the same blue tree, then there exists a cycle with blue edges and edge $[u, v]$. In this case the edge $[u, v]$ can be colored red. If, however, the vertices u and v belong to different blue trees T_1 and T_2 (where single vertices without adjacent blue edge also count as a blue tree), then there is a cut without blue edge. Let V_1 be the vertex set of T_1. Then $(V_1, V \setminus V_1)$ is such a cut without blue edges. In this case we can color edge $[u, v]$ blue. This completes the proof. ∎

If we only use the blue coloring rule, we get Prim's algorithm for constructing a minimum spanning tree:

Prim's algorithm

1. Start with an arbitrary vertex v_1. Set $A := \{v_1\}$.

2. Consider the cut $(A, V \setminus A)$. Color the shortest edge $e = [u, v]$ in the cut blue.

3. Assume $u \in A$. Set $A := A \cup \{v\}$.
   If $A \neq V$, return to Step 2.
   Otherwise stop. The blue edges form a minimum spanning tree.

Prim's algorithm grows the minimum spanning tree from the root vertex v_1. Since all blue edges have always both endpoints in the set A, the cut $(A, V \setminus A)$ never contains a blue edge. The algorithm yields the minimum spanning tree in $n - 1$ rounds. In every round we have to find an edge in the current cut with minimum length. This requires at most $O(n)$ operations. Thus the total time complexity of Prim's algorithm is $O(n^2)$. By using the advanced data structure of so-called Fibonacci heaps, the complexity can be reduced to $O(m + n \log n)$, see Fredman and Tarjan [10].

Another application of Theorem 2.1 is Kruskal's algorithm:

Kruskal's algorithm

1. Sort the edges according to increasing lengths:
$$c(e_1) \leq c(e_2) \leq c(e_3) \leq \ldots \leq c(e_m). \tag{2}$$
   Initially all edges are uncolored.

2. Start with edge e_1.

3. If the curent edge forms no cycle with the blue edges, color it blue. Otherwise color it red.

4. If $|V| - 1$ edges are already colored blue, terminate. These blue edges form a minimum spanning tree.
   Otherwise proceed with the next edge.

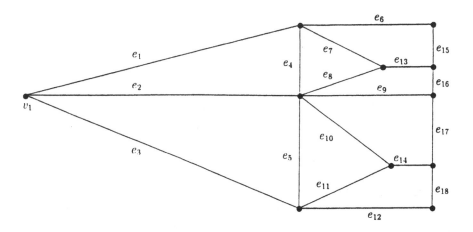

Figure 2.7: Example for Prim's and Kruskal's algorithm

Let us illustrate Prim's and Kruskal's algorithm by means of a small example. Consider the graph as represented by Fig. 2.7.

Starting from vertex v_1 Prim's algorithm finds the edges of a minimum spanning tree in the following sequence:

$$e_2, e_4, e_7, e_{13}, e_{16}, e_{15}, e_{17}, e_{14}, e_{18}, e_{11}.$$

Kruskal's algorithm determines the edges of a minimum spanning tree in the following sequence:

$$e_{16}, e_{15}, e_{14}, e_{18}, e_{13}, e_{17}, e_4, e_7, e_{11}, e_2.$$

Kruskal's algorithm can be implemented to run in $O(|E| \log |V|)$ time. It is the prototype of a so-called greedy algorithm which is addressed in the next section.

Note that neither Prim's nor Kruskal's algorithm uses the fact that we sum up the lengths of the edges of the spanning tree. They solve at the same time the *bottleneck problem*

$$\min_{T} \max_{e \in E(T)} c(e).$$

Hence there exist optimal solutions for the spanning tree problem which minimize at the same time the sum objective function $\sum_{e \in E(T)} c(e)$ and the bottleneck objective function $\max_{e \in E(T)} c(e)$. Note, moreover, that both algorithms find immediately also a spanning tree of *maximum* length: just reverse the order relation in the algorithms.

Kruskal's algorithm can be used for *clustering* points: in the first steps of Kruskal's algorithm those points are connected which are close to each other. Thus we get an initial forest. Every tree in this forest represents clusters of points which differ not much from each other. We can run the algorithm until a prescribed number of clusters (trees) is reached.

Another application of minimum spanning trees occurs in *system reliability*, see e.g. Hunter [13] or Worsley [21]. Consider a technical system with n components. Let us suppose that all components can perform the same functions. Thus the system fails if all of its components fail. Usually component failures are not independent events, for example, if the whole system is exposed to heat or dust. Thus an exact formula for the probability that the system works can be given by the inclusion-exclusion principle. But this requires that the probabilities of all joint events are known, which is usually not the case. Let E_i denote the event that the i-th component is operable. A trivial upper bound for the probability that the system works is given by

$$\mathbf{Pr}(\bigcup_{i=1}^{n} E_i) \leq \sum_{i=1}^{n} \mathbf{Pr}(E_i).$$

Sometimes, however, $\mathbf{Pr}(E_i \cap E_j) =: \mathbf{Pr}(E_{ij})$ may be known. This knowledge enables us to improve the upper bound. We define a graph G in the following way: the vertex i of G corresponds to the event E_i. We have an edge $[i, j]$ in G, if and only if $E_i \cap E_j \neq \emptyset$. Moreover, let $\mathbf{Pr}(E_{ij})$ be the length of such an edge. Now let T be any spanning tree in G. Using the length of T we find the following probabilistic bounds that the system works:

$$\sum_{i=1}^{n} \mathbf{Pr}(E_i) \quad - \sum_{i=1}^{n} \sum_{j=i+1}^{n} \mathbf{Pr}(E_{ij}) \leq \mathbf{Pr}(\text{system operates})$$

$$\leq \sum_{i=1}^{n} \mathbf{Pr}(E_i) - \sum_{[i,j] \in E(T)} \mathbf{Pr}(E_{ij}).$$

(3)

The *best* upper bound will be obtained, if the length of T is as large as possible. Thus we get a good upper bound by determining a *maximum* spanning tree in G.

2.3   The greedy algorithm and matroids

The greedy algorithm is one of the most basic algorithms in combinatorial optimization. Let us consider a general combinatorial optimization problem of the following form. Let $E = \{e_1, e_2, ..., e_m\}$ be a finite ground set. Let S be a nonempty class of subsets of E. Every subset $S \in S$ is called a *feasible solution.* Moreover, every element e_i of the ground set has a certain weight $c(e_i)$. The weight of a feasible solution $S \neq \emptyset$ is defined by

$$c(S) := \sum_{e \in S} c(e).$$

Our combinatorial optimization problem consists in finding a maximal feasible set (*basis*) with minimum weight, where maximal refers here to set inclusion. Let us denote the class of maximal sets in S by $\mathcal{B}$.

$$\min_{S \in \mathcal{S}} \sum_{e \in B} c(e). \tag{4}$$

Let us give an example. Let E be the edge set of a connected graph G and let $c(e)$ be the weight (length) of edge e. Now let S be the class of all edge sets corresponding to forests in G. All maximal elements of S have the same cardinality and correspond one to one to the spanning trees of G. Thus the combinatorial optimization problem described above is nothing else than the minimum spanning tree problem. Kruskal's algorithm applied to problem (4) has the following form:

Greedy algorithm

1. Sort the elements $e \in E$ according to increasing weights:

$$c(e_1) \leq c(e_2) \leq c(e_3) \leq \ldots \leq c(e_m). \tag{5}$$

2. Let $T := \emptyset, \quad i := 1$.

3. If $T \cup \{e_i\} \in S$, set $T := T \cup \{e_i\}$.
   Otherwise go to Step 4.

4. If $i < m$, set $i := i + 1$ and goto Step 3. Otherwise terminate.

Now we can ask: for which combinatorial optimization problems does this greedy algorithm yield an optimal solution? The answer is closely related to a combinatorial structure called *matroids* which was introduced by Whitney [20] as generalization of independent sets in linear algebra.

Given a finite set E together with a nonempty class S of subsets of E, the pair (E, S) is called an *independence system* if the following two conditions hold:

1. $\emptyset \in \mathcal{S}$.

2. $A \in \mathcal{S}, B \subseteq A$ implies $B \in \mathcal{S}$.

The sets of the class S are called *independent sets*. A maximal independent set (with respect to set inclusion) is called a *basis*. If the independence system (E, S) fulfills in addition the exchange property

3. $A \in S$, $B \in S$ and $|A| < |B|$ implies $\exists e \in B \setminus A: \; A \cup \{e\} \in S$.

then (E, S) is called a *matroid*. Let us give two examples for matroids:

Example 1. Let E consist of the column vectors of some given matrix. A subset $S \subseteq E$ lies in the class S if the column vectors of S are linearly independent. Property 2 above is valid due to the extension lemma for independent sets. Property 3 is nothing else than Steinitz' Exchange Theorem.

Example 2. Let E be the edge set of some undirected graph. A subset S belongs to the class S if the edges in S do not contain a cycle. The proof that this system fulfills also Property 2 and Property 3 of a matroid, is left to the reader. In this case, (E, S) is called the *graphic matroid*.

The exchange property 3 is equivalent with the fact that all maximal sets (with respect to set inclusion) of the class S have the same cardinality. Therefore all bases of a matroid have the same number of elements. This property is well known from linear algebra where all bases of a certain subspace have the same cardinality (the dimension of this subspace), and in graph theory where all spanning trees of a connected graph with n vertices have $n - 1$ edges. The connection between matroids and the greedy algorithm is exhibited by the following theorem.

Theorem 2.2 *(Rado, 1957) An independence system (E, S) is a matroid if and only if the greedy algorithm computes a basis with minimum weight for all weight functions c.*

This theorem shows that the structure of a matroid can be defined via an algorithm, namely the greedy algorithm. Such a situation occurs very rarely in mathematics.

Sketch of the proof. Let (E, S) be a matroid. The assumption that the optimal solution found by the greedy algorithm is not a basis with minimum weight, can easily be reduced to a contradiction by considering lexicographically ordered solutions: Let the elements of the ground set be ordered according to (5), and let $B_G = (e_{i_1}, e_{i_2}, ..., e_{i_r})$ be the solution found by the greedy algorithm and $B_o = (e_{j_1}, e_{j_2}, ..., e_{j_r})$ be the lexicographically smallest optimal solution. Suppose we have $i_1 = j_1, ..., i_{s-1} = j_{s-1}$, but $i_s \neq j_s$. Due to the greedy algorithm $j_s > i_s$ must hold which implies $c_{j_s} \geq c_{i_s}$. The exchange property 3 allows us now to exchange some element e_{j_t} with $j_t \geq j_s$ against the element e_{i_s}. This yields a solution whose weight is not greater than $c(B_o)$ and which is lexicographically smaller than B_o. But this is a contradiction.

Conversely, let (E, S) be not a matroid. Then there exist maximal independent sets with different cardinality. Let O be a base with largest cardinality, say cardinality r. We define a weight function $c(e)$, $e \in E$ by

$$c(e) := \begin{cases} r, & \text{if } e \in O \\ r + 1, & \text{if } e \in E \setminus O \end{cases}$$

Thus the greedy algorithm will find the solution O with $c(O) = r^2$, whereas there existsan other basis having weight $\leq (r+1)(r-1) = r^2 - 1$. This means that the greedy algorithm does not find a basis with minimum weight. ∎

In combinatorial optimization, the greedy algorithm is often used as a fast heuristic for finding a good suboptimal solution. Consider for example the **knapsack problem**:

n items can be packed in a knapsack, each of them having value c_j and weight w_j, $j = 1, 2, ..., n$. The weight of the knapsack should not exceed b units. Which items should be packed in the knapsack such that its value becomes as large as possible?

Formally, we can state this problem as follows

$$\text{Maximize} \quad c_1 x_1 + c_2 x_2 + ... + c_n x_n$$
$$\text{s.t.} \quad w_1 x_1 + w_2 x_2 + ... + w_n x_n \leq b$$
$$x_j \in \{0, 1\} \qquad \text{for } j = 1, 2, ..., n.$$

In order to apply the greedy algorithm, we order the items with respect to their *relative value* c_j / w_j. Then we apply the following algorithm:

Greedy heuristic for the knapsack problem

1. Sort the items according to decreasing relative weights:

$$\frac{c_1}{w_1} \geq \frac{c_2}{w_2} \geq ... \geq \frac{c_n}{w_n}.$$

2. $W := 0, \quad i := 1$.

3. If $W + w_i \leq b$, pack item i in the knapsack and set $W := W + w_i$. Otherwise go to Step 4.

4. If $i < n$, set $i := i + 1$ and goto Step 3. Otherwise terminate.

Example. Consider the following knapsack problem:

$$\text{Maximize} \quad 10x_1 + 5x_2 + 8x_3 + 4x_4 + 5x_5$$
$$\text{s.t.} \quad 2x_1 + 3x_2 + 8x_3 + 5x_4 + 10x_5 \leq 12$$
$$x_j \in \{0, 1\} \qquad \text{for } j = 1, 2, ..., 5.$$

The greedy solution for this problem is

$$x_1 = 1, \quad x_2 = 1, \quad x_3 = 0, \quad x_4 = 1, \quad x_5 = 0.$$

Knapsack problems occur in practice for example in the forest industry in connection with cutting stock problems. The quality of greedy heuristics has been analysed for example by Fisher and Wolsey [8].

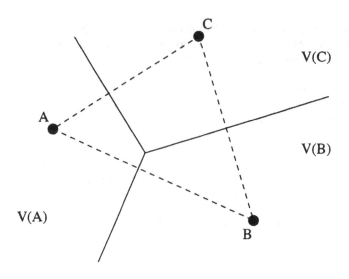

Figure 2.8: Voronoi diagram for three points

2.4   Shortest spanning trees in the plane

If we want to find a shortest spanning tree for points in the Euclidean plane,
geometric considerations can help us to design an algorithm for this problem
which is faster than the previously discussed algorithms. The basic considerations are independent of the used distance measure. They are equally valid
for the Euclidean distance, Manhattan distance and maximum distance. We
shall use in our formulation a general distance function d, in the figures,
however, the Euclidean distance.

Let n points $P_1, P_2, ..., P_n$ in the plane be given. The *Voronoi region*
$V(P_i)$ of point P_i consists of all points X in the plane which fulfill

$$d(P_i, X) \leq d(P_j, X) \quad \text{for all } j \neq i.$$

The Voronoi regions of three points in the plane is shown in Fig. 2.8, the
so-called *Voronoi diagram* of these points. In order to exclude pathological
cases we assume in the following that no four or more of the given points
lie on a circular line. Starting from the Voronoi diagram of a point set S
we define the *Delaunay triangulation* $D(S)$ of S as undirected graph with
vertex set S. $D(S)$ has the edge $[P_i, P_j]$, if and only if $V(P_i) \cap V(P_j) \neq \emptyset$.
See Figure 2.9 for a Voronoi diagram and Figure 2.10 for the corresponding
Delaunay triangulation.

An immediate consequence of this definition is the following lemma:

Lemma 2.3 *Let S be a finite point set. If in the Delaunay triangulation
$D(S)$ the points P_1 and P_2 are not connected by an edge, then there exists a
third point P_3 in S which is nearer to both P_1 and P_2, i.e.,*

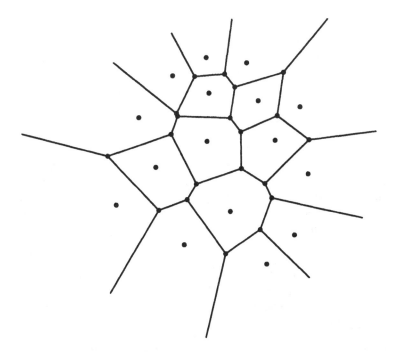

Figure 2.9: Voronoi diagram

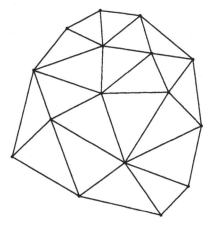

Figure 2.10: Delaunay triangularization

$$\max\left(d(P_1, P_3), d(P_2, P_3)\right) < d(P_1, P_2).$$

The graph $D(S)$ is *planar*, i.e. it can be drawn in the plane such that no edges cross. Planar graphs have the important property that they have only "few" edges as the following lemma shows:

Lemma 2.4 *A planar graph with n vertices $(n \geq 3)$ has at most $3n-6$ edges.*

Proof. We prove this lemma by induction. For a graph G with 3 vertices the lemma is obviously true. Therefore we may assume that the lemma holds for any planar graph with k, $3 \leq k \leq n$ vertices. Now we consider a planar graph with $n + 1$ vertices. We delete one vertex v and all edges incident with v. So we obtain a new planar graph G'. If the degree $d(v)$ of vertex v is at most 3, then we can immediately apply the induction hypothesis and conclude that G has no more than $3n - 6 + d(v) \leq 3(n + 1) - 6$ edges. If $d(v) > 3$, then all neighboring vertices of v lie around one face and form the vertices of a polygon. Thus we could add at least $d(v) - 3$ new edges (diagonals in the polygon) without violating the planarity of graph G'. This yields that G' has at most $3n - 6 - (d(v) - 3)$ edges. Adding the $d(v)$ edges incident with v, we get that G has again at most $3(n + 1) - 6$ edges. ■

Next we prove the following rsult on minimum spanning trees for point sets in the plane.

Theorem 2.5 *There exists a minimum spanning tree for the point set S such that all edges of this tree are edges of the Delaunay triangulation $D(S)$.*

Proof. Let us consider the complete graph with vertex set S. The corresponding edge lengths are $d(P_i, P_j)$ for $1 \leq i < j \leq |S|$. Now we apply Kruskal's algorithm to this graph. If all edges which this algorithm selects for the minimum spanning tree are also edges of $D(S)$, we are done. So, let $e = [P_i, P_j]$ be the first edge which does not belong to $D(S)$, but which according to Kruskal's algorithm should be an edge of the minimum spanning tree. According to Lemma 2.3 there exists a point $P_k \in S$ with $\max\left(d(P_i, P_k), d(P_j, P_k)\right) < d(P_i, P_j)$. Therefore these edges have been investigated in an earlier phase of Kruskal's algorithm which implies that both vertices P_i and P_j lie in the same blue tree. By adding edge e we would get a cycle. Thus edge e cannot belong to the minimum spanning tree. ■

Since it is possible to construct the Delaunay triangulation $D(S)$ of S in $O(n \log n)$ time and since due to Lemma 2.4 $D(S)$ has only $O(n)$ edges, we can solve the minimum spanning tree problem in the plane using Kruskal's algorithm in $O(n \log n)$ time.

2.5   Steiner tree problems

In Section 2.1 we have seen that the shortest connection of n points in the plane is in general not given by a minimum spanning tree, but by a Steiner tree. The difference between these two notions is that in Steiner trees additional points are inserted, the so-called *Steiner points*. No additional points are allowed in minimum spanning trees.

A Steiner minimum tree problem (SMT) for points in the plane can be described in the following way: Let a finite set V of points in the plane be given. Fix a metric in the plane, e.g. the Euclidean or the Manhattan metric. Find a set S of additional points, the Steiner points, such that the minimum spanning tree of the point set $V \cup S$ has minimum length. The Steiner minimum tree problem is an $\mathcal{NP}$-hard problem. Thus no polynomial solution algorithm is known for (SMT).

In a more formal graph theoretical setting the Steiner minimum tree problem can be stated as follows: Let an undirected, connected graph $G = (V, E)$ be given. The edges of G have the lengths $c(e)$, $e \in E$. Let T be a subset of the vertices of G, the so-called *terminals* which have to be connected. Find a subset S of vertices $V \setminus T$ such that the minimum spanning tree on the vertices $T \cup S$ has minimum length. This problem is once again $\mathcal{NP}$-hard. It occurs frequently in VLSI design, where the terminals have to be joined by "wires". In this context one has to deal even with a much more complicated problem, the so-called *Steiner tree packing problem,* see Fig. 2.11. In a Steiner tree packing problem the terminals belong to different nets. We have to find a Steiner minimum tree for every single net under the additional constraint that edges of different nets do not intersect. This is a very important and hard problem in practice which has attracted much attention recently, see e.g. Uchoa and Poggi de Aragão [19].

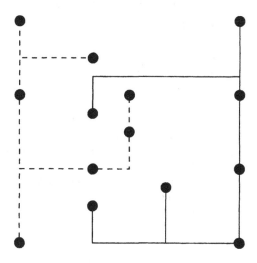

Figure 2.11: A packing of two Steiner trees

3 Shortest paths

3.1 Shortest paths problems

Shortest path problems play a crucial role in many applications. They are important in traffic planning, in data compression (see e.g. Békési, Galambos, Pferschy and Woeginger [2]), in communication networks, for location problems and in project management.

One of the most obvious applications arises in *traffic planning* where a shortest or fastest way from an origin s to a terminal t has to be found. This is right now of special interest where by the use of geomatic information systems such a decision can be achieved onboard a car. The underlying structure is a hierarchical road network: local roads provide access to the highway and freeway network.

In communication networks one can ask for the most reliable path between a source s and a sink t. In this context the arcs represent communication links and their weights are probabilities of their breakdown.

As a third introductory example we can consider a heavy transport which has to avoid narrow passages. In order to find a feasible way from s to t we can model bridges as arcs of an underlying directed graph. The width of the bridges is the weight of the arcs. The problem is to find a path from s to t such that the minimum weight is as large as possible. Several further examples will be mentioned later in this chapter.

The *shortest path problem* can be modeled in the following way. We consider a directed graph $G = (N, A)$ where every arc $e = (x, y) \in A$ has some length $w(e) = w(x, y)$. The length of the path $P = (e_1, e_2, ..., e_k)$ is defined by

$$w(P) := w(e_1) + w(e_2) + ... + w(e_k).$$

In the *single pair problem* we want to find a shortest path from a source s to a sink t. In the *single source problem* we want to find shortest paths from the source s to all other nodes in G. In the *single sink problem* we want to find shortest paths from all nodes to the sink t. Finally, in the *all pairs problem* we want to find shortest paths between all pairs of nodes.

Let us recall that a simple path is a path where no node is repeated. If there is a path from node i to node j, then there is also a simple path from i to j. This is immediate to see by the following "shortcutting" argument: if a node is repeated, the path contains a cycle which can be deleted.

A cycle C is called a *negative cycle*, if $w(C) < 0$. If we pass through a negative cycle again and again, we can make the length of a path arbitrarily small: in this case the shortest path problem does not have an optimal solution. So we get:

Proposition 3.1

1. *There is a shortest path from s to t if and only if no path from s to t contains a negative cycle.*

2. *If there is a shortest path from s to t, then there exists also a simple shortest path from s to t.*

We shall see in the following that the algorithmic solution of shortest path problems depends on the weights $w(e)$ being nonnegative or not. If all weights $w(e)$ are nonnegative, then a straightforward implementation of Dijkstra's algorithm solves the single pair problem in $O(n^2)$ steps. In the presence of negative weights, one can use Bellman-Moore's algorithm which takes $O(n^3)$ steps.

3.2 Dijkstra's algorithm

One of the fastest algorithms for computing shortest paths is due to Dijkstra [6]. This algorithm determines a shortest path from a single source to one or all other nodes of the network under the assumption that all arc-weights are nonnegative.

The underlying idea of Dijkstra's algorithm is the following. Let X be a proper subset of the nodes which contains the source s. We assume that we already know the shortest distance $d(x)$ from the source s to all other nodes in X. This is surely the case when we start with $X = \{s\}$. For any node $y \in N \setminus X$ we define $d(y)$ as shortest length of a path $P = (s, x_1, ..., x_k, y)$, where all intermediate nodes $x_1, x_2, ..., x_k$ lie in the set X. Now we determine a node $\bar{y}$ with

$$d(\bar{y}) := \min\{d(y)| \ y \notin X\} \qquad (6)$$

and add the node $\bar{y}$ to the set X. $d(\bar{y})$ is the shortest distance from s to $\bar{y}$: consider any path from s to $\bar{y}$. Such a path must cross the cut $(X, N \setminus X)$. Thus it has the form $\hat{P} = (s, x_1, ..., x_r, y_1, ..., \bar{y})$, where $x_1, ..., x_r$ are nodes in X and y_1 is the first node of the path which does not lie in X. Due to (6) we have $d(y_1) \geq d(\bar{y})$ and the length of the path from node y_1 to the node $\bar{y}$ is nonnegative due to the assumption of nonnegative arc lengths. Therefore we get that path $\hat{P}$ has at least the length $d(\bar{y})$ which we wanted to prove.

Algorithmically, we start with $X = \{s\}$ and determine in each step the labels $d(y)$ for $y \notin X$. At the begining we set $d(y) := w(s, y)$, if y is a neighbour of s, and $d(y) := \infty$, otherwise. Let $N(s)$ denote the set all all neighbours of s. The set $\bar{X}$ contains those nodes which are not in X, but have a neighbour in the set X. After determining $d(\bar{y}) := \min\{d(y)| \ y \in \bar{X}\}$ we can update the labels $d(y)$, $y \in \bar{X} \cap N(\bar{y})$ by comparing $d(y)$, the length of a shortest path to y via nodes in X with $d(\bar{y}) + w(\bar{y}, y)$, the length of the path which is composed of shortest path from s to $\bar{y}$ and the arc $(\bar{y}, y)$. If $d(y) > d(\bar{y}) + w(\bar{y}, y)$, then $d(y)$ has to be replaced by $d(\bar{y}) + w(\bar{y}, y)$. Since we are not only interested in the shortest path length, but also in the corresponding path, we store for any $x \neq s$ a label $p(x)$ which finally will be the predecessor of x on a shortest path from s to x. Summarizing we get the following algorithm:

Dijkstra's algorithm for determining a shortest path from s to t.

Start: $X := \{s\}$

1. $d(x) := \infty$ for all $x \notin X$.

2. Define for all neighbors of s
   $d(x) := w(s, x)$,
   $p(x) := s$.

3. Determine $d(\bar{y}) := \min\{d(y)| \, y \notin X\}$.

4. $X := X \cup \{\bar{y}\}$.

5. If $t \in X$, terminate. Otherwise go to Step 6.

6. Set $d(y) := \min\{d(y), d(\bar{y}) + w(\bar{y}, y)\}$,
   $p(y) := \bar{y}$, if the minimum is attained by $d(\bar{y}) + w(\bar{y}, y)$.
   Return to Step 3.

Let $G = (N, A)$ be a network with n nodes and m arcs. In this case Dijkstra's algorithm performs at most $n - 1$ iterations and every iteration needs at most $O(n)$ arithmetic operations. Thus the time complexity of this algorithm is $O(n^2)$. This complexity can be improved to $O(n \log n + m)$ by the use of *Fibonacci heaps*, see Fredman and Tarjan [10].

Let us illustrate Dijkstra's algorithm by means of the example shown in Figure 3.1. We start with $X = \{s\}$ and get

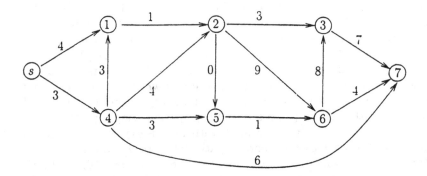

Figure 3.1: Example for Dijkstra's algorithm

node x	1	2	3	4	5	6	7
$d(x)$	4	∞	∞	3	∞	∞	∞
$p(x)$	s	$-$	$-$	s	$-$	$-$	$-$

The minimum d-value is 3 and is attained for node 4. Thus $X = \{s, 4\}$, $p(4) = s$ and we get as new table

node $x \notin X$	1	2	3	5	6	7
$d(x)$	4	7	∞	6	∞	9
$p(x)$	s	4	$-$	4	$-$	4

The minimum d-value is now attained for $x = 1$. Thus $X = \{s, 4, 1\}$, $p(1) = s$ and an update of the table yields

$x \notin X$	2	3	5	6	7
$d(x)$	5	∞	6	∞	9
$p(x)$	1	$-$	4	$-$	4

Now we get $\min\limits_{x \notin X} d(x) = 5$ for $x = 2$ and $X = \{s, 4, 1, 2\}$ with $p(2) = 1$. Thus we have

$x \notin X$	3	5	6	7
$d(x)$	8	5	14	9
$p(x)$	2	2	2	4

The next step yields $\min\limits_{x \notin X} d(x) = 5$ for $x = 5$. Thus we obtain $p(5) = 2$, $X = \{s, 4, 1, 2, 5\}$ and

$x \notin X$	3	6	7
$d(x)$	8	6	9
$p(x)$	2	5	4

The minimum d-value is now 6, attained for node $x = 6$. Thus $X = \{s, 4, 1, 2, 5, 6\}$, $p(6) = 5$ and

$x \notin X$	3	7
$d(x)$	8	9
$p(x)$	2	4

So we get $\min\limits_{x \notin X} d(x) = 8$ for $x = 3$, $p(3) = 2$ and finally $d(7) = 9$ with $p(7) = 4$. Now we are finished. The solution can be represented by the so-called *shortest path-tree* (see Fig. 3.2), taking into account that we know for every node $x \in N \setminus \{s\}$ the predecessor $p(x)$ on a shortest path from the

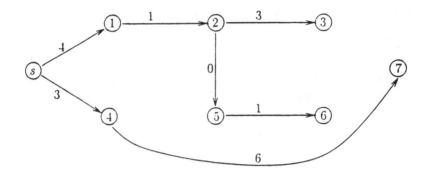

Figure 3.2: Shortest path tree

source s to node x. This tree consists of the arcs $(p(x), x)$ for all $x \in N \setminus \{s\}$.

Dijkstra-like algorithms is used for computing shortest routes in traffic networks. An efficient adaption to handle real-world situations which have millions of arcs (links) can be found in Ertl [7].

3.3   Different objective functions

The discussion of Dijkstra's algorithm shows that it is applicable for computing shortest paths whenever adding some arcs to a path does not decrease its length. This remark enables us to use Dijkstra's algorithm in a much more general framework as in the previous section. First we might define the length of a path in a different way. Let us consider some examples:

In the introductory example of a heavy transport we can define the length of a path $P = (e_1, e_2, ..., e_k)$ by

$$w(P) := \min_{1 \leq i \leq k} w(e_i).$$

In the most reliable path problem we might define the length of a path $P = (e_1, e_2, ..., e_k)$ by

$$w(P) := \prod_{i=1}^{k} w(e_i).$$

More generally, we can consider the weights as elements of a commutative, ordered semigroup $(H, *, \preceq)$ with semigroup operation $*$ and order relation $\preceq$. The operation $*$ should be compatible with the order relation, i.e.,

$$a \preceq b \text{ implies } a * c \preceq b * c \text{ for all } c \in H.$$

In such a semigroup we can define the length of a path $P = (e_1, e_2, ..., e_k)$ by

$$w(P) := w(e_1) * w(e_2) * ... * w(e_k).$$

This leads to the so-called *algebraic path probem:*

Find a path P from s to t with minimum length $w(P)$.

In such a framework an element $a \in H$ is called nonnegative, if

$$c * a \succeq c \quad \text{for all} \quad c \in H.$$

Thus the heavy transport example can be modeled in the semigroup $(\mathbb{R}, \min, \geq)$. In this system obviously all elements are nonnegative, since $\min(a, c) \leq c$ for all $c \in H$. Thus a Dijkstra-like algorithm can be applied. Just replace in the classical Dijkstra algorithm of Section 3.2 every $\leq$ by $\geq$ and every $+$ by min. A similar model is $(\mathbb{R}, \max, \leq)$ which leads to a bottleneck path problem: make the largest weight of a path as small as possible. The reliability problem can be modeled in the system $([0, 1], \cdot, \leq)$. More about algebraic path problems can be found in the book of Zimmermann [22].

3.4 Bellman-Moore algorithm

In Section 3.2 we saw that Dijkstra's algorithm essentially needs that all arc weights are nonnegative. There are, however, path problems which need negative weights. For example, the problem to find a longest path in a network is equivalent with finding a shortest path in a network where all arc length are negative. Longest path problems play an important role with so-called project networks. Planning a large project requires a careful analysis to decide when every single activity has to start at the latest so that the project can be completed in time. Such an analysis can be carried out by means of a so-called *project network* whose arcs correspond to activities. The project network models exactly the logical interdependence of the activities: all activities corresponding to arcs which enter a node u must be completed before the activities are started which correspond to arcs leaving this node u, see Figure 3.3 for an example. The logical structure of the project implies that project networks are *acyclic*, i.e., they do not contain cycles. Every arc in a project network is associated with a positive length, the duration of the corresponding activity. The question of when has the first activity to be started in order to complete the whole project in time, leads to the problem of finding a *longest path* in the project network. The critical path problem can be solved in $O(m)$ time.

The problem of finding a longest path in the network $G = (N, A)$ with arc lengths $w(e)$, $e \in A$, is equivalent to finding a shortest path in the same network, but now with arc lengths $-w(e)$, $e \in A$. If $G(N, A)$ has no negative cycle, this problem has an optimal solution due to Theorem 3.1. But Dijkstra's algorithm is not applicable, since the arc weights are negative.

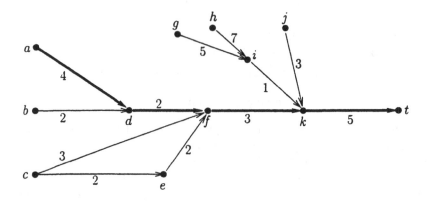

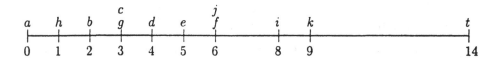

Figure 3.3: Project network and latest starting times of the activities. The thick line in the upper digraph shows the longest path. The lower figure shows on a time scale when the single activities have to be started at latest to assure that the project is completed within 14 units of time..

Let an acyclic digraph $G = (N, A)$ with arbitrary real arc weights $w(e)$, $e \in A$, be given. In order to find a longest path from the source s to the sink t we apply an algorithm due to Bellman [3] and Moore [14]. This algorithm proceeds in passes. After pass k, $k = 1, 2, ..., n - 1$ the longest paths from the source s to the other nodes with at most k arcs are known. The algorithm uses the data structure of a *queue:* In this context a queue is a sequence of nodes. The first element of the queue is taken to be examined next. If we add elements to the queue, they are added at the end (tail) of the queue. In the following, $d(x)$ is the current distance of node x from s and $p(x)$ is the immediate predecessor on a path of length $d(x)$ from s to x.

Bellman-Moore algorithm

Start:
Initially, let all queues $Q_1, Q_2, \ldots$ be empty.
If x is a neighbour of s do

$d(x) := w(s, x)$,
$p(x) := s$,
add node x to queue Q_1

else

$d(x) := -\infty$.

Recursion (pass $i + 1$):

As long as the queue Q_i is nonempty, take the first element x of Q_i,
check for each $(x, y) \in A$:

if $d(y) < d(x) + w(x, y)$, then

$d(y) := d(x) + w(x, y)$,
$p(y) := x$,
store y in queue Q_{i+1}, if it is nor already there.

If Q_i is empty, set $i := i + 1$ and start the next iteration.

The correctness of this algorithm relies on the fact that at the end of pass i the value $d(x)$ is the length of a longest path from s to x with no more than i arcs. Since in the recursion step $O(|A|)$ operations are performed and a longest path has at most $n - 1$ arcs (due to the absence of positive cycles), the total complexity of this algorithm is $O(nm)$.

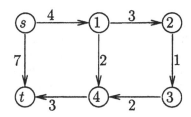

Figure 3.4: Longest path example

Let us illustrate this algorithm on the small example of Figure 3.4. We start with the table

x	1	2	3	4	t
$d(x)$	4	∞	∞	∞	7
$p(x)$	s	–	–	–	s

and obtain as first queue $Q_1 = (1, t)$. In the second pass we obtain

x	1	2	3	4	t
$d(x)$	4	7	∞	6	7
$p(x)$	s	1	–	1	s

and the queue $Q_2 = (2, 4)$. The third pass yields

x	1	2	3	4	t
$d(x)$	4	7	8	6	9
$p(x)$	s	1	2	1	4

and the queue $Q_3 = (3, t)$. The next iteration yields

x	1	2	3	4	t
$d(x)$	4	7	8	10	9
$p(x)$	s	1	2	3	4

and the queue $Q_4 = (4)$. Finally, we get the following optimal solution of the given longest path problem

x	1	2	3	4	t
$d(x)$	4	7	8	10	13
$p(x)$	s	1	2	3	4

The length of a longest path from s to t is 13. The path can be reconstructed by means of the labels $p(x)$: The immediate predecessor of t on this path is node $p(t) = 4$, the predecessor of node 4 is $p(4) = 3$ and so on. Thus the longest path is $(s, 1, 2, 3, 4, t)$.

Goldberg and Radzik [12] designed a heuristic improvement of the Bellman-Moore algorithm relying on a topological scan of the underlying network which is faster in practice and has the same worse case complexity as Bellman-Moore's algorithm. For acyclic networks there is a version of Goldberg and Radzik's algorithm, which runs in $O(m)$-time.

3.5   The all-pairs problem

Sometimes, e.g. in connection with location problems, one is interested in finding the shortest distances between all pairs of nodes. This can obviously be done by applying Dijkstra's or Bellman-Moore's algorithm for any node x as source, yielding routines of the complexity $O(n^3)$ and $O(n^2 m)$, respectively. An alternative direct approach dates back to Floyd [9] and uses dynamic programming.

We start with a directed graph G with n nodes which has no negative cycles. Let w_{ij} be the weight (length) of arc (i, j). We compute $n \times n$ matrices $D(k) := (d_{ij}(k))$ for $k = 0, 1, ..., n$. $d_{ij}(k)$ is the length of a shortest path from node i to node j, where only the nodes $1, 2, ..., k$ are allowed as intermediate nodes. In order to reconstruct the shortest path, we also define an $n \times n$ array $P(i, j)$. After these preparations we can state Floyd's algorithm.

Floyd's algorithm

Start:

For all $i = 1$ to n and all $j = 1$ to n define

$$d_{ij}(0) := w_{ij},$$
$$p_{ij} = 0.$$

Recursion:

For $k := 0$ to $n - 1$ do

for $i := 1$ to n do

for $j := 1$ to n do

if $d_{ij}(k) \leq d_{i,k+1}(k) + d_{k+1,j}(k)$ then

$d_{ij}(k + 1) := d_{ij}(k)$

else

$d_{ij}(k+1) := d_{i,k+1}(k) + d_{k+1,j}(k),$
$p_{ij} := k + 1.$

By induction can easily be shown that $d_{ij}(k)$ is the length of a shortest path from node i to node j where only the nodes $1, 2, ..., k$ are allowed as intermediate nodes. This proves the correctness of the algorithm. The complexity of the algorithm is $O(n^3)$.

Let us illustrate Floyd's algorithm by means of the following example (see Fig. 3.5). We start with the weighted *adjacency matrix* $D(0)$ and $P = 0$. The elements of the weighted adjacency matrix are 0 on the main diagonal, equal to w_{ij}, if $(i, j) \in A$, and are equal to ∞, otherwise.

$$D(0) = \begin{pmatrix} 0 & 4 & 2 & 8 & \infty \\ 7 & 0 & \infty & \infty & \infty \\ \infty & 1 & 0 & 2 & 7 \\ 3 & \infty & 1 & 0 & 9 \\ \infty & 8 & 3 & 1 & 0 \end{pmatrix} \qquad P = \begin{pmatrix} 0 & 0 & 0 & 0 & 0 \\ 0 & 0 & 0 & 0 & 0 \\ 0 & 0 & 0 & 0 & 0 \\ 0 & 0 & 0 & 0 & 0 \\ 0 & 0 & 0 & 0 & 0 \end{pmatrix}$$

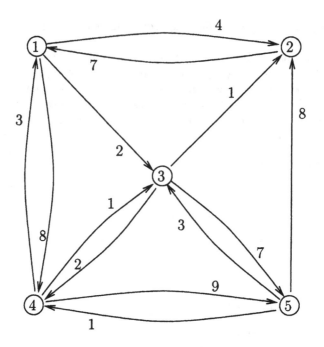

Figure 3.5: Example for the all pairs algorithm

In the first iteration we insert node 1 and compare the direct arc lengths with the lengths of the paths via node 1. We get the following improvements:

$$d_{23}(1) = 9, \quad d_{24}(1) = 15, \quad d_{42}(1) = 7.$$

This leads to

$$D(1) = \begin{pmatrix} 0 & 4 & 2 & 8 & \infty \\ 7 & 0 & 9 & 15 & \infty \\ \infty & 1 & 0 & 2 & 7 \\ 3 & 7 & 1 & 0 & 9 \\ \infty & 8 & 3 & 1 & 0 \end{pmatrix} \qquad P = \begin{pmatrix} 0 & 0 & 0 & 0 & 0 \\ 0 & 0 & 1 & 1 & 0 \\ 0 & 0 & 0 & 0 & 0 \\ 0 & 1 & 0 & 0 & 0 \\ 0 & 0 & 0 & 0 & 0 \end{pmatrix}$$

Next we compare the length of shortest paths inf $D(1)$ with those using node 2 as intermediate node as well. We get the following improvements:

$$d_{31}(2) = 8, \quad d_{51}(2) = 15.$$

This leads to

$$D(2) = \begin{pmatrix} 0 & 4 & 2 & 8 & \infty \\ 7 & 0 & 9 & 15 & \infty \\ 8 & 1 & 0 & 2 & 7 \\ 3 & 7 & 1 & 0 & 9 \\ 15 & 8 & 3 & 1 & 0 \end{pmatrix} \qquad P = \begin{pmatrix} 0 & 0 & 0 & 0 & 0 \\ 0 & 0 & 1 & 1 & 0 \\ 2 & 0 & 0 & 0 & 0 \\ 0 & 1 & 0 & 0 & 0 \\ 2 & 0 & 0 & 0 & 0 \end{pmatrix}$$

Now we insert node 3. We get the following improvements:

$$d_{12}(3) = 3, \quad d_{14}(3) = 4, \quad d_{15}(3) = 9,$$
$$d_{24}(3) = 11, \quad d_{25}(3) = 16,$$
$$d_{42}(3) = 2, \quad d_{45}(3) = 8,$$
$$d_{51}(3) = 11, \quad d_{52}(3) = 4.$$

Thus we get

$$D(3) = \begin{pmatrix} 0 & 3 & 2 & 4 & 9 \\ 7 & 0 & 9 & 11 & 16 \\ 8 & 1 & 0 & 2 & 7 \\ 3 & 2 & 1 & 0 & 8 \\ 11 & 4 & 3 & 1 & 0 \end{pmatrix} \qquad P = \begin{pmatrix} 0 & 3 & 0 & 3 & 3 \\ 0 & 0 & 1 & 3 & 3 \\ 2 & 0 & 0 & 0 & 0 \\ 0 & 3 & 0 & 0 & 3 \\ 3 & 3 & 0 & 0 & 0 \end{pmatrix}$$

Next we insert node 4 which leads to the following improvements:

$$d_{31}(4) = 5,$$
$$d_{51}(4) = 4, \quad d_{52}(4) = 3, \quad d_{53}(4) = 2.$$

Therefore we get

$$D(4) = \begin{pmatrix} 0 & 3 & 2 & 4 & 9 \\ 7 & 0 & 9 & 11 & 16 \\ 5 & 1 & 0 & 2 & 7 \\ 3 & 2 & 1 & 0 & 8 \\ 4 & 3 & 2 & 1 & 0 \end{pmatrix} \qquad P = \begin{pmatrix} 0 & 3 & 0 & 3 & 3 \\ 0 & 0 & 1 & 3 & 3 \\ 4 & 0 & 0 & 0 & 0 \\ 0 & 3 & 0 & 0 & 3 \\ 4 & 4 & 4 & 0 & 0 \end{pmatrix}$$

In the last step, inserting node 5, we get no further improvement. Thus $D = D(4)$ is the matrix of the shortest path lengths. In order to construct now the shortest path, say, from node 5 to node 2 we use the matrix P.

Since $p(5, 2) = 4$, node 4 is an intermediate node on this path. Thus the shortest path can be split in the path from node 5 to node 4 and in the path from node 4 to node 2. Now $p(5, 4) = 0$ which says that we go directly from node 5 to node 4. Furthermore $p(4, 2) = 3$ says that from node 4 to node 2 we pass node 3. Since $p(4, 3) = p(3, 2) = 0$ we get for the shortest path $(5, 4, 3, 2)$.

Dantzig [5] proposed another $O(n^3)$ algorithm for the all pairs shortest path problem which works by successive extensions of the graph. If all arc weights are positive, an algorithm due to Spira [17] can be applied which has a better average time complexity than Floyd's algorithm. Spira's algorithm runs in $O(n^2 \log^2 n)$ average time. Another alternative for solving the all pairs shortest path problem are *matrix algorithms* which use the following recursion:

Matrix algorithm

Start:

Let $D(1)$ be the weighted adjacency matrix of the given digraph.

Recursion:

For $l := 2$ to $n - 1$ do

for $i := 1$ to n do

for $j := 1$ to n do

$d_{ij}(l)$ $:=$

$\min_{1 \leq k \leq n} (d_{ij}(l-1), d_{ik}(l-1) + d_{kj}(1))$

In this recursion $d_{ij}(l)$ is the length of a shortest path from node i to node j with no more than l arcs. The recursion above relies on the fact that a path with no more than l arcs has either no more than $l - 1$ arcs or exactly l arcs. In the latter case the path has an arc (k, j) as last arc. If we replace in the recursion min by $+$ and $+$ by the multiplication $\cdot$, we get the classical formula for matrix multiplication:

$$d_{ij}(l) := \sum_{k=1}^{n} d_{ik}(l-1) \cdot d_{kj}(1).$$

Due to this analogy it is possible to design the lay-out of special chips, so-called *systolic arrays*, which perform matrix multiplication and shortest path computations and differ only by the arithmetic operations used. See Rote [16] for more details. By properly redefining the necessary algebraic operations also the all-pairs version of algebraic path problems can be solved in this way.

A straightforward implementation of the matrix algorithm has the time complexity $O(n^4)$. It is, however, not necessary to compute *all* products $D(2), D(3), D(4), ...$, since computing $D(4)$ by $D(2)$ and $D(2)$ yields also all shortest paths of a length up to 4 arcs. Thus it is enough to compute $D(2), D(4), D(8), ...$ which leads to an algorithm of complexity $O(n^3 \log n)$.

3.6 An application: 1-median problem

One of the basic problems in location theory is the 1-median problem which focusses on the question where to locate a facility which serves several customers such that the sum of all weighted distances between the facility and the customers becomes as small as possible. Figure 3.6 shows an example in the plane where an optimal location for the distribution center y is asked. Mathematically, we can describe this *1-median problem* as

$$\min_{y} \sum_{x} w(x)d(x, y) \tag{7}$$

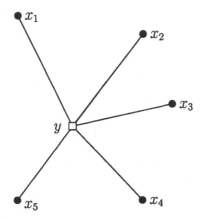

Figure 3.6: 1-median problem in the plane. The distribution center y serves the clients $x_1,...,x_5$.

where $w(x)$ is the amount of goods customer x has to get from the distribution center y and $d(y, x)$ is the distance between customer x and facility y. If the customers and the possible locations for the new facility are represented as vertices of a graph, we get the 1-median problem in a graph, see Fig. 3.7. In this case $d(x, y)$ is the shortest distance between vertex x and vertex y and (7) asks for a 1-median in a graph.

Recently facility location problems were considered where the customers have positive as well as negative weights. For example, a new container terminal is important for the transportation industry and certain factories around, but is not attractive in the neighbourhood of appartment buildings, hospitals and schools. We can model this situation by giving some customers a positive weight and other customers a negative weight. If all possible new locations are nodes of the considered digraph, the corresponding 1-median problem (7) can be solved in the following way:

1. Compute the pairwise shortest distances between all nodes of the underlying digraph. This yields a matrix $D = (d_{ij})$.

2. Multiply every column of D with the weight $w(x)$ and sum up the rows.

3. Find the smallest entry among these n sums.

Obviously, this procedure for finding a 1-median in a digraph with positive and negative node weights has the complexity $O(n^3)$. In a cactus, see Fig. 3.8, this problem can even be solved in linear time, see Burkard and Krarup [4].

Acknowledgement I thank Bettina Klinz for her careful reading of various drafts of this article and her many constructive comments.

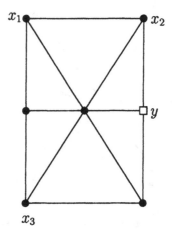

Figure 3.7: 1-median problem in a graph. The distribution center y serves the clients x_1, x_2 and x_3

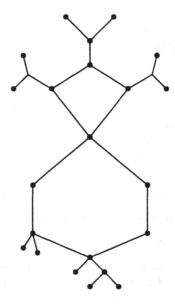

Figure 3.8: Cactus

References

[1] R. K. Ahuja, T.L. Magnanti and J. B. Orlin, *Network flows. Theory, algorithms, and applications* Englewood Cliffs, NJ: Prentice Hall, 1993

[2] J. Békési, G. Galambos, U. Pferschy and G. J. Woeginger, Greedy algorithms for on-line data compression. *J. Algorithms* **25**, 1997, 274–289.

[3] R. E. Bellman, On a routing problem. *Quart. Appl. Math.* **16**, 1985, 87–90.

[4] R. E. Burkard and J. Krarup, A linear algorithm for the pos/neg weighted 1 -median problem on a cactus. *Computing* **60**, 1998, 193–215.

[5] G. B. Dantzig, All shortest routes in a graph. *Theory of Graphs (Internat. Sympos., Rome, 1966)* New York: Gordon and Breach, 1967, pp. 91–92.

[6] E. W. Dijkstra, A note on two problems in connection with graphs. *Numerische Mathematik* **1**,1958, 269–271.

[7] G. Ertl, Shortest path calculation in large road networks. *Operations Research Spektrum* **20**, 1998, 15–20.

[8] M.L. Fisher and L.A. Wolsey, On the greedy heuristic for covering and packing problems. *SIAM Journal on Algebraic and Discrete Methods* **3**, 1982, 584–591.

[9] R. W. Floyd, Algorithm 97: Shortest Path. *Communication ACM* **5**, 1962, 345.

[10] M. L. Fredman and R. E. Tarjan, Fibonacci heaps and their uses in improved network optimization algorithms. *J. of the ACM* **34**, 1987, 596–615.

[11] M. R. Garey and D. S. Johnson, *Computers and Intractability: A Guide to the Theory of NP-Completeness.* San Francicso: W. H. Freeman and Co., 1979.

[12] A. V. Goldberg and T. Radzik, A heuristic improvement of the Bellman-Ford algorithm. *Appl. Math. Lett.* **6**, 1993, 3–6.

[13] D. Hunter, An upper bound for the probability of a union. *Journal of Aplied Probability* **13**, 1976, 597–603.

[14] E. F. Moore, The shortest path through a maze. In *Proc. of the Int. Symp. on the Theory of Switching*, Harvard University Press, 1959, pp. 285–292.

[15] R. Rado, Note on independence functions. *Proceedings of the London Mathematical Society* **7**, 1957, 300–320.

[16] G. Rote, A systolic array algorithm for the algebraic path problem. *Computing* **34**, 1985, 191–219.

[17] P. M. Spira, A new algorithm for finding all shortest paths in a graph of positive arcs in average time $O(n^2 \log^2 n)$. *SIAM J. Comput.* **2**, 1973, 28–32.

[18] R. E. Tarjan, *Data Structures and Network Algorithms*. SIAM: Philadelphia, 1983.

[19] E. Uchoa and M. Poggi de Aragão, Vertex-disjoint packing of two Steiner trees: polyhedra and branch-and-cut. In *IPCO '99*, G. Cornuéjols, R. E. Burkard and G. J. Woeginger, eds., Springer Lecture Notes in Computer Science Vol. **1610**, pp. 439–452. Berlin, Heidelberg: Springer, 1999.

[20] H. Whitney, On the abstract properties of linear dependence. *American Journal of Mathematics* **57**, 1935, 509–533.

[21] R. J. Worsley, An improved Bonferroni inequality. *Biometrika* **69**, 1982, 297–302.

[22] U. Zimmermann, *Linear and Combinatorial Optimization in Ordered Algebraic Structures*, Annals of Discrete Mathematics Vol. 10, Amsterdam: North-Holland, 1981.

Mathematical Models for Polymer Crystallization Processes

Vincenzo Capasso

MIRIAM - Milan Research Centre for Industrial and Applied Mathematics,
Università degli Studi di Milano, via Saldini 50, I-20133 Milano, Italy
e-mail: vincenzo.capasso@mat.unimi.it

Contents

1   Introduction

Polymer industry raises a large amount of relevant mathematical problems with respect to the *quality* of manufactured polymer parts. These include in particular questions about the crystallization kinetics of the polymer melt, in presence of a temperature field.

The *final morphology* of the crystallised material is a fundamental factor in the physical properties of the solidified part. Also the long term behaviour of such properties (dimensional stability, physical ageing, ...) is strongly influenced by the microstructure of the crystallized material. Furthermore polymers are used in surgical application as glue for the implant of femoral prostheses. In this case the polymer is crystallized in situ and it is very important to obtain a complete crystallization since the uncrystallized material is toxic for human beings.

Figure 1: A schematic representation of a spherulite and an impingement phenomenon

Crystallization is a mechanism of phase change in polymeric materials. If an experiment is started with a liquid (the polymer melt) and the temperature is decreased subsequently below a certain point (the melting point of the material), crystals appear *randomly in space and time* and start to grow. Growth processes may be very complicated, but usually, with polymers we have growth of either spherical crystallites (*spherulites*) or of cylindrical crystallites (so called *shish-kebabs*). In the following the restriction to the case of spherulitic growth is made, which is also a good assumption for the crystallization of relaxed polymer melts. But also in the situation of flowing melts, as occurring in injection moulding, the central core of solidified part shows spherulitic morphologies.

The spherulites grow until they hit another crystal, which abruptly stops

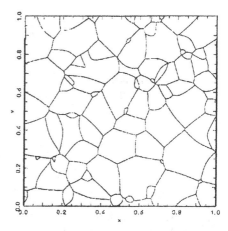

Figure 2: The final Johnson-Mehl tessellation of a crystallization process, both from real and simulated experiments

the growth at the interface (Fig. 1). This phenomenon, called *impingement*, causes the final morphology of the crystallized material; the resulting division of space into cells is called *Johnson-Mehl tessellation.* (Fig. 2). Because of the randomness in time and location of the birth of crystals, the final morphology is random. A mathematical theory is needed to describe the mean theoretical properties of the process (and possibly the *variability* around the mean behaviour too), to predict the final properties of the crystallized material. Tuning of these mathematical models with respect to experimental data may bring polymer industry to optimize the solidification process(even at the large scale industrial level) so to obtain materials with the best mechanical properties.

In order to model the kinetics of the crystallization process in an heterogeneous temperature field we have to introduce the basic mathematical structures which will represent nucleation and growth of crystals. The nucleation process, being random both in time and space, will be modelled as a *marked point process*, that is a marked counting process, the mark being the (random) spatial location of the germ; this process will be the basis of the building of a *dynamical Boolean model*, representing the growth of the crystals in absence of impingement. The dynamical Boolean model, coupled with the concept of causal cone in presence of spatial heterogeneities, provides a generalization of the well known Kolmogorov-Avrami-Evans formula to compute the evolution of the *degree of crystallinity*, that is the mean volume fraction of the space occupied by crystals (a rigorous definition of degree of crystallinity will be provided later).

2   The stochastic modelling of the crystallization process in a given temperature field

In this section we describe the crystallization process in a deterministic given field of temperature, neglecting, at the moment, the influence on temperature of crystallization due to the release of latent heat during the phase change.

Crystallization is a stochastic process in time and space due to the combination of birth (nucleation) and growth of crystals. The first one is modelled as a marked point process, and the whole process by a dynamical Boolean model.

2.1   The nucleation process

The nucleation process is described by a stochastic spatially marked point process (MPP) N on an underlying probability space $(\Omega, \mathcal{A}, P)$, with marks in the physical space $E \in \mathbb{R}^d, d = 1, 2, 3$ (we shall denote by $\mathcal{B}_{\mathbb{R}_+}$, resp. $\mathcal{E}$, the σ-algebra of Borel sets on $\mathbb{R}_+$, resp. on E). The counting process N is a random measure given by

$$N = \sum_{n=1}^{\infty} \epsilon_{T_n, X_n}$$

where

- T_n is an $\mathbb{R}_+$-valued random variable representing the time of birth of the $n-$th nucleus,

- X_n is an E-valued random variable representing the spatial location of the nucleus born at time T_n,

- $\epsilon_{t,x}$ is a measure on $\mathcal{B}_{\mathbb{R}_+} \times \mathcal{E}$ such that for any $t_1 < t_2$ and $B \in \mathcal{E}$,

$$\epsilon_{t,x}([t_1, t_2] \times B) = \left\{ \begin{array}{ll} 1 & \text{if } t \in [t_1, t_2], x \in B, \\ 0 & \text{otherwise.} \end{array} \right.$$

In a crystallization process with nucleation events

$$\{(T_n, X_n) \,|\, 0 \le T_1 \le T_2 \le \ldots\}$$

the crystalline phase at time $t > 0$ is given by a random set

$$\Theta^t = \bigcup_{T_j \le t} \Theta_j^t,$$

that is the union of all crystals born at time T_j at location X_j and freely grown up to time t. Note that impigement is obscured in this representation.

The integer valued random variable $N([0,t] \times B)$ counts the (random) number of nucleation events occurred in the time-space region $[0,t] \times B \in \mathcal{B}_{\mathbb{R}_+} \times \mathcal{E}$. We introduce the following

Definition 1. *The* **crystallinity** *at a point $x \in E$ and time $t > 0$, $\xi(t,x)$ is the probability that, at time t, x is covered (or "captured") by some crystal, i.e.*

$$\xi(t,x) := P(x \in \Theta^t) = E[I_{\Theta^t}(x)].$$

It is possible to show [3] that, under general conditions, a compensator ν of N exists, such that, $\forall B \in \mathcal{E}, t > 0$

$$N([0,t] \times B) = \nu([0,t] \times B) + \mathcal{M}(t,B)$$

where $\forall B \in \mathcal{E}$, $\{\mathcal{M}(t,B)\}_{t \in \mathbb{R}_+}$ is a zero mean martingale. The (random) measure $\nu(dt \times dx)$ is known as the "stochastic intensity" of the process. It provides the probability that a new nucleation event occurs during the infinitesimal time interval $[t, t+dt[$, in the infinitesimal volume dx, given the "history" of the process before time t.

For the MPP that models the birth process of crystal nuclei, we assume that the random measure ν is given by

$$\nu(dt \times dx) = \alpha(t,x)(1 - I_{\Theta^{t-}}(x))dt \, dx. \tag{2.1}$$

where α, known as "nucleation rate", is a real-valued measurable function on $\mathbb{R}_+ \times E$ such that $\forall t \in \mathbb{R}_+ : \alpha(t,\cdot) \in \mathcal{L}^1(E)$.

The (deterministic) measure defined by the expected values

$$\Lambda([0,t] \times B) = E[N([0,t] \times B)],$$

for $t \geq 0, B \in \mathcal{E}$, is known as the "intensity measure" of N.

The following holds

$$
\begin{aligned}
\Lambda(dt \times dx) &= E[N(dt \times dx)] \\
&= E[\nu(dt \times dx)] \\
&= E[\alpha(t,x)(1 - I_{\Theta^{t-}}(x))dt \, dx] \\
&= \alpha(t,x)E[(1 - I_{\Theta^{t-}}(x))]dt \, dx \\
&= \alpha(t,x)(1 - \xi(t,x))dt \, dx.
\end{aligned}
$$

If we define $\tilde{\nu}(dt) := \nu([t, t+dt[\times E)$, it is always possible [17] to factorize the measure ν in the following way

$$\nu(dt \times dx) = k(t,dx)\tilde{\nu}(dt) \tag{2.2}$$

where the kernel $k(t, \cdot)$ denotes, for any $t \in \mathbb{R}_+$, a probability measure on $(E, \mathcal{E})$ such that

$$k(t, E \setminus \Theta^t) = \int_{E \setminus \Theta^t} k(t, dx) = 1. \tag{2.3}$$

By comparing Equations (2.2) and (2.1) and imposing Condition (2.3), we get

$$\tilde{\nu}(dt) \;=\; \int_E \nu(dt \times dx) = \left(\int_E \alpha(t,y)(1 - I_{\Theta^{t-}}(y))dy \right) dt$$

$$=\; \left(\int_{\overline{\Theta^{t-}}} \alpha(t,y)dy \right) dt \tag{2.4}$$

$$k(t,dx) \;=\; \frac{\alpha(t,x)(1 - I_{\Theta^{t-}}(x))}{\displaystyle\int_{\overline{\Theta^{t-}}} \alpha(t,y)dy} dx \tag{2.5}$$

where $\overline{\Theta^{t-}}$ denotes the complement of Θ^{t-}.

Remark 1. From Eq.(2.4) and (2.5) we can give a physical meaning to the disintegration of the compensator.

- $\tilde{\nu}(dt)$ is the compensator of the (univariate) counting process

$$\tilde{N}(t)\colon\; = N([0,t] \times E), t \geq 0.$$

$\tilde{N}$ is called *underlying counting process* associated with N and at each instant t it counts the number of nuclei born in the whole region E up to time t. The process $\tilde{\nu}$ decreases to zero as far as all the material is crystallized

$$\lim_{\Theta^t \to E} \tilde{\nu}(dt) = 0.$$

- $k(t,dx)$ gives the spatial probability distribution of a new nucleus born during $[t, t+dt[$. Note that with the increase of the crystallized volume, the available space for new nuclei is reduced.

The average number of spherulites born up to time t per unit volume at point $x \in E$ is

$$E(N(t,x)) = \int_0^t \alpha(s,x)(1 - \xi(s,t))ds.$$

2.2   Growth of crystals

Experimental results and numerical simulations show that the crystallization process can be represented by a dynamical Boolean model. It has been introduced in [7, 8, 9] to model isothermal crystallization processes, where nucleation and growth rates are time but not space dependent (spatially homogeneous).

The dynamical Boolean model is a dynamical version of the classical definition of Boolean model [4, 18]. We give here the definition of an "heterogeneous version" of the dynamical Boolean model, that may represent a crystallization process with nucleation rate $\alpha(t, x)$ and growth rate $G(t, x)$.

The set representing the crystalline phase Θ^t, at any time $t \in \mathbb{R}_+$, is completely characterized by its hitting functional [18], defined as

$$T_{\Theta^t}(K) := P(\Theta^t \cap K \neq \emptyset),$$

for every K compact subset of E.

In particular for $K = \{x\}$, $x \in E$, and $t \in \mathbb{R}_+$, we have the crystallinity

$$T_{\Theta^t}(\{x\}) \;\; = \;\; P(x \in \Theta^t)$$

$$= \;\; E[I_{\Theta^t}(x)] \equiv \xi(t, x).$$

The classical Kolmogorov-Avrami-Evans theory for isothermal crystallization heavily used the fact that crystals $\Theta_s^t(x)$ are of spherical shape if the growth rate is constant; the same is true whenever the growth rate depends upon time only. In the case of non-homogeneous growth, i.e., if the growth rate G depends on both space and time, the shape of a polymeric crystal (in absence of impingement) is no longer a ball centered at the origin of growth. In the case of a growth rate with constant gradient it has been verified that the growing crystal is the union of facets (the *growth lines*) which lead to growth in minimal time (cf.[21]). This principle can be adapted for the case of arbitrary growth rates (cf.[23]).

Assumption 1 *Minimal-time Principle.*[4] The curve along which a crystal grows, from its origin to any other point, is such that the needed time is minimal.

The minimal-time principle is obviously satisfied for homogeneous growth, since there the growth lines are just straight lines. The growth of a crystal in $\mathbb{R}^2$ between its origin (x_0, y_0) and another point (x_1, y_1) due to Assumption 1 may be formulated as follows:

$$t_1 = \min_{(x,y,\phi)}$$

subject to

$$\dot{x}(t) = G(x(t), y(t), t) \cos \phi(t), \qquad t \in (t_0, t_1)$$

$$\dot{y}(t) = G(x(t), y(t), t) \sin \phi(t), \qquad t \in (t_0, t_1)$$

$$x(t_0) = x_0, \; y(t_0) = y_0$$

$$x(t_1) = x_1, \; y(t_1) = y_1$$

The necessary first order conditions for this control problem (cf.e.g.[15]) lead to the following equation for the control variable ϕ:

$$\dot{\phi} = \langle \nabla G(x, y, t), (-\sin\phi, \cos\phi)^T \rangle$$

For the growth of a crystal in $\mathbb{R}^3$ we obtain the control problem

$$t_1 = \min_{(x, y, \phi, \theta)}$$

subject to

$$\dot{x}(t) = G(x(t), y(t), z(t), t) \cos\phi(t) \cos\theta(t)$$

$$\dot{y}(t) = G(x(t), y(t), z(t), t) \sin\phi(t) \cos\theta(t)$$

$$\dot{z}(t) = G(x(t), y(t), z(t), t) \sin\theta(t)$$

$$x(t_0) = x_0, \ y(t_0) = y_0, \ z(t_0) = z_0$$

$$x(t_1) = x_1, \ y(t_1) = y_1, \ z(t_1) = z_1$$

which leads to the necessary conditions

$$\dot{\phi}(t) = \langle \nabla G(x(t), y(t), z(t), t), (-\sin\phi(t), \cos\phi(t), 0)^T \rangle$$

$$\dot{\theta}(t) = \langle \nabla G(x(t), y(t), z(t), t),$$

$$(\quad -\cos\phi(t)\sin\theta(t), -\sin\phi(t)\sin\theta(t), \cos\theta(t))^T \rangle$$

By eliminating the angles we may deduce a second order ODE for the growth lines given by

$$\frac{d}{dt}\left(\frac{\dot{\mathbf{x}}}{G(\mathbf{x}, t)}\right) = -\nabla G(\mathbf{x}, t) + \left\langle \nabla G(\mathbf{x}, t), \frac{\dot{\mathbf{x}}}{G(\mathbf{x}, t)} \right\rangle \frac{\dot{\mathbf{x}}}{G(\mathbf{x}, t)}, \qquad (2.6)$$

where $\mathbf{x}$ denotes the vector $(x, y)^T$ in $\mathbb{R}^2$ and $(x, y, z)^T$ in $\mathbb{R}^3$, respectively. The crystal at time t is now given as the union of all growth lines, i.e.

$$\Theta_0^t = \{\mathbf{x}(\tau) \,|\, \mathbf{x} \text{ solves } (2.6), \mathbf{x}(t_0) = \mathbf{x}_0, \tau \in (t_0, t)\}.$$

Each growth line is determined uniquely by the origin of growth $\mathbf{x}(t_0) = \mathbf{x}_0$ and the initial speed of growth, which may be written as

$$\dot{\mathbf{x}}(t_0) = G(\mathbf{x}_0, t_0)\nu_0,$$

where ν_0 is an arbitrary vector in $\mathbb{R}^d$ with $||\nu_0|| = 1$. Thus, we may introduce a parametrization for the crystal based on the initial direction, namely, in $\mathbb{R}^2$

$$\Theta_0^t = \{\mathbf{x}(\tau, \gamma) \,|\, \tau \in (t_0, t), \gamma \in [0, 2\pi)\}, \tag{2.7}$$

where $\mathbf{x}(\tau, \gamma)$ denotes the solution of (2.6) with initial values

$$\begin{aligned}
\mathbf{x}(t_0) &= \mathbf{x}_0, \\
\dot{\mathbf{x}}(t_0) &= G(\mathbf{x}_0, t_0)(\cos \gamma, \sin \gamma)^T,
\end{aligned}$$

and in $\mathbb{R}^3$

$$\Theta_0^t = \{\mathbf{x}(\tau, \gamma_1, \gamma_2) \,|\, \tau \in (t_0, t), \gamma_1 \in [0, 2\pi), \gamma_2 \in [-\frac{\pi}{2}, \frac{\pi}{2}]\}, \tag{2.8}$$

where $\mathbf{x}(\tau, \gamma_1, \gamma_2)$ denotes the solution of (2.6) with initial values

$$\begin{aligned}
\mathbf{x}(t_0) &= \mathbf{x}_0, \\
\dot{\mathbf{x}}(t_0) &= G(\mathbf{x}_0, t_0)(\cos \gamma_1 \cos \gamma_2, \sin \gamma_1 \cos \gamma_2, \sin \gamma_2)^T.
\end{aligned}$$

Equation (2.6) yields a description of the crystal growth based on growth lines, which are computed independently. The parameterizations introduced in (2.7) or (2.8) also provide another view upon the growing crystal. For the sake of simplicity we concentrate on the case of a crystal in $\mathbb{R}^2$, but similar reasoning is possible in higher dimensions. By fixing the time t we obtain the set

$$\partial\Theta_0^t = \{\mathbf{x}(t, \gamma) \,|\, \gamma \in [0, 2\pi)\}, \tag{2.9}$$

which is called the *growth front*.

In particular (2.6) implies that at any point $\mathbf{x}(t, \gamma)$ of the growth front at time t (initial direction γ) we have

$$\begin{aligned}
\dot{\mathbf{x}}(t, \gamma) &= G(\mathbf{x}(t, \gamma), t)\nu(t, \gamma), & (2.10) \\
\dot{\nu}(t, \gamma) &= -\nabla G(\mathbf{x}(t, \gamma), t) + \langle\nabla G(\mathbf{x}(t, \gamma), t), \nu(t, \gamma)\rangle\nu(t, \gamma), & (2.11)
\end{aligned}$$

which clearly shows that the growth is determined by the actual normal direction of the growth front as well as by the growth rate and its gradient. The initial values are given by

$$\begin{aligned}
\mathbf{x}(t_0, \gamma) &= \mathbf{x}_0, & (2.12) \\
\nu(t_0, \gamma) &= (\cos \gamma, \sin \gamma)^T. & (2.13)
\end{aligned}$$

In $\mathbb{R}^3$ the initial conditions are given by (2.12) and by

$$\nu(t_0, \gamma) = (\cos \gamma_1 \cos \gamma_2, \sin \gamma_1 \cos \gamma_2, \sin \gamma_2)^T. \tag{2.14}$$

As opposed to the original minimal-time principle, the derivated system (2.10), (2.11) needs only information about the shape of the crystal at the

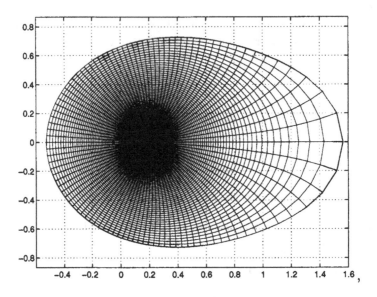

Figure 3: A crystal shape in a typical temperature field.

actual time, but not about the history of growth. Hence, this description seems to be suitable not only for the case of growth in a given field, but also for growth in interaction with the field.

Fig. 3 shows a simulation of non-homogeneous crystal growth we used a typical parabolic temperature profile (i.e., the solution of the heat equation without source term) and data for the growth rate obtained by measurements of i-PP. The result, presented in Fig. 3, shows the growth front in the first time steps. The deviation from the spherical shape obviously increases with time, nevertheless the crystals still remain convex and do not produce exotic shapes.

2.3 The causal cone

From the definition of crystallinity and of Boolean model, we have

$$\xi(t, x) = P(x \in \Theta^t)$$

$$= P(x \in \bigcup_{s_n \leq t} \Theta^t_{s_n}(x_n))$$

$$= P(\exists (s_n, x_n) \in [0, t] \times E \mid x \in \Theta^t_{s_n}(x_n)). \qquad (2.15)$$

Eq.(2.15) justifies the following definition

Definition 2. *The **causal cone** $A(t,x)$ of a point x at time t is the set of the couples (s,y) such that a crystal born in y at time s covers the point x by time t*

$$A(t,x) := \{(s,y) \in [0,t] \times E \mid x \in \Theta_s^t(y)\}$$

where we have denoted by $\Theta_s^t(y)$ the crystal born at $y \in E$ at time $s \in \mathbb{R}_+$, and observed at time $t \geq s$.

We denote by $C_s(t,x)$ the section of the causal cone at time $s < t$,

$$
\begin{aligned}
C_s(t,x) &:= \{y \in E \mid (s,y) \in A(t,x)\} \\
&= \{y \in E \mid x \in \Theta_s^t(y)\}.
\end{aligned}
$$

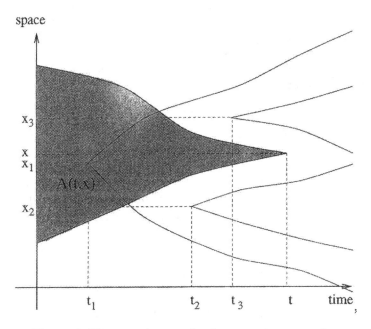

Figure 4: The causal cone of point x at times s and t.

Let us suppose that the growth rate of a crystal is depending only upon the point (t,x) that we are considering and not, for example, upon the age of the crystal. Then the probability $\xi(t,x)$ that the point x is covered at time t may be substituted by the probability that no nucleation occurs inside the causal cone [4, 8]

$$P(x \notin \Theta^t) = P(\text{no nucleation occurs in } A(t,x)). \qquad (2.16)$$

In order to verify (2.16), we have to show that any nucleation in the causal cone leads to coverage independently of the amount of crystalline phase present at that time. More precisely, we have to show that even if

$(t_1, x_1) \in A(t, x)$ is already covered by another crystal, the point x will be covered at time t.

This can be done by comparing two crystals - the one already existing before (born at (t_0, x_0) denoted by Θ_0^t) and the (virtual) new-born crystal Θ_1^t. We show that the later born will always stay 'inside' the other one. Defining the *cone of influence* by

$$\mathcal{I}(t_0, x_0) := \{(t, x) \mid x \in \Theta_0^t\}, \qquad (2.17)$$

we can express this statement mathematically in the following lemma [4]

Lemma 1. *If* $(t_1, x_1) \in \mathcal{I}(t_0, x_0)$, *then* $\mathcal{I}(t_1, x_1) \subset \mathcal{I}(t_0, x_0)$.

From Lemma 1, there follows that the event "no nucleation takes place into the causal cone $A(t, x)$" has the same probability as the event

$$F_{t,x}^0 = \{\Theta^s \cap C_s(t, x) = \emptyset, \ s \le t\} \in \mathcal{F}_t.$$

Conditional upon $F_{t,x}^0$, the expression (2.1) for the compensator in the causal cone reduces to

$$\nu(ds \times dy \mid F_{t,x}^0) = \alpha(s, y)ds \, dy, \qquad (s, y) \in A(t, x) \qquad (2.18)$$

so that we have a deterministic compensator (the "free space" nucleation rate).

It is possible to prove [16] that a Poisson process is characterized by a deterministic compensator. Hence, in this case, in the causal cone, we have a space-time Poisson Process. In particular the number of nuclei that fall into the causal cone $N(A(t, x))$ is a Poisson random variable with intensity

$$\Lambda_0(A(t, x)) = \nu(A(t, x) \mid F_{t,x}^0). \qquad (2.19)$$

So, since

$$1 - \xi(t, x) = P(N(A(t, x)) = 0)$$

we have that

$$\xi(t, x) = 1 - e^{-\Lambda_0(A(t,x))}.$$

Considering the factorization of the compensator (2.2) and Eqs.(2.18) and (2.19), we may write

$$
\begin{aligned}
\Lambda_0(A(t, x)) &= \int_0^t \int_{C(s;t,x)} \alpha(s, y)ds \, dy \\
&= \int_0^t \tilde{\nu}(ds) \int_{C(s;t,x)} k_0(s, dy) \\
&= \tilde{\nu}([0, t]) \int_0^t \frac{\tilde{\nu}(ds)}{\tilde{\nu}([0, t])} k_0(s, C(s; t, x)) \\
&= \tilde{\nu}([0, t]) E_S[k_0(S, C(S; t, x))] \qquad (2.20)
\end{aligned}
$$

where $\tilde{\nu}([0, t])$, defined in (2.5) is the mean number of nuclei born in the whole space E up to time t; $k_0(s, dy) = k(s, dy \mid F^0_{t,x})$ (cf. (2.5) is the distribution of the nuclei at time s, knowing that no point is born in the causal cone $A(t, x)$ and S is a random variable with distribution

$$\frac{\tilde{\nu}(ds)}{\tilde{\nu}([0, t])} I_{[0,t]}(s).$$

If we denote by $a(t, x) = E_S[k_0(S, C(S; t, x))]$,

$$\xi(t, x) = 1 - e^{-\tilde{\nu}([0,t])a(t,x)} \tag{2.21}$$

may be seen as an extension of the Kolmogorov-Avrami-Evans formula to the heterogeneous case.

3 Interaction with latent heat

The analysis is much more complicated if we wish to include in the modelling of crystallization processes the interaction with temperature due to latent heat. This implies randomness in the temperature field due to the intrinsic randomness of the nucleation process.

Indeed nucleation and growth rates are temperature dependent (cf. [22] for an experimental estimation of the functional dependence of the growth rate from the temperature). This causes problems with the definition of the causal cone and consequently with the derivation of the degree of crystallization. As a first approach to circumvent these problems we shall focus on the fact that in a typical industrial situation we face a multiple-scale phenomenon. Under these circumstances averaging is possible by applying laws of large numbers.

3.1 A random evolution equation

In an experimental situation, where spatial heterogeneities are caused only by the heat transfer in the material, we may assume that growth and nucleation rates depend upon space and time via suitable functions of the temperature only [13]

$$G(t, x) = \tilde{G}(T(t, x)),$$

$$\alpha(t, x) = \frac{\partial}{\partial t}\tilde{N}(T(t, x)) + \alpha_1(t, x).$$

Usually it is assumed that the two terms forming the nucleation rate satisfy the following condition:

$$\alpha(x,t) = \frac{\partial \tilde{N}}{\partial T} \cdot \frac{\partial T}{\partial t} + \alpha_1$$

$$= \alpha_0(t,x)\frac{\partial T}{\partial t} + \alpha_1(t,x).$$

In typical industrial situations the first term largely dominates; nucleations are numerous when the sample is being cooled (that is when $\frac{\partial T}{\partial t} \neq 0$), even though in various materials some rare nucleations may still be observed when the cooling is stopped (i.e. when $\frac{\partial T}{\partial t} = 0$) at a temperature between the melting and the glass transition point.

It is also often assumed that $\tilde{G}$ and $\tilde{N}$ are exponential functions of the temperature [1, 13]

$$\tilde{G}(T) = G_{ref} \exp(-\beta_G(T - T_{ref})), \tag{3.22}$$

$$\tilde{N}(T) = N_{ref} \exp(-\beta_N(T - T_{ref})), \tag{3.23}$$

where G_{ref}, N_{ref}, T_{ref}, β_G and β_N are constants depending on the material.

Viceversa, the growing of the crystalline phase influences the heat transfer process in the material because of the release of latent heat,

$$(\rho c T)_t = \nabla.(\kappa \nabla T) + (h I_{\Theta^t})_t, \qquad \text{in } \mathbb{R}^+ \times E, \tag{3.24}$$

$$T_n = \beta(T - T_{out}), \qquad \text{on } \mathbb{R}^+ \times \partial E. \tag{3.25}$$

Here $(I_{\Theta^t})_t$ denotes the time derivative of the indicator function I_{Θ^t} of the crystalline phase at time t. Equation (3.24) has to be understood in a weak sense. ρ denotes the density, c the heat capacity, κ the heat conductivity, β the heat transfer coefficient and h the latent heat due to the phase change.

The parameters in the heat equation may depend upon the phase, i.e., if ρ_1, c_1, κ_1 and β_1 denote the parameters of the crystallized material and ρ_2, c_2, κ_2 and β_2 the ones of the non-crystallized, we may write

$$\rho = I_{\Theta^t}\rho_1 + (1 - I_{\Theta^t})\rho_2,$$
$$c = I_{\Theta^t}c_1 + (1 - I_{\Theta^t})c_2,$$
$$\kappa = I_{\Theta^t}\kappa_1 + (1 - I_{\Theta^t})\kappa_2,$$
$$\beta = I_{\Theta^t}\beta_1 + (1 - I_{\Theta^t})\beta_2.$$

This heat transfer model is a random differential equation, since all parameters depend upon the random variable I_{Θ^t}.

A direct consequence is the stochasticity of the temperature field, whose evolution influences (via (3.22) and (3.23)) the crystallization process itself.

3.1.1   Stochasticity of the Causal Cone

As we have seen in Sec.2.3, a way of deducing model equations is based on the investigation of the causal cone $A(t, x)$, i.e., the set of all points y and times s, such that nucleation at y at time s would lead to coverage of x at time t. Lemma 1 can be applied to any situation, where the growth rate $G(t, x)$ is a given field.

In the case of non-isothermal crystallization, this is a very delicate point, since the growing crystals influence temperature and thus nucleation and growth rates, so that, in the case of a substantial non-isothermal situation, they depend upon temperature, and via (3.24), (3.25), the temperature depends upon the crystalline phase. In a situation like that, the causal cone itself cannot be defined independently of the process anymore, because, for the reasons described above, it always depends on the specific history of the crystallization process.

Even the numerical simulation of the crystallization process coupled, via (3.22) and (3.23), with (3.24), (3.25) raises relevant problems of numerical instabilities. A typical approximation in the literature is the so-called hybrid model. In the heat equation (3.24), the indicator function of the crystalline phase is replaced with its mean value, the degree of crystallinity

$$I_{\Theta^t}(x) \longrightarrow \xi(t, x).$$

In this way we obtain a modified heat equation, that gives a deterministic temperature field. Coupling this equation with the extension of the Kolmogorov-Avrami-Evans equation (2.21), solves our problem.

Our approach tends to make this kind of approximation more rigorous by a multiple scales arugument.

3.2   The multiple scales

Under typical industrial conditions, the following assumptions can be made

- the nucleation rate and consequently the number of crystals is very large;

- the growth rate and consequently the size of crystals are very small;

- the typical scale for diffusion of temperature (macroscale) is larger than the typical crystal size (microscale).

If t_0 is a typical time scale for the process, the typical scale of the heat transfer problem is given by

$$x_T = \sqrt{\frac{k_0 t_0}{\rho_0 c_0}},$$

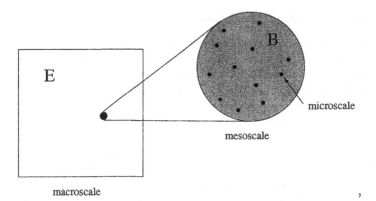

Figure 5: Schematic rapresentation of the scales in polymer crystallization.

where t_0 is the length of the considered time interval, k_0, ρ_0 and c_0 are typical scales for k, ρ and c.

The typical scale for the growth of a nucleus is given by

$$x_G = G_0 t_0,$$

where G_0 is a typical value for the growth rate G. In applications it turns out that $x_T >> x_G$, which is due to the fact that crystal growth is very slow, whereas the heat conduction is rather fast. This means [4] that there exist two significant scales in the problem, i.e.

- microscale x_G for growth.

- macroscale x_T for heat conduction.

It is a direct consequence that if one is interested only in local microscopic effects, the temperature variation can be neglected, whereas for a pure macroscopic description the growth effects are not important.

A "mesoscale" may be introduced, sufficiently small with respect to the macroscale of heat conduction so that temperature at that scale may be considered approximately constant, but large enough with respect to the typical scale of the size of individual crystals so that it contains a large number of them, making a "law of large number" applicable.

In the subsequent sections we will derive a model on such a mesoscale.

3.3 Averaging

Consider the indicator function

$$f(t, x) \quad = \quad I_{\Theta^t}(x) \qquad t \in \mathbb{R}^+, x \in E, \tag{3.26}$$

so that the crystallinity degree is given by

$$\xi(t,x) = E[f(t,x)], \qquad t \in \mathbb{R}^+, x \in E. \tag{3.27}$$

If we denote by f^j the indicator function of the single crystal Θ_j^t, we may write

$$f(t,x) \;=\; 1 - \prod_{T_j \le t}(1 - f^j(t,x)), \qquad t \in \mathbb{R}^+, x \in E, \tag{3.28}$$

having denoted by T_j the (random) time of birth of the j-th crystal.

Besides the volume distribution f^j of a crystal, we may also define the surface density u^j and the surface oriented density v^j, which are generalized functions, via

$$\langle u^j, \phi\rangle \;=\; \int_{\partial\Theta_j^t}\phi(x)\,d\sigma(x), \qquad \forall\,\phi \in C_0^\infty(E,\mathbb{R}) \tag{3.29}$$

$$\langle v^j, \phi\rangle \;=\; -\int_{\partial\Theta_j^t}\phi(x)\nu(x)\,d\sigma(x), \qquad \forall\,\phi \in C_0^\infty(E,\mathbb{R}). \tag{3.30}$$

Due to the evolution equations in Section 2.2, we have, for $t > T_j$

$$f_t^j \;=\; (1 - f^j)Gu^j,$$

$$\nabla f^j \;=\; (1 - f^j)v^j,$$

$$v^j \;=\; \nabla(Gu^j). \tag{3.31}$$

Eqns (3.31) may be extended to any time $t > 0$, by including the birth of the j-th crystal via the time-derivative of u^j:

$$u_t^j = \nabla \cdot (Gv^j) + S_j^d, \quad t > 0, \tag{3.32}$$

where

$$S_j := \frac{\partial}{\partial t}\langle u^j, \psi\rangle\big|_{t=T_j}, \quad \text{for } \psi \in C_0^\infty(\mathbb{R}_+ \times E, \mathbb{R}),$$

describes the event of birth of a new crystal at (T_j, X_j), so that it is a stochastic quantity.

It can be shown that for

$$\begin{aligned}
\langle S_j^1, \psi\rangle &= 2\psi_t(T_j, X_j) \quad \text{for } d = 1,\\
\langle S_j^2, \psi\rangle &= 2\pi G(T_j, X_j)\psi(T_j, X_j) \quad \text{for } d = 2,\\
\langle S_j^3, \psi\rangle &= 8\pi T_j G(X_j, T_j)^2\psi(X_j, T_j) \quad \text{for } d = 3, \text{ etc.}
\end{aligned}$$

We get the corresponding global quantities as

$$u := \sum_{T_j < t} u^j, \quad v := \sum_{T_j < t} v^j, \tag{3.33}$$

together (3.28). The corresponding evolution equations coupled with temperature will be [6]

$$(\rho c T)_t = \nabla(k\nabla T) + (h\ f)_t, \tag{3.34}$$

$$f_t = \tilde{G}(T)(1 - f)u, \tag{3.35}$$

$$u_t = \nabla(\tilde{G}(T)v) + \sum_{T_j = t} S_j, \tag{3.36}$$

$$v_t = \nabla(\tilde{G}(T)u), \tag{3.37}$$

in $\mathbb{R}_+ \times E$. If we assume that the whole crystallization process starts at time $t = 0$ the initial conditions are given by

$$T = T^0, \qquad\qquad f = 0,$$
$$u = 0, \qquad\qquad v = 0,$$

in $\{0\} \times E$ and, because of the cooling from outside, the boundary condition

$$T_\nu = \gamma_0(T - T_{out}), \tag{3.38}$$

must be satisfied on $\mathbb{R}_+ \times \partial E$, where γ_0 denotes the heat transfer coefficient and T_{out} the temperature of the cooling material.

Note that the only source of stochasticity in (3.34)-(3.37) is explicitily given by the random measures S_j in 3.36.

Under the multiple scales assumptions given above, if we take a sufficiently large volume C, then the number of crystals produced in C in the time interval $(t, t + \Delta t)$ can be approximated by its expected value,i.e.(by neglecting overlapping of these "small" crystals),

$$N([t, t + \Delta t[\times C) =\simeq \int_t^{t+\Delta t} dt \int_C dx \alpha(t, x). \tag{3.39}$$

/ If (3.39) is valid, on a typical volume for this mesoscale, we may apply a law of large numbers to approximate the random source by its expected value,

$$\int_C dx \left(\sum_{T_j = t} S_j \right) \simeq \int_C dx \left[\sum_{T_j = t} S_j \right].$$

More precisely we get, for $d = 1$

$$\int_C dx \left(\sum_{T_j = t} S_j^1 \right) \simeq \int_C 2\alpha(t, x)dx$$

$$= \int_C 2\frac{\partial}{\partial t}\tilde{N}(T(t, x))dx, \tag{3.40}$$

and for $d = 2$

$$\int_C dx \left(\sum_{T_j = t} S_j^2 \right) \simeq \int_C 2\pi \int_C G(t, x) \int_0^t \alpha(\tau, x) d\tau dx$$

$$= \int_C 2\pi dx \tilde{G}(T(t, x)) \left[\tilde{N}(T(t, x)) - \tilde{N}(T(0, x)) \right]. \qquad (3.41)$$

By introducing the above approximations we obtain an averaged model

$$T_{\bar{t}} = \text{div }_{\bar{x}}(D\nabla_{\bar{x}}T) + \frac{h}{c}\xi_{\bar{t}}, \qquad (3.42)$$

$$\xi_{\bar{t}} = (1 - \xi)\tilde{G}(T)\bar{u}, \qquad (3.43)$$

$$\bar{u}_{\bar{t}} = \nabla_{\bar{x}}(\tilde{G}(T)\bar{v}) + 2\pi\tilde{G}(T)\tilde{N}(T), \qquad (3.44)$$

$$\bar{v}_{\bar{t}} = \nabla_{\bar{x}}(\tilde{G}(T)\bar{u}), \qquad (3.45)$$

in $(0, t_*) \times E$, supplied by the boundary conditions on $(0, t_*) \times \partial E$,

$$\frac{\partial T}{\partial \bar{\nu}} = \gamma_0(T - T_{out}),$$

$$\bar{v} + \bar{w}^T \nu = 0,$$

and the initial conditions in $\{0\} \times E$

$$T = T^0, \qquad\qquad\qquad \xi = 0, \qquad (3.46)$$

$$\bar{u} = 0, \qquad\qquad\qquad \bar{v} = 0. \qquad (3.47)$$

Here T denotes the temperature, ξ the degree of crystallinity and $\bar{v}$, $\bar{w}$ represent the mean free surface distributions of the crystals if they would grow freely. The boundary condition (3.46) holds only if it is assumed that no nucleation event occurs on the boundary of E. The capital $(T, \tilde{G}, \tilde{N})$ and overlined characters $(\bar{v}, \bar{w}, \bar{t}, \bar{x})$ represent dimensional quantities, only the degree of crystallinity ξ is dimensionless.

Relevant inverse problems for the identification of the nucleation rate have been analyzed in [5]; see also [14].

3.3.1 Numerical simulations

For the numerical simulation of the crystallization one has to solve the coupled system (3.42)-(3.47), consisting of nonlinear hyperbolic and parabolic equations. In our numerical simulations we considered crystallization in a rectangular domain, whose length is twice the width ($\Omega = (0, L) \times (0, 2L)$). We performed simulations with two different choices for the temperature of the cooling material, using a uniform temperature $T_{out} = 45^0$ C (Figs 6 and 7 - Example1) and a temperature T_{out} which is lower on two consecutive boundary segments ($T_{out} = 10^0$C) and higher on the opposite ones

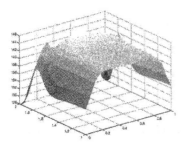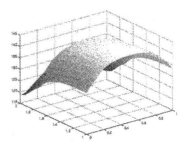

Figure 6: Temperature in Example 1 after 5 and 10 minutes. Because of the symmetry, only the upper part of the rectangle $((0, L) \times (L, 2L))$ is plotted.

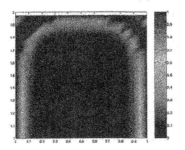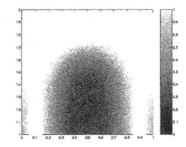

Figure 7: Degree of crystallinity in Example 1 after 5 and 15 minutes.

$(T_{out} = 80^0 \text{C})$ ((Figs 8 and 9 - Example 2). For all material parameters we used measurements for isotactic polypropylene [2].

The heat transfer coefficient and the outer temperature were chosen such that the assumptions about nucleation and growth are satisfied, but cooling is still slow compared to industrial conditions.

The results (which where performed without boundary nucleation) clearly show the boundary layer in ξ (Figures 7 and 9), but there is no such effect in the temperature, which is due to the fact that cooling at the boundary is much stronger than the reheating effect caused by the latent heat. A comparison of the left and right hand side in Figure 6 shows that the reheating effect in the interior is more significant, the temperature is not necessarily monotone decreasing in time there.

In Example 2 the temperature after 6 and 9 minutes is shown in Figure 8, the evolution of the degree of crystallinity (after 6 and 9minutes) is plotted in Figure 9. Here the crystallization process seems to behave almost like a two-phase problem, the crystalline phase propagates from the corner with lowest temperature to the interior. This corresponds very well to earlier modelling

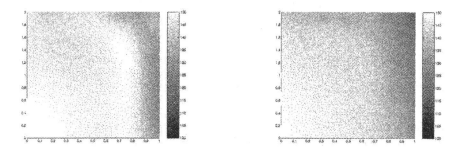

Figure 8: Temperature in Example 2.

approaches, approximating crystallization by a two-phase process [19].

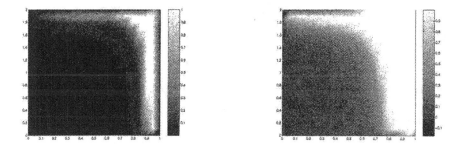

Figure 9: Degree of crystallinity in Example 2.

3.4 A particle model

A mathematically rigorous approach for averaging is possible via an interacting particles model as the one proposed in [11].

By assuming that nucleation occurs much faster than growth, we have modelled crystals as points in $\mathbb{R}^d$ ($d = 1, 2, 3$) with a growing (crystallized) mass at that point. In this way an explicit geometric growth is lost a priori; smoothing of the pointwise mass distribution will anyway recover this aspect. This model allows to take into account perfection, or secondary crystallization, that had been neglected in previous models.

A scale parameter $M \in \mathbb{N}$ is introduced rapresenting the order of magnitude of the total number of crystal nuclei in the system at a sufficiently large time. The behavior of the system is analyzed for $M \to \infty$. Given the underlying counting process $\{\tilde{N}(t), t \in \mathbb{R}_+\}$, defined as in Sect. 2 the system is described in terms of

- $X_M^j \in \mathbb{R}^d$, $j = 1, \cdots, \tilde{N}(t)$, (random) positions of crystals at time $t \geq 0$,

- $Y_M^j(t) \in \mathbb{R}_+$, $j = 1, \cdots, \tilde{N}(t)$, (random) masses of the crystals at time $t \geq 0$.

The macroscopic description of the process is given in terms of the random empirical distribution of crystals $\mathbb{X}_M(t)$ and the random empirical mass distribution of crystals $\mathbb{Y}_M(t)$

$$\mathbb{X}_M(t) := \frac{1}{M} \sum_{j=1}^{\tilde{N}(t)} \epsilon_{X_M^j}, \quad t \geq 0, \tag{3.48}$$

$$\mathbb{Y}_M(t) := \frac{1}{M} \sum_{j=1}^{\tilde{N}(t)} Y_M^j(t) \epsilon_{X_M^j}, \quad t \geq 0, \tag{3.49}$$

where ϵ_x is the localizing measure at x: $\forall B \in B_{\mathbb{R}^d}$,

$$\epsilon_x(B) = \begin{cases} 1, & \text{if } x \in B, \\ 0, & \text{otherwise.} \end{cases}$$

Given a spatial distribution of crystals up to some time t a new crystal nucleates during the time interval $[t, t + dt[$ in a volume dx centered at $x \in E$ at a rate

$$\alpha_M(t, x) = M\kappa((\mathbb{Y}_M(t) * \phi_M)(x)) b_b(T_M(t, x)). \tag{3.50}$$

Here in addition to the usual dependence upon temperature the nucleation rate is assumed to depend on the mass distribution of crystallized material in a suitable neighborhood of the relevant point $x \in E$. Such dependence is introduced to describe possible speed up and/ or saturation effects, for suitable choices of the function k. In this way we may recover a completely analogous modelling of the intensity of the birth process as in (2.1).

The convolution kernel ϕ_M is chosen as a scaled probability density on E

$$\phi_M(y) = M^\delta \phi(M^{\delta/d} y), \quad 0 < \delta < 1, \tag{3.51}$$

given a probability density ϕ on E; observe that by (3.51)

$$\lim_{M \to \infty} \phi_M = \delta_0. \tag{3.52}$$

Because of the scaling, the range of the neighborhood that will influence the nucleation rate at $x \in E$ will decrease with M, and the strength of this dependence will increase. The choice $\delta \in (0, 1)$ is made so that the birth rate at

x is influenced by existing crystals standing within a range which is macroscopically small, but microscopically large (coeherently with our mesoscale assumption). The range of interaction of each crystal becomes smaller as M tends to infinity, but will still be large enough to contain a number of crystals sufficient to make a law of large numbers applicable (moderate limit) [20].

As far as the growth of crystallized mass is concerned, suppose the j-th nucleus was born at time $T_M^j > 0$, at location $X_M^j = x \in E$, with initial mass $Y_M^j(T_M^j) = v_0 > 0$. We assume that for $t > T_M^j$ the process $Y_M^j(t)$ jumps from $v > 0$ to $v + 1/\lambda_M$ with a rate (per unit time)

$$G_M(t, x, v) = \lambda_M \beta((\mathbb{Y}_M(t) * \phi_M)(x), v) b_g(T_M(t, x)), \qquad (3.53)$$

where $\lambda_M >> M$ is again a scale parameter.

Besides the usual dependence upon temperature, here we may assume ,via the function β, a dependence upon the crystallized mass distribution (via the convolution kernel ϕ_M) in a neighborhood of x, and the crystallized mass v at x. The first one may describe saturation effects, while the second one may take into account that larger crystals may grow faster (surface reaction).

As before the evolution equation of the temperature field will be coupled to the birth and growth process via the production of latent heat.

$$\frac{\partial}{\partial t} T_M(t, x) = \sigma \Delta T_M(t, x) + a_g \, b_g(T_M(t, x))$$

$$\times \quad \frac{1}{M} \sum_{k=1}^{\tilde{N}(t)} \tilde{\beta}((\mathbb{Y}_M(t) * \phi_M)(X_M^k)) \phi_M(x - X_M^k) \tilde{g}(Y_M^k(t)) \quad (3.54)$$

As a technical simplification here we have assumed a spatially homogeneous heat diffusion coefficient σ.

3.4.1 Evolution equations for the empirical distributions

Under typical regularity assumptions on the rates α_M and G_M, the evolution equations for $\{X_M(t), t \in \mathbb{R}_+\}$ and $\{Y_M(t), t \in \mathbb{R}_+\}$ are the following ones; for any $f \in C_b^\infty(\mathbb{R}^d) \cap L^2(\mathbb{R}^d)$,

$$\int_{\mathbb{R}^d} f(x)\mathbb{X}_M(t)(dx) = \int_{\mathbb{R}^d} f(x)\mathbb{X}_M(0)(dx)$$

$$+ \int_0^t ds \int_{\mathbb{R}^d} dx f(x)\kappa((Y_M(s) * \phi_M)(x))b_b(T_M(x,s))$$

$$+ \mathcal{M}_{1,M}(f,t), \tag{3.55}$$

$$\int_{\mathbb{R}^d} f(x)\mathbb{Y}_M(t)(dx) = \int_{\mathbb{R}^d} f(x)\mathbb{Y}_M(0)(dx)$$

$$+ \int_0^t ds \left[\frac{1}{M}\langle \mathbb{Y}_N(t), \tilde{\beta}(\mathbb{Y}_M(s) * \psi_N)b_g(T_M(t,\cdot))f \rangle \right.$$

$$\left. + \int_{\mathbb{R}^d} v_0 f(x)\kappa((\mathbb{Y}_M(s) * \phi_M)(x))b_b(T_M(s,x))dx \right]$$

$$+ \mathcal{M}_{2,M}(f,t), \tag{3.56}$$

where $\mathcal{M}_{i,M}(f,t), i = 1,2$ are suitable zero mean martingales.

Both martingales vanish in probability for M tending to infinity. This is the substantial reason for the asymptotically deterministic dynamics of the whole system for a sufficiently large number of spherulites per unit volume (which may also mean a sufficiently large time).

In the limit we obtain [11] a deterministic system for the mass density y coupled with the temperature field T

$$\begin{cases} \frac{\partial}{\partial t}y(x,t) = \tilde{\beta}(y(x,t))b_g(T(x,t))y(x,t) + v_0\kappa(y(x,t))b_b(T(x,t)) \\ \frac{\partial}{\partial t}T(x,t) = \sigma\Delta T(x,t) + a_g\tilde{\beta}(y(x,t))b_g(T(x,t))y(x,t). \end{cases} \tag{3.57}$$

3.4.2 Simulations

We calculated the numerical solutions of system (3.57) with an initial condition linear in space and time-constant boundary conditions.

We performed also stochastic simulations of the many-particle system so to compare qualitatively the results.

The PDE system

System (56) is solved in a domain $E = [0,1] \times [0,1] \subset \mathbb{R}^2$, by finite difference methods, with initial condition

$$\begin{aligned}
\rho(0, x_1, x_2) &= 0, \\
y(0, x_1, x_2) &= 0, \qquad\qquad\qquad\qquad x \in E \qquad (3.58)\\
T(0, x_1, x_2) &= T_{\max} - x_1(T_{\max} - T_{\min}),
\end{aligned}$$

and boundary condition for the temperature constant in time

$$T(t, x_1, x_2) = T_{\max} - x_1(T_{\max} - T_{\min}), \qquad t > 0, \ (x_1, x_2) \in \partial E. \quad (3.59)$$

The functions of the temperature are chosen in the following way:

$$\begin{aligned}
b_b(T) &= N_{\text{ref}} \exp(-\beta_N(T - T_{\text{ref}})), \\
b_g(T) &= G_{\text{ref}} \exp(-\beta_G(T - T_{\text{ref}})),
\end{aligned}$$

and , since we have chosen κ and $\tilde{\beta}$ as saturating functions of y ,

$$\kappa(y) = \begin{cases} y^2 - 2y + 1 & 0 \leq y \leq 1 \\ 0 & y > 1 \end{cases}$$

$$\tilde{\beta}(y) = \begin{cases} 1 - y & 0 \leq y \leq 1 \\ 0 & y > 1 \end{cases}$$

The results are shown in the following pictures. Crystallization starts in the colder side of the sample and from there diffuses in all the available space (Fig.(10)-(11)). The crystallization rate becomes smaller getting close to saturation: we may see, in Fig.(12)-(13), that its effect on temperature, via enthalpy, is quite strong at the beginning of the process and becomes almost negligible when the process is close to the end.

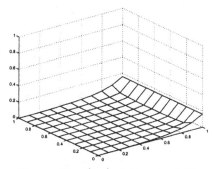

Figure 10: $y(t, x)$ at time $t = 100$

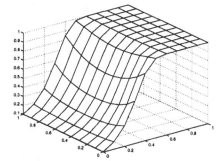

Figure 11: $y(t, x)$ at time $t = 10000$.

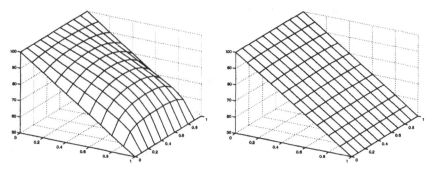

Figure 12: $T(t,x)$ at time $t = 100$ **Figure 13:** $T(t,x)$ at time $t = 10000$.

In Fig.(14) and (15) we have the profile respectively of temperature and crystallization in the middle of the sample for $t =$100, 200, 300, 400, 500, 1000, 2000, 3000, 4000, 5000, 10000, 20000, 30000 sec.

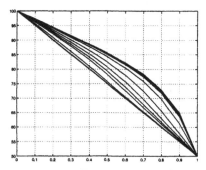 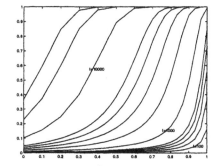

Figure 14: Temperature profile.   **Figure 15:** Crystallization profile.

The SDE system

We simulated the crystallization process by a many-particle system of size $M = 10^4$, the boundary-initial condition being the same as in (3.58) and (3.59). With 10000 nuclei we are able to simulate only the initial part of the process, but we may see that, according with the solution of the corresponding PDEs system, the crystallization starts in the coolest part of the sample and from here it expands in the remaining available space.

In Fig.16, we have the position of the nuclei. We note that the right part of the window, corresponding to the coolest temperature, is almost filled with the spots, while they become rarer going to the hotter part.

The temperature is shown in Fig.17. We may see the effect of latent heat of crystallization, that causes a re-heating in the middle of the sample (in absence of the latent heat, the temperature would be a sloping plane).

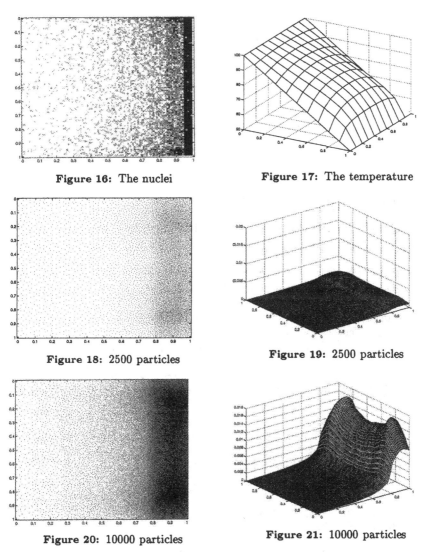

Figure 16: The nuclei

Figure 17: The temperature

Figure 18: 2500 particles

Figure 19: 2500 particles

Figure 20: 10000 particles

Figure 21: 10000 particles

Fig.18-21, show the density $h_M = \mathbb{Y}_M * W_M$ of the process $\mathbb{Y}_M$ represented in a 2-D and in a 3-D plot, for 2500, 5000, 7500 and 10000 particles.

Acknowledgements.

It is a pleasure to acknowledge relevant and stimulating discussions with all members of the ECMI Special Interest Group on "Polymers" (www.mat.-unimi.it/ ~ miriam). Particular thanks are due to Dr. S. Mazzullo of the "G.Natta" Research Centre of MONTELL-Italia and to my students and collaborators M.Burger (Linz), A.Micheletti, D.Morale and C. Salani (MIRIAM-Milan Research Centre for Industrial and Applied Mathematics). Financial support from the Italian Space Agency (ASI) and the EU-TMR Programme DEIC is gratefully acknowledged.

References

[1] G. Alfonso, Polimeri cristallini, in *Fondamenti di Scienza dei Polimeri*, (M. Guaita et al. Eds.), Pacini, Pisa, 1998.

[2] G. Alfonso, Private Commmunications, 1999.

[3] P.Brémaud, *Point Processes and Queues, Martingale Dynamics*, Springer-Verlag, New York, 1981.

[4] M. Burger, V. Capasso, C. Salani, Modelling multidimensional crystallization of polymers in interaction with heat fransfer. (1999). Submitted.

[5] M. Burger, V. Capasso, H.W. Engl, Inverse problems related to crystallization of polymers, *Inverse Problems*, 15 (1999) 155-173

[6] M. Burger, V. Capasso, Mathematical modelling and simulation of non-isothermal crystallization of polymers. *Technical Report 1, Industrial Mathematics Institute, J.Kepler Universität, Linz, 2000.*

[7] V.Capasso, A.Micheletti, Stochastic geometry of birth-and-growth processes, *Quaderno n. 20/1997*, Dip. di Matematica, Università di Milano, 1997.

[8] V.Capasso, A.Micheletti, Birth-and-Growth stochastic processes modelling polymer crystallization *Technical Report 14, Industrial Mathematics Institute, J.Kepler Universität, Linz, 1997.*

[9] V.Capasso, A.Micheletti, M.De Giosa, R.Mininni, Stochastic modelling and statistics of polymer crystallization processes, *Surv. Math. Ind.*, 6, (1996), pp.109-132.

[10] V.Capasso, C.Salani, Stochastic birth and growth processes modelling crystallization of polymers with spatially heterogeneous parameters, *Quad.Dip.Matematica* 6 (1999).To appear on *Journal of Nonlinear Analysis*.

[11] V.Capasso, D. Morale, K.Oelschläger, C.Salani, An interacting particle approach for the crystallization of polymers, To appear on Proceedings " La Matematica nelle Scienze della Vita e nelle Applicazioni" (M. Fabrizio and C. Vettori, Eds.) Bologna, Academy od Sciences, 1999.

[12] N.A.C. Cressie, *Statistics for Spatial Data*, Wiley, New York, 1993.

[13] G.Eder, H.Janeschitz-Kriegl, Structure development during processing: crystallization, in *Materials Science and Technology* , Vol.18 (H.Meijer, Ed.), Verlag Chemie, Weinheim, 1997.

[14] H. W. Engl, Inverse Problems and Their Regularization, This Volume.

[15] R.V. Gamkrelidze, *Principles of Optimal Control Theory* Plenum Press, N.Y., London, 1976.

[16] B.G.Ivanoff, E.Merzbach, A martingale characterization of the set-indexed Poisson Process, *Stochastics and Stochastics Report* **51** (1994), 69-82.

[17] G.Last, A.Brandt, *Marked Point Processes on the Real Line. The Dynamic Approach*, Springer, New York, 1995.

[18] G.Matheron, *Random Sets and Integral Geometry*, Wiley, New York, 1975.

[19] S.Mazzullo, M.Paolini, C.Verdi, *Polymer crystallization and processing: free boundary problems and their numerical approximation*, Math. Engineering in Industry **2** (1989), 219-232.

[20] K.Oelschläger, *Many-Particle Systems and the continuum Description of their Dynamics* Habilitationsschrift, Faculty of Mathematics, University of Heidelberg, Germany, 1989.

[21] G.E.W.Schulze, T.R.Naujeck, A growing 2D spherulite and calculus of variations, *Colloid & Polymer Science* **269** (1991), 689-703.

[22] C.Salani, Crystallization of polymers with thermal heterogeneities, ECMI Thesis, Linz (1997).

[23] J.E.Taylor, J.W.Cahn, C.A.Handwerker, Geometric models of crystal growth, *Acta metall. mater.* **40** (1992), 1443-1475.

Differential Equations in Technology and Medicine: Computational Concepts, Adaptive Algorithms, and Virtual Labs.

P. Deuflhard

Konrad-Zuse-Zentrum Berlin (ZIB) and Free University Berlin
URL: http://www.zib.de/deuflhard

Contents

Introduction

This series of lectures deals with a variety of challenging real life problems selected from clinical cancer therapy, communication technology, polymer production, and pharmaceutical drug design. All of these problems from rather diverse application areas share two common features: (a) they have been modelled by various *differential equations* – elliptic, parabolic, or Schrödinger–type partial differential equations, countable ordinary differential equations, or Hamiltonian systems, (b) their numerical solution has turned out to be a real challenge to computational mathematics.

Therefore, *before* diving into actual computation, the *computational concepts* to be applied need to be carefully considered. To start with, any numerical analyst must be prepared to totally remodel problems coming from science or engineering – see e.g. Sections 3 and 4 below. The computational problems to be treated should be well–posed, important features of any underlying continuous model should be passed on, if at all possible, to the discrete model, and the computational resources employed (computing time, storage, graphics) should be adequate.

Speaking in mathematical terms, the solutions to be approximated live in appropriate infinite dimensional function spaces, e.g. in Sobolev spaces in Sections 1 and 2, in discrete weighted sequence spaces in Section 3, or in certain statistically weighted function spaces in Section 4. The *mathematical paradigm* advocated throughout this paper is that – already due to mathematical aesthetics – any *infinite* dimensional space should *not* be represented by just a *single finite* dimensional space (with possibly large dimension), but by a well–designed *sequence of finite dimensional spaces*, which successively exploit the asymptotic properties characterizing the original function space. The fascinating, but (for a mathematician) not really surprising experience is that *mathematical aesthetics go directly with computational efficiency*. In other words, a careful and sufficiently ingenious realization of the above paradigm will lead to efficient algorithms that actually work in hard real life problems. The reason for this coincidence of aesthetics and efficiency lies in the fact that function spaces describe some *data redundancy* that can be exploited in the numerical solution process. In order to do so, *adaptivity* of algorithms is one of the leading construction principles. Typically, wherever adaptivity has been successfully realized, algorithms with a computational complexity close to the (unavoidable) complexity of the problem emerge – a feature of extreme importance especially in challenging problems of science, technology, or medicine.

In traditional industrial environments, however, new efficient mathematical algorithms are not automatically accepted – even if they significantly supercede already existing older ones (if not old–fashioned ones) within long established commercial software systems. Exceptions do occur where simulation or optimization is the dominating competition factor. Due to this experience the author's group has put a lot of effort in the design of *virtual labs.* These specialized integrated software systems permit a fast and convenient switch between numerical code and interactive visualization tools (that we also develop, but do not touch here). Sometimes only such virtual labs open the door for new mathematical ideas in hospitals or industrial labs.

1 Partial Differential Equations in Cancer Therapy Planning

The present section deals with partial differential equation (PDE) models arising in medicine (example: cancer therapy hyperthermia) and high frequency electrical engineering (example: radio wave absorption). In this type of application the 3D geometry – say, of human patients – motivates the choice of *tetrahedral* finite element methods (FEM). The clinical setting requires the robust computational solution of problems to prescribed accuracy at highest possible speed on local workstations. Reliability plays the dominant role in medicine, which is a nice parallelism with the intentions of mathematics. Numerical speed is required to permit a fast simulation of different scenarios for different patients. In other words: the situation both requires and deserves the construction of highly efficient algorithms, numerical software, and visualization tools.

1.1 Multilevel Finite Element Methods Revisited

The presentation of this section focusses on elliptic or parabolic PDEs and Maxwell's equations. Mathematically speaking, the stationary solutions of these PDEs live in some Sobolev space like H^α or H_{curl} depending on the prescribed boundary conditions, whereas the time dependent solutions live in some scale of these spaces. In view of the above mentioned *paradigm* and the expected *computational complexity*, these spaces are approximated by a sequence of finite element spaces in the frame of multigrid (MG) or multilevel (ML) methods. Before going into the technical details of the real life problems to be presented below, a roadmap of several advanced computational concepts will be given first that have turned out to be important not only for the herein selected applications.

Optimal multigrid complexity. Classical MG methods have been first advocated for actual computation in the 70's by A. BRANDT [26] and HACKBUSCH [52]. The latter author has paved the success path for MG methods by first proving an optimal computational complexity estimate $\mathcal{O}(N)$ for the so–called W–cycle, where N was understood to be the number of nodes in a *uniform* space grid. The same attractive feature was observed in suitable implementations of the simpler V–cycle and later proved by BRAESS AND HACKBUSCH [23] under certain regularity assumptions and for uniform grids.

The subsequent development was then characterized by a successive extension of MG methods from the originally only elliptic problems to larger and larger problem classes.

Adaptive multilevel methods. In quite a number of industrially relevant problems rather localized phenomena occur. In this case, uniform grids are by no means optimal, which, in turn, also means that the classical MG methods on uniform grids could not be regarded as optimal. For this reason, multigrid methods on *adaptive* grids have been developed quite early, probably first by R. BANK [8] in his code PLTMG in the context of problems arising from semiconductor device modelling where sharp local boundary layers arise naturally. Later adaptive MG implementations are the code family KASKADE [14] by the author's group and the code family UG [11] by WITTUM, BASTIAN, and co-workers. UG in particular pays careful attention to parallelization issues [12]. The proof of an optimal computational complexity estimate $\mathcal{O}(N_{ad})$, where N_{ad} is now understood to be the often much smaller adaptive number of nodes, turned out to need more sophisticated proof techniques; for the well-known V–cycle, this challenging task has been performed by BRAMBLE, PASCIAK, WANG, AND XU [25] – see also XU [90]. His rather elegant theoretical tools came from the interpretation of MG methods as abstract *multiplicative* Schwarz methods (equivalent to abstract Gauss–Seidel methods) based on an underlying multilevel splitting in function space.

Hierarchical bases finite element methods. Independent of the classical MG methods, a novel multilevel method based on conjugate gradient iteration with some *hierarchical basis* (HB) *preconditioning* had been suggested in the mid 80's for elliptic PDEs by YSERENTANT [91]. From the scratch, this new type of algorithm turned out to be competitive with classical MG in terms of computational speed. An *adaptive* 2D version of the new method had been designed and implemented in the late 80's by DEUFLHARD, LEINEN, AND YSERENTANT [39] in the code KASKADE. On top of that first realization, a more mature version including also 3D has been worked out by BORNEMANN, ERDMANN, AND KORNHUBER [16]. The present version of KASKADE [14] contains the original HB–preconditioner for 2D and the more recent BPX–preconditioner due to XU [89, 24] for 3D. For an account of its performance see Section 1.2.

Additive versus multiplicative multigrid methods. After the theoretical milestone paper by XU [90], the hierarchical basis type methods are now interpreted as abstract *additive* Schwarz methods (equivalent to abstract Jacobi methods) also based on a multilevel decomposition in function space. By construction additive Schwarz methods provide some preconditioning. In this interpretation, which the author prefers to adopt, the classical multigrid methods are then called *multiplicative MG methods*, whereas the HB–

or BPX–preconditioned CG methods are called *additive MG methods*. In particular, the BPX–preconditioning and the V–cycle are just the additive and multiplicative counterparts. From theoretical analysis, multiplicative MG methods might require less iterations – which, however, need not imply less computing time (see e.g. the Maxwell MG solvers in Section 1.2 below, Table 1.1). Moreover, if an additive MG involves only one iteration on the finest grid, then multiplicative MG methods cannot gain too much. In the subsequently described elliptic problems, the bulk of computing time is anyway spent in the evaluation of the stiffness matrix elements and the right hand side elements. Summarizing, the question of whether additive or multiplicative MG methods should be preferred, appears to be less important than other conceptual issues – see below. For the orientation of the reader: UG is strictly multiplicative, PLTMG is dominantly multiplicative with some additive options, KASKADE is dominantly additive with some multiplicative code e.g. for eigenvalue problems and the harmonic Maxwell's equations. A common software platform of UG and KASKADE is in preparation.

Cascadic multigrid methods. These rather recent MG methods can be understood as some confluence of additive and multiplicative MG methods. From the additive point of view, cascadic multigrid (CMG) methods are characterized by the simplest possible preconditioner: either no or just a diagonal preconditioner is applied; as a distinguishing feature, coarser levels are visited more often than finer levels – to serve as preconditioning substitutes. From the multiplicative side, CMG methods may be understood as MG methods with an increased number of smoothing iterations on coarser levels, but without any coarse grid corrections. As a first algorithm of this type, a *cascadic conjugate gradient method* (CCG) had been proposed by the author in [30]. The general CMG class with arbitrary smoothers beyond CG has been presented by BORNEMANN AND DEUFLHARD [20]. In their paper they analyzed CMG in terms of convergence and computational complexity in an adaptive setting – based on first much more restrictive convergence results due to SHAIDUROV [82]. These CMG methods exhibit good convergence properties only in H^1, but not in L^2 – unlike additive (with appropriate preconditioning) or multiplicative MG methods. Therefore, though being certainly easiest to implement among all MG methods, CMG methods – the youngest members of the MG family – are still in the process of maturing. Just to avoid mixing terms: CMG is different from the code KASKADE, which predominantly realizes additive MG methods.

Local error estimators. Any efficient implementation of *adaptive* MG methods (additive, multiplicative, cascadic) must be based on cheap *local error estimators* or, at least, *local error indicators*. In the best case, these are derived from theoretical *a–posteriori error estimates*. These estimates will be local only, if local (right hand side) perturbations in the given problem remain local – i.e. if the Greens' function of the PDE problem exhibits local behavior. As a consequence of this elementary insight, *adaptive* MG methods will be essentially applicable to linear or nonlinear elliptic or parabolic problems. As for a comparative assessment of the different available local error estimators, there is a beautiful paper by BORNEMANN, ERDMANN, AND KORNHUBER [17] that gives a unified theoretical framework for most of the popular 2D and 3D error estimators. For orientation: PLTMG uses the triangle oriented estimator of BANK AND WEISER [10], KASKADE the edge oriented estimator of DEUFLHARD, LEINEN, YSERENTANT, and UG the estimator of BABUSKA AND MILLER [4].

Adaptive grid refinement. Within adaptive ML methods simplicial grids play a dominant role, since they behave nicely in local refinement processes. In connection with any selected error estimator, the local extrapolation method due to BABUSKA AND RHEINBOLDT [5] can be applied to determine some threshold value, above which a geometrical element (tetrahedron, triangle, edge) is marked for local refinement. Once this marking has been done, well–designed strategies need to be applied to produce a complete FE grid on the next refinement level. The art of refinement is quite established in 2D (see the "red" and "green" refinements due to BANK ET AL. [9] or the "blue" refinement due to KORNHUBER AND ROITZSCH [62]). In 3D there is still work left to be done, even though successful strategies due to RIVARA [73], ONG [71], or BEY [15] have been around for quite a time.

Multilevel methods for nonlinear elliptic problems. For nonlinear elliptic problems there are two basic lines of MG methods: (I) the *nonlinear MG method*, sometimes also called *full approximation scheme* (FAS), wherein nonlinear residuals are evaluated within MG cycles, and (II) the *Newton MG method*, wherein *linear residuals* are evaluated within the MG method for the solution of the linear systems for the Newton corrections. In [42, 44] DEUFLHARD AND WEISER proposed an adaptive version for the second MG approach based on an *affine conjugate* characterization of nonlinearity via the special Lipschitz condition

$$\|F'(x)^{-1/2}\big(F'(y) - F'(x)\big)(y - x)\| \ \leq \ \omega\|F'(x)^{1/2}(y - x)\|. \qquad (1.1)$$

This type of condition enters into certain affine invariant convergence results for both local and global *inexact* Newton methods in the function spaces $W^{p,q}$. The associated code Newton–KASKADE realizes a theoretically backed optimal balance between outer Newton iterations with possible adaptive damping, multilevel discretization, and inner preconditioned CG iterations; its performance is exemplified in Section 1.2 below.

Method of lines for parabolic PDEs. For time dependent PDEs, the most popular approach is still the so–called *method of lines* (MOL), which

realizes a *first space / then time* discretization. After space discretization
a typically large block–structured system of ordinary differential equations
(ODEs) arises, which is then solved by any stiff ODE integrator: in the
simplest (but often inefficient) case by an implicit or backward Euler with
constant timestep, in advanced versions by some implicit multistep code (like
DASSL), some implicit Runge–Kutta code (like RADAU 5), or some linearly
implicit extrapolation code (like LIMEX) with *adaptive* control of timestep
and possibly time discretization order. However, if one aims at dynamically
adapted non–uniform space grids in 2D or 3D with MG methods to be applied, which is the typical case in *parabolic* PDEs, then the MOL approach
will lead into some mass.

Adaptive Rothe method for linear parabolic PDEs. Starting 91,
BORNEMANN [18, 19] suggested to abandon the MOL for parabolic PDEs
and to use the so–called ROTHE method instead, which realizes a *first time /
then space* discretization. His first papers dealt with initial boundary value
problems for linear scalar parabolic PDEs such as

$$u_t = -\Delta u + f(x),\ u(x,0) = \varphi(x),\ u(x,t)\,|_{\partial\Omega} = \psi(t),\ t \geq 0,\ x \in \Omega \subset \mathbb{R}^d\ .$$
(1.2)

Upon incorporating the boundary conditions into a linear elliptic operator
$\mathcal{A}$ some function U is defined by virtue of the *abstract Cauchy problem*

$$U'(t) = \mathcal{A}U + F,\ U(0) = U_0\ .$$
(1.3)

Note that U represents a spatial function living in some *scale of Hilbert
spaces* $H^\alpha(\Omega)$. The above abstract ordinary differential equation (ODE) may
now be formally discretized for time step τ by some stiff integration scheme.
For simplicity, we choose the implicit Euler method, which generates an equation of the type

$$(I - \tau\mathcal{A})\Delta U = \tau F\ .$$
(1.4)

This equation represents some (τ–dependent) elliptic boundary value problem, which can be solved by any adaptive multilevel method. Moreover, the
available advanced ODE technology may also enter, but now in function space
– which means that any error control devices known from finite dimensional
ODEs are realized via spatial approximations using an adaptive MLFEM.
Summarizing, a substantial advantage of this reversed order of discretization
turns out to be that dynamic space grid adaptation and adaptive MG methods within each time layer are, in principle, easy to apply. Moreover, this
approach nicely reflects the underlying theoretical structure.

Adaptive Rothe method for nonlinear parabolic PDEs. The above
algorithmic approach can be extended to the nonlinear parabolic case. A

rather direct extension is obtained on the basis of some abstract stiffness theory presented by the author in [31]. In this paper stiff time discretization of a nonlinear ODE initial value problem, say

$$U' = F(U), \quad U(0) = U_0 \tag{1.5}$$

has been interpreted as a *simplified Newton iteration* for the evolution problem *in function space*. This Newton iteration, in turn, may be formally understood as a Picard iteration for the slightly rewritten ODE

$$U' - \mathcal{A}U = F(U) - \mathcal{A}U, \quad U(0) = U_0 \tag{1.6}$$

wherein $\mathcal{A} \approx F'(U_0)$ – i.e. in finite dimension $\mathcal{A}$ is just an approximate Jacobian (n, n)–matrix of the right hand side. From this theoretical insight *linearly implicit* stiff integration methods appear naturally – as opposed to nonlinear stiff discretization schemes like BDF or implicit RK methods. The concept directly carries over to infinite dimension when equations (1.5) and (1.6) are any abstract Cauchy problem. Upon applying, for simplicity, the linearly implicit Euler discretization to (1.5), we arrive at some linear boundary value problem of the kind

$$(I - \tau\mathcal{A})\Delta U = \tau\big(F(U) - \mathcal{A}U\big) . \tag{1.7}$$

Following this line, LANG [63, 64] developed the adaptive multilevel code **KARDOS** that realizes a linearly implicit (embedded) Runge–Kutta method of low order on each discretization level. In its present form, this most recent code from the **KASKADE** family is applicable to 3D nonlinear systems of reaction–diffusion equations with mild convection. Generally speaking, since the adaptive Rothe method is fully adaptive in both time and space, it is able to resolve extreme *multiscales* in time and space that often arise in hard real life problems – like e.g. in chemical combustion. An early comparison of the new approach with the more traditional MOL approach (both adaptive 1D implementations) can be found in the survey paper [38]. The Rothe method will play a role in Section 1.2, Section 2, and Section 3.

1.2   Clinical Therapy Planning by Virtual Patients

The so–called *regional hyperthermia* is a rather recent promising cancer therapy based on the *local* heating of tumor tissue to above a threshold value of about 42 °C. At present this therapy is applied in combination with chemotherapy or radiotherapy. The idea is that heated tumor cells are more sensitive to extinction by either rays or drugs. For the medical treatment, the cancer patient is put into an applicator, which essentially consists of a set of 83 (old) or 24 (new) radiofrequency antennas and a water bolus to allow for a low reflection passage of the radio waves into the body – see Fig. 1.1.

The antennas emit radiation at a frequency of about 100 MHz corresponding to a wave length in water of about 30 cm, which – physically speaking –

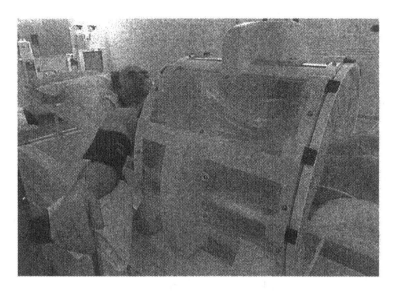

Figure 1.1: Real patient in hospital (Sigma–60 applicator).

means that *wave optics* and interference phenomena rather than ray optics must be modelled. Heat within the body is produced by absorption of the radio waves and distributed by blood circulation in the tumor as well as in sane tissue. Mathematically speaking, the whole system (*patient, water bolus, applicator, surrounding air*) is modelled by the time harmonic Maxwell's equations in inhomogeneous media and a so–called bio–heat transfer (BHT) partial differential equation describing the heat distribution in the body. The task is to tune the set of radiofrequency antennas optimally such that the heat will concentrate within the tumor of a patient, but not at any hot spots elsewhere.

In the project to be reported here, we have been collaborating with internationally renowned oncologists at one of the large Berlin hospitals, the Rudolf–Virchow–Klinikum at the Charité of the Humboldt University. Our task is to support the patient–specific planning of individual therapies. In order to make the method at all useful in a clinical environment, the computational results must be obtained within hours (at most) on a workstation in hospital *to medical reliability*. In addition, any numerical results are to be presented in visual form so that they can be directly interpreted and conveniently handled by medical staff. These requirements made the development of an integrated software package necessary that combines efficient 3D interaction tools with both numerical and computer graphical algorithms. As a prerequisite for the PDE solvers, a rather detailed *virtual patient* needs to be built up from medical imaging input (at present computed tomograms). The system as it stands now is already able to decide about the question whether a given patient can be expected to be successfully treated by hyperthermia using a given applicator. The presentation herein essentially follows the articles [41, 40].

Electric field simulation. We model the antennas by a fixed (angular) frequency ω and the human tissues so that Ohm's law holds. Let the electric field have a representation of the form $Re\,\mathbf{E}(\mathbf{x})e^{i\omega t}$ with a complex amplitude $\mathbf{E}(\mathbf{x})$ defined on a computational domain $\Omega \subset \mathbb{R}^3$. Then the time harmonic Maxwell's equations in terms of the electric field $\mathbf{E}$ and the magnetic field $\boldsymbol{H}$ read

$$\operatorname{curl} \boldsymbol{H} = \ i\,\omega\epsilon\,\boldsymbol{E}\,, \quad \operatorname{curl} \boldsymbol{E} = -i\,\omega\mu\,\boldsymbol{H}\,, \qquad (1.8)$$

where μ is the permeability and $\epsilon = \epsilon' - i\sigma/\omega$ is defined via the generic dielectric constant ϵ' and the conductivity σ. The two equations in (1.8) are combined with the well–known double–curl equation

$$\operatorname{curl}\,(\frac{1}{\mu}\operatorname{curl} \boldsymbol{E}) - \omega^2\epsilon\,\boldsymbol{E} \ = \ 0, \qquad (1.9)$$

which will be the basis for the subsequent FE model. An appropriate function space for the differential operator in (1.9) and Dirichlet boundary conditions on the boundary Γ_D is

$$H_{curl} := \{\mathbf{w} \in (L^2(\Omega))^3\,;\ \mathbf{curl}\,\mathbf{w} \in (L^2(\Omega))^3,\ \mathbf{w}_t = \mathbf{E}_t^0 \text{ on } \Gamma_D\}.$$

The function space $H_{curl;0}$ is used for homogeneous boundary conditions $\mathbf{w}_t = 0$. We are now ready to give a variational formulation for the desired field $\mathbf{E}$ in the form: Determine $\mathbf{E} \in H_{curl}$ such that for all $\mathbf{w} \in H_{curl;0}$

$$\int\limits_{\Omega} \{\frac{1}{\mu}\mathbf{curl}\,\mathbf{E}\,\mathbf{curl}\,\mathbf{w} - \omega^2\epsilon\mathbf{E}\,\mathbf{w}\}\,d\Omega \ - \int\limits_{\Gamma_{ext}} \beta(\mathbf{n} \times \mathbf{E})\,(\mathbf{n} \times \mathbf{w})\,d\Gamma \ = \ 0.$$

$$(1.10)$$

Herein the second integral describes a contribution on the exterior boundary Γ_{ext}. The above bilinear form is coercive for non–vanishing σ, which implies that the problem has a unique solution. Note that the negative part of the integrand plays an important role especially for *high frequency* ω. In the positive semi–definite part, the ample nullspace of the **curl**–operator is rather undesirable, since all standard iterative methods (including MG) are known to preserve nullspace components. This causes a slowing down of convergence, once these nullspace components are present.

For the FE discretization of (1.10) we employ NÉDÉLEC's **curl**–conforming finite elements of lowest order [70] on a tetrahedral triangulation $\mathcal{T}_h$ of the domain – also called Whitney 1–forms or *edge elements*. These elements are easy to refine, which is a necessary prerequisite for any adaptive FEM. They are divergence–free and inherit continuity of the tangential electrical field components from the physical equations so that unwanted spurious discrete solutions [21] are suppressed. Moreover, as pointed out recently by HIPT-MAIR [55], they permit a *discrete Helmholtz decomposition*, which turned out to be crucial for the construction of an adaptive multigrid method with so–called hybrid smoothing [13]. As exemplified in Table 1.1 for Gauss–Seidel

Ref. Depth	Nodes	#Iter			CPU [min]		
		Std	M–Hyb	A–Hyb	Std	M–Hyb	A–Hyb
0	128 365	4250	354	413	150	24	20
1	373 084	4832	265	277	800	76	60
2	1 085 269	> 10000	186	194	> 2000	215	160

Table 1.1: Multilevel solvers: standard smoothing (Std), multiplicative hybrid smoothing (M-Hyb), and additive hybrid smoothing (A–Hyb).

smoothing, both the multiplicative and the additive hybrid MG versions exhibit optimal multigrid complexity: the number of iterations does not increase with increasing refinement levels; this had not been the case for the former standard (Std) versions. Note that in this case the additive compared to the multiplicative version turns out to require slightly more iterations but slightly less computing time.

Heat Transfer Model. Our present model for the dissipation of heat in the human body (cf. PENNES [72]) assumes *potential flow for the blood* within the various tissues including the tumor. This leads to the so–called *bio–heat transfer equation* (BHT)

$$\rho_t c_t \frac{\partial T}{\partial t} = \text{div}(k \text{ grad} T) - W \rho_b \rho_t c_b (T - T_a) + Q \qquad Q = \frac{\sigma}{2} |E|^2 . \quad (1.11)$$

Herein ρ_t, ρ_b denotes the density of tissue and blood, c_t, c_b the specific heat of tissue and blood, T, T_a the temperature of tissue and arterial blood, k the thermal conductivity of tissue, W the blood perfusion, Q the power deposition within the tissue, and σ the electric conductivity. The thermal effects of *strong* blood vessels are excluded in this simplified model – but will be included in a future stage of the project. For $W = W(x)$, this parabolic PDE is *linear*. We solve it by the adaptive multilevel method KASTIO as implemented within the tool box KASKADE. A more realistic model takes into account that blood flow depends on tissue temperature: experiments in [85] have shown that the blood flow in normal tissues, e.g., skin and muscle, increases significantly when heated up to $41 - 43°C$, whereas in the tumor zone the blood flow decreases with temperature. On this experimental basis, we chose $W = W(T)$ monotonically increasing in muscle and fat tissue, but monotonically decreasing in tumor tissue. The arising *nonlinear* PDE has been solved by the code KARDOS [65] for the time dependent case and by the code Newton–KASKADE [43] in the stationary case – both within the package KASKADE.

Initial Grid Generation. The multilevel FEMs just described require an initial coarse grid, which captures the essential geometric features of the stated problem including a subdomain characterization for the different materials (bone, fat, muscle, ...). The total number of elements should be as large as necessary to state the problem correctly, but as small as possible in order to reduce computational costs. Starting point is a stack of plane CT images (about 60 per patient) containing only density information, which first need to be *segmented* according to physiological and oncological knowledge; this is done by the medical staff. The task then is to construct 3D grids from this type of input. It has turned out in the course of the project, that we had to develop our own fast and robust techniques for grid generation. These techniques include: (a) *extraction of compartment interfaces from segmentation results* by a proper generalization of the marching cubes algorithm [66] to non–binary classifications [54]: a significant speed–up is obtained via lookup–tables; (b) *coarsening of compartment surfaces* to allow for initial grids with as few elements as possible [83]; (c) *tetrahedral mesh generation*: each compartment is filled with tetrahedra starting from its surface by using a *3D–advancing front* method [84].

At present, the whole grid generation process can be performed automatically within about 15 minutes CPU time on a UNIX workstation. A typical coarse grid patient model consists of 40,000 – 60,000 tetrahedra and 8,000 – 10,000 vertices.

Optimization Algorithm. In therapy planning, the antenna parameters for each channel $j = 1, ..., k$ (equivalent to k pairs of coupled antennas) must be computed. We parametrize the complex amplitudes z_j by their real amplitudes a_j and their phases θ_j according to $z_j = a_j \exp(-i\theta_j)$. Then parameters $p = \{\Re z_j, \Im z_j\}$ must be determined such that the following therapeutic goals are achieved:

- within the tumor a therapeutic temperature level $T_t \approx 43°$C is maintained,

- regions of healthy tissue are not heated above $T_h \approx 42°$C.

For most patients both requirements cannot be fulfilled simultaneously. In searching for a compromise we avoid destruction of healthy tissue by the additional constraint that temperature in healthy tissue must not exceed certain limits which depend on the tissue type: 42°C for more sensitive tissue compartments (like bladder or intestine) and 44°C otherwise.

From these goals we arrive at the following objective function

$$f(p) = \int\limits_{\substack{x \in V_{tumor} \\ T(x,p) < T_t}} (T_t - T(x,p))^2 \, dx \; + \int\limits_{\substack{x \notin V_{tumor} \\ T(x,p) > T_h}} (T(x,p) - T_h)^2 \, dx \qquad (1.12)$$

to be minimized subject to the constraints

$$T(x,p) \leq T_{lim}(x), \quad x \notin V_{tumor}.$$

In the *linear* heat transfer model, simple superposition of the electric field $\mathbf{E}$ into k modes can be employed, which in $Q \sim |\mathbf{E}|^2$ leads to k^2 basic modes to be computed in advance, plus one further mode for the basal temperature T_{bas}. For the *nonlinear* bioheat transfer model, we constructed some fixed point iteration [40] that converges at an average contraction rate of $\theta \approx 0.3$. This algorithm exploits the fact that the Maxwell solves are considerably more expensive than the **BHT** solves.

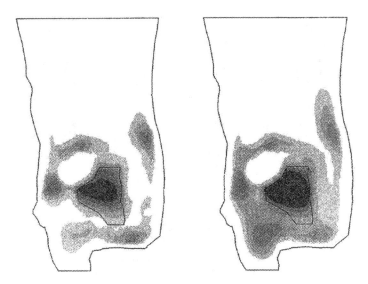

Figure 1.2: Optimized temperature distributions: linear (left) versus nonlinear model (right). Black lines: body outline and tumor contour. Light grey to dark grey: regions from 39°C to 43°C.

The total computational cost for the *nonlinear* case (with n iterations) can be counted to be

$$cost_{total} = k * cost_{Maxwell} + \\ n * (cost_{nlBHT} + (k^2 + 1) * cost_{lBHT} + cost_{Opt}) \qquad (1.13)$$

where the notation is certainly self–explaining. We observed $n \approx 6$. The total cost for the *linear* case is obtained by inserting $cost_{nlBHT} = 0$ and $n = 1$ above.

Upon comparing linear versus nonlinear perfusion models, significant differences show up. As can be seen in Fig. 1.2, the nonlinear model predicts a tumor heating, which from the therapeutic point of view is slightly preferable. The nonlinear model also influences the choice of optimal parameters for the k channels.

Old versus new applicator. Our earlier computations have led to considerable improvements over the old applicator (Sigma–60, $k = 4$ channels, circular cross section with larger water bolus) in the form of some new applicator (Sigma–Eye, $k = 12$ channels, eye shaped cross section with smaller water bolus), see Fig. 1.3.

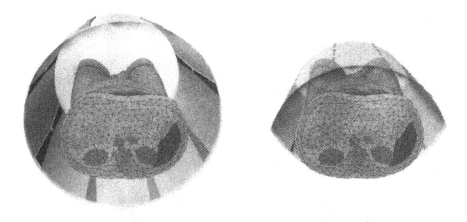

Figure 1.3: Virtual patient in Sigma–60 (left) and Sigma–Eye (right) applicator.

The therapeutic improvement can be seen in Table 1.2, which condenses the information obtained from simulation results for three virtual patients with different tumor locations. In order to illustrate the relative computational weights for the different algorithmic parts, we document some comparative results for both the old and the new applicator in Table 1.3 for the linear heat transfer model. The field computation times per channel of the old Sigma–60 appeared to be ~ 20 minutes as compared to ~ 10 minutes for the new Sigma–Eye, an effect due to the smaller bolus volume. As expected, the temperature computation times roughly scale with k^2.

Virtual Lab. The whole integrated software environment HyperPlan now consists of about 300.000 lines of code, wherein only about 120.000 lines are numerical code, the other parts are segmentation algorithms, grid generation methods, and visualization tools. This virtual lab has been recently sold

Virtual patient	part of tumor volume heated to above 43°C	
	Sigma–60	Sigma–Eye
distal (supraanal) rectal carcinoma	17.5%	62.5
highly presacral rectal carcinoma	0.7%	18.4
cervical carcinoma at pelvic wall	24.8%	49.1

Table 1.2: Therapeutic improvement of new (Sigma–Eye) over old (Sigma–60) hyperthermia applicator.

	Sigma–60 $k = 4$	Sigma–Eye $k = 12$
Segmentation	2 – 4 hours*	
Grid Generation	15 min**	
Field Calculations	80 min**	120 min**
Temperature Calculations	2 min**	20 min**
Optimization	6 sec**	1 min**
	* interactive ** CPU time (SUN UltraSparc)	

Table 1.3: Computation times per patient.

to industry and will be worldwide distributed together with the applicator hardware – increasing the applicator's efficiency significantly.

2 Partial Differential Equations in Optical Chip Design

Every netsurfer now and then tends to have the impression that the abbreviation www means *world wide waiting* rather than *world wide web* – despite the tremendous information propagation speed of modern glass fibres. A drastically better performance rate – by many orders of magnitude! – can be expected by future so–called *optical networks*. In such networks all active components (like microlasers) or passive components (like couplers or tapers) are assembled on integrated optical chips. The technological aim is that signal *processing* on such a chip should reach a speed comparable to that of signal *propagation* along the fibre. In the project to be reported here

the author's group at ZIB has been collaborating with the Heinrich–Hertz–
Institute (HHI) in Berlin and with an industry research lab. As an example,
Fig. 2.1 shows a patented optical chip that has been designed by HHI with
parameters carefully specified on the basis of ZIB simulations. Its schematic
representation is given in Fig. 2.2.

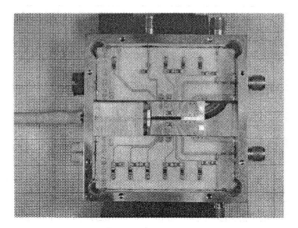

Figure 2.1: Integrated optical chip (central black stripe) mounted on ceramics substrate.

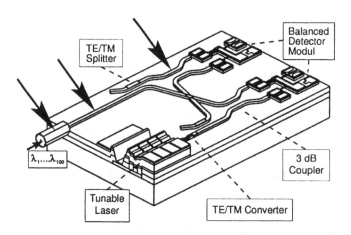

Figure 2.2: Schematic representation of the chip in Fig. 2.1.

The design of integrated optical components is presently based on two
different simplified mathematical models. Their efficient simulation requires
the construction of two types of computational methods, the *beam propagation methods* (initial boundary value problems) and the *guided mode methods*
(Helmholtz eigenproblems in selected cross section planes). For both of them
we have made suggestions to be described now. Typical features coming from

the technological problem are its *geometric complexity*, its *multiscale structure*, and the requirement of quite *stringent error tolerances* to control the behavior of the signals over long distances.

2.1   Beam Propagation Analysis

When modelling the *signal propagation* along a glass fibre, the fibre axis naturally arises as a time–like coordinate z. In order to derive the so–called FRESNEL approximation, the electric field E is written in terms of a slowly varying amplitude u as

$$E = ue^{-in_0k_0z} \quad , \tag{2.14}$$

with k_0 the vacuum wave number and n_0 some *effective refraction index* to be specified below. Assume that we start from Maxwell's equations in the form (1.9). Let u depend on the propagation variable z and, for simplicity, only on one cross section variable x. For ease of writing we redefine $x := k_0x, z := k_0z, g := n^2 - n_0^2, c := 2in_0$. The specification of n_0 comes in by some projection argument to take energy conservation in the FRESNEL approximation at least to some extent into account. This leads to

$$n_0^2 = \frac{(n^2u, u) - (\nabla u, \nabla u)}{(u, u)} \quad . \tag{2.15}$$

We thus end up with the normalized *paraxial wave equation* for the transversal electrical mode (TE) in the form

$$\Delta u + g \cdot u = cu_z \quad , \tag{2.16}$$

which obviously is some complex SCHRÖDINGER–type equation. For pure beam propagation, we may even neglect the term u_{zz} so that Δu here means just u_{xx}. This initial boundary value problem has been solved numerically in several technological projects by F. SCHMIDT [76] using an *adaptive Rothe method*. As already described in Section 1.1 above, this technique offers simultaneous adaptivity in both time and space together with multilevel speed. However, the desirable adaptivity cannot be fully exploited, unless suitable boundary conditions have been constructed, which are to be discussed next.

Discrete transparent boundary conditions. The idea behind the construction of transparent boundary conditions (TB) for wave type equations is to restrict the computations to some region of interest choosing boundary conditions such that waves touching the boundaries just pass these boundaries without any reflections. For some time the canonical approach has been to start from a set of TB derived from the *continuous* model, i.e. from the wave equation itself; these (non–local) boundary conditions were then discretized. However, proceeding like that will often induce discretized reflected waves and even instability [68]. For this reason, we derived a different approach in [78] that we called *discrete* transparent boundary conditions (DTB) – directly based on the Rothe method. Just like in the continuous case, these DTB are also of nonlocal Cauchy type. In addition each linear implicit discretization scheme induces its own DTB.

In order to exemplify the approach, we return to the above PDE (2.16). In the Rothe method the discretization for the time–like variable z, i.e. the direction of propagation, goes first. We deliberately apply the *implicit midpoint rule*, which has the selective feature that it conserves energy also in the discrete case. Consequently, any energy jumps observed in the course of the simulations must originate from the FRESNEL approximation – a convenient and cheap monitor for the validity of the employed model. After z–discretization of (2.16) neighboring time layers $(i, i + 1)$ will be related according to (note: $j = \sqrt{-1}$)

$$\frac{\partial^2 u_{i+1}}{\partial x^2} - \lambda_{i+1}^2 u_{i+1} \;=\; -\frac{\partial^2 u_i}{\partial x^2} + \kappa_{i+1}^2 u_i \tag{2.17}$$

$$\lambda_{i+1}^2(x) \;:=\; \frac{4jn_0(z_i + \tfrac{1}{2}\Delta z_{i+1})}{\Delta z_{i+1}} - g(x, z_i + \tfrac{1}{2}\Delta z_{i+1})$$

$$\kappa_{i+1}^2(x) \;:=\; -\frac{4jn_0(z_i + \tfrac{1}{2}\Delta z_{i+1})}{\Delta z_{i+1}} - g(x, z_i + \tfrac{1}{2}\Delta z_{i+1})$$

$$\sigma_{i+1}^2(x) \;:=\; 2g(x, z_i + \tfrac{1}{2}\Delta z_{i+1}).$$

This relation defines a nested sequence of 1D *boundary value problems* for successive solutions $u_i(x)$ within some finite region of interest; let $x = a$ denote one of the boundaries. Note that we need not restrict the time steps Δz_i to be constant. Following the lines of [78], the above non–local pattern can be taken into account in terms of certain LAPLACE transforms $U_i(p)$, which can be defined via the recurrence relations

$$U_{i+1}(p) = \frac{u_{i+1}(a) - u_i(a)}{p + \lambda_{i+1}} + U_i(p) - \sigma_{i+1}^2 \frac{U_i(p) - U_i(\lambda_{i+1})}{p^2 - \lambda_{i+1}^2} \,. \tag{2.18}$$

Once this can be solved, the boundary values of the solution u_{i+1} at $x = a$ are defined by

$$\left.\frac{\partial(u_{i+1} - u_i)}{\partial x}\right|_{x=a} + \lambda_{i+1}(u_{i+1}(a) - u_i(a)) = \sigma_{i+1}^2 U_i(\lambda_{i+1}). \tag{2.19}$$

Note that the new boundary conditions at time layer $i + 1$ require the old boundary conditions from time layer i and the term $U_i(\lambda_{i+1})$ so that the recurrence (2.18) should be evaluated at $p = \lambda_{i+2}$. Hence, whenever $\lambda_{i+1} = \lambda_{i+2}$ – typically when locally constant stepsizes $\Delta z_{i+1} = \Delta z_{i+2}$ and homogeneous materials occur – then both the denominator and the numerator in the third right hand term of (2.16) vanish: so some limit needs to be taken. Numerical trouble will already arise for nearly zeroes. For this reason, the above recurrence relation turned out to be hard to stabilize numerically. Once this has been achieved (see [78]), the obtained algorithm was easy to realize. Summarizing, this type of DTB goes perfectly together with full adaptivity in space and time.

Remark. An even more elegant derivation of DTB by SCHMIDT AND YEVICK [79] applies some shift operator calculus. An inspired extension of DTB to the 2D Helmholtz equation can be found in the recent paper by SCHMIDT [77], who derives and exploits some type of discrete MIKUSIŃSKI operator calculus.

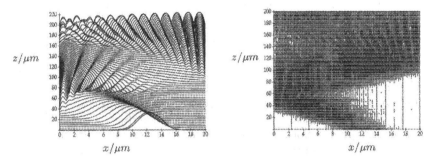

Figure 2.3: Gaussian peak propagation with metallic boundary conditions. Left: solution with interference pattern. Right: corresponding nodal flux.

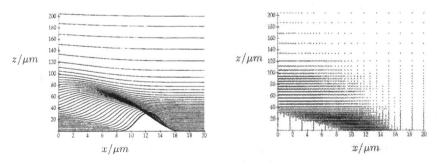

Figure 2.4: Gaussian peak propagation with discrete transparent boundary conditions for the implicit midpoint rule as z–discretization. Left: solution without any reflections. Right: corresponding nodal flux.

Numerical illustration. In order to illustrate both the adaptive Rothe method for the paraxial wave equation and the role of the discrete transparent boundary conditions, we give a numerical comparison from [78]. In Fig. 2.3 (left) a Gaussian peak with slight axial deviation is shown to hit homogeneous Dirichlet conditions (metallic boundary); the corresponding nodal flux obtained from the adaptive scheme is presented in Fig. 2.3 (right). Obviously, due to multiple reflections and the associated interference pattern the spatial grids fill up after few time steps making adaptivity dispensable. In contrast to this undesirable behavior, the numerical results obtained with our discrete transparent boundary conditions for the implicit midpoint discretization are pleasing: see Fig. 2.4 (left) and the corresponding nodal flux pattern in Fig. 2.4 (right). Recall that the computational amount is roughly proportional to the number of occurring nodes.

2.2   Guided Mode Analysis

As an alternative tool in chip design, engineers study so–called guided modes. For this purpose they select certain cross sections orthogonal to the propagation direction (see e.g. the arrows in Fig. 2.2). In the thus defined planes Maxwell's equations are simplified to (scalar or vectorial) 2D Helmholtz eigenvalue problems. Recall the already mentioned typical features coming from the technological problem like geometric complexity and multiscale structure. In addition, close *eigenvalue clusters* naturally occur that must be resolved to *high accuracy*. Therefore *subspace iteration methods* play a dominant role to assure a proper condition of the numerical problem (compare [50]) and, at the same time, reasonable convergence rates. In what follows we present adaptive FEMs for the complex Helmholtz eigenproblem, which have been recently developed by the author's group and successfully applied to the challenging problem class from integrated optics.

In order to derive the associated mathematical model, we look for solutions that are translation invariant along the propagation direction. They may be found through the ansatz

$$E(x, y, z) = u(x, y)e^{-i\mu z}. \tag{2.20}$$

Upon inserting this form into the paraxial wave equation (2.16) and introducing the notation $\lambda = -\mu^2$, we arrive at the *Helmholtz eigenvalue problem*

$$-\Delta u(x, y) - g(x, y) \cdot u(x, y) = \lambda u(x, y), \quad (x, y) \in \Omega \subset \mathbb{R}^2 . \tag{2.21}$$

Eigenfunctions to $\mathrm{Re}(\lambda) \geq 0$ are called *evanescent modes*: they die out along the fibre due to exponential damping. Eigenfunctions to $\mathrm{Re}(\lambda) < 0$ are called *guided modes*: they live along the whole fibre as long as spatial dependencies are ignored. As already stated above, the latter are the objects of interest for the design of integrated optical chips. In weak formulation the above eigenvalue problem reads

$$a(u, v) = \lambda (u, v), \quad \forall v \in H, \tag{2.22}$$

with the sesquilinear form $a(u, v) = (\nabla u, \nabla v) - (gu, v)$ and some Sobolev space H chosen according to either Dirichlet or Neumann boundary conditions. For g real (no material losses assumed) the inner product $(\cdot, \cdot)$ is just the L^2–product, $a(\cdot, \cdot)$ is a symmetric bilinear form, and the eigenproblem is *selfadjoint*. If material losses are included into the model – as they should! – then the eigenproblem is generally *non–selfadjoint*.

Selfadjoint Helmholtz eigenproblems. In this case all eigenvalues λ are real and the eigenfunctions form an orthonormal basis. Technological interest focusses on clusters of the q lowest negative eigenvalues, which give rise to the q largest z–frequencies μ in the ansatz (2.20) and thus – via the dispersion relation – to the q lowest (x, y)–frequencies, i.e. to the q smoothest spatial modes. Multigrid methods for the solution of selfadjoint eigenproblems have been around for quite a time – see e.g. the cokernel projection method due to HACKBUSCH [52] or the *Rayleigh quotient minimization method* (RQM) due to MANDEL/MCCORMICK [67]. Asymptotically both approaches are equivalent. For the present challenging technological problem class, however, we nevertheless had to develop our own code, which is based on the latter MG approach. Let in finite dimension the invariant subspace associated with the eigenvalue cluster be represented by some orthogonal matrix U. Then RQM means

$$R(U) = \min_V R(V) \text{ with } R(V) = \text{ trace} \left((V^*BV)^{-1}(V^*AV) \right) . \qquad (2.23)$$

In addition to adaptivity, we carefully studied and selected the smoother: we found out that (nonlinear) conjugate gradient methods are less sensitive to clustering of eigenvalues than e.g. Gauss–Seidel or Jacobi methods; moreover, cg–methods are better suited to start the subspace iteration – for details we refer to [34]. For coarse grid correction the canonical interpolation would do. By construction, global monotonicity of the MG method with respect to the Rayleigh quotient has been achieved – which, in turn, led to a high robustness of the algorithm even for rather poor initial coarse grids. A simple illustrating 1D example showing this increased robustness of the RQM MG method [67] over the projected MG method [52] can be found in [34].

Non–selfadjoint Helmholtz eigenproblem. In this case the eigenvalues λ are complex lying in a left bounded half stripe region of the complex plane such that $\text{Re}\lambda_1 \leq \text{Re}\lambda_2 \leq \ldots \to \infty$. The eigenvalues with lowest negative real part are those of technological interest. After proper discretization a non–Hermitian eigenvalue problem would arise. Looking over the fence into numerical linear algebra (cf. [50]), we expect to solve such eigenvalue problems via orthogonal transformations to a Schur normal form. As it turns out, we can actually follow this line of thought also in our present operator case, since the *complex* Helmholtz equation differs from its real counterpart only via the complexity of g. In other words, the complex Helmholtz operator is nearly selfadjoint – up to only a compact perturbation. In this situation a completeness result due to KATSNELSON [60] states that (a) the corresponding spectrum is discrete and (b) there exists a *Schur basis* $\{u_j\}_{j=1}^{\infty}$ of $L^2(\Omega)$ such that

$$a(v, u_j) = \lambda_j \,(v, u_j) + \sum_{k=1}^{j-1} \tau_{kj} \,(v, u_k) \quad \forall v \in H_0^1(\Omega). \qquad (2.24)$$

With this result in mind, we developed a generalization of the above MG method for the selfadjoint case. Of course, the solution of our problem now is no longer a minimum of the Rayleigh quotient, but still a stationary point. In the selfadjoint case, discretization by finite elements had led to a set of nested discrete eigenvalue problems of the type

$$\tilde{A}\Theta = \tilde{B}\Theta\hat{\Lambda}$$

with the orthogonally projected matrices

$$\tilde{A} = \left(\tilde{U} \ \tilde{P}\right)^* A \left(\tilde{U} \ \tilde{P}\right) \quad \text{and} \quad \tilde{B} = \left(\tilde{U} \ \tilde{P}\right)^* B \left(\tilde{U} \ \tilde{P}\right) \quad.$$

Herein Θ denotes the eigenmode approximations, while the matrices $\tilde{P}$ represent the simultaneous subspace iteration. The essential extension idea to the non–selfadjoint case now is to replace these projected eigenvalue problems in each smoothing step and each coarse grid correction by projected Schur problems

$$\tilde{A}\Theta = \tilde{B}\Theta\hat{T} \quad,$$

wherein the $\hat{T}$ are now triangular matrices. Numerical experience confirms that in this way some smoothing can actually be realized. As in the simpler case the canonical coarse grid correction (interpolation) is taken. The thus modified algorithm resembles a block ARNOLDI method [74]. The arising coarse grid problems are of the same type as the original problem but of smaller dimension. Therefore a recursive construction leads to a multigrid algorithm, which is schematically written down here [48].

Algorithm.

$[U_l, T_l] = \text{MGM}(A_l, B_l, U_l, T_l, l)$

1. presmoothing: $U_l \to \tilde{U}_l, \ T_l \to \tilde{T}_l$

2. coarse grid correction: $\tilde{U}_l \to \hat{U}_l, \ \tilde{T}_l \to \hat{T}_l$

    *   compute $A_{l-1} = V_l^* A_l V_l$ and $B_{l-1} = V_l^* B_l V_l$, where in case

        * $l = l_{\max}: V_l = \left(\ \tilde{U}_l \quad P_l \ \right)$

$$* \ l < l_{\max} : \ V_l = \left(\begin{array}{cc} \tilde{U}_l & 0 \\ & P_l \end{array} \right)$$

- if

  $$* \ l > 1 : \ [U_{l-1}, T_{l-1}] = \mathrm{MGM}(A_{l-1}, B_{l-1}, \left(\begin{array}{c} I \\ 0 \end{array} \right), \tilde{T}_l, l-1)$$

  $$* \ l = 1 : \ \text{solve } A_0 U_0 = B_0 U_0 T_0 \, , \ U_0^* B_0 U_0 = I$$

- set $\hat{U}_l = V_l U_{l-1}$, $\hat{T}_l = T_{l-1}$

3. postsmoothing: $\hat{U}_l \to U_l$, $\hat{T}_l \to T_l$

Herein the matrices A_l and B_l are the (augmented) system and mass matrices corresponding to FE spaces S_l. The matrices U_l with q columns and the upper triangular matrices T_l represent the unknowns. The prolongation matrices P_l perform the interpolation from the finer spaces S_l to the coarser spaces S_{l-1}.

The above algorithm is just a working algorithm in the sense that it has proved to be efficient in rather difficult application problems with multiscale structure and that experience has confirmed optimal MG complexity. A thorough theoretical investigation, however, is still missing.

Multi Quantum Well (MQW) Laser. For illustration purposes we here give the results of computations for a special integrated optical component, a MQW layer as part of some MQW laser. The cross section of the MQW layer is depicted in Fig. 2.5: the left hand structure has some length scale of about $10\mu m$, whereas the right hand zoom shows several monomolecular layers of about $10nm$ thickness.

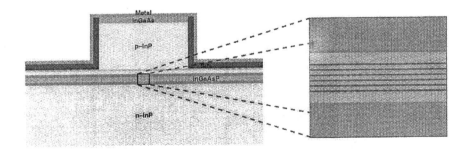

Figure 2.5: Cross section of MQW layer with zoom.

The scale factor of $1000 \approx 2^{10}$ must be spanned by *adaptive* grids – uniform grids would give rise to unreasonably large numbers of nodes. In our adaptive setting we started with coarse grids of about 2.500 nodes and ended up with finest grids of about 15.000 nodes – uniform grid refinement would have produced some estimated 2.500.000 nodes. In this problem $g(x,y) = k_0^2 n^2(x,y)$ with k_0 the vacuum wave number of light and $n(x,y)$ the different refractive indexes, which are complex valued within the MQW layer and the metal layers, but real valued otherwise. The exact design parameters must be kept confidential. In what follows we describe a typical technologically relevant parameter study about the dependence of the eigenvalues upon the imaginary part of some refractive index. We computed the four eigenvalues with lowest real parts for five parameter values – see Fig. 2.6. We also computed the corresponding invariant subspaces. Logarithmic contour plots of the finest mesh approximation of the two Schur modes u_1, u_2 associated with eigenvalues λ_1, λ_2 (see arrows in Fig. 2.6) are given in Fig. 2.7. Observe that u_1 is the symmetric fundamental mode, while u_2 is the anti–symmetric first order mode.

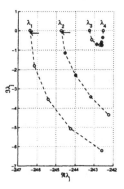

Figure 2.6: Dependence of eigenvalue cluster on imaginary part of refractive index.

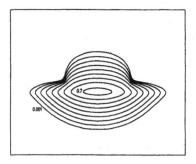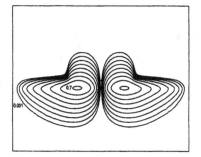

Figure 2.7: Logarithmic contour plots of $|u_1|^2$ and $|u_2|^2$.

3 Countable Ordinary Differential Equations in Polymer Industry

Chemically speaking, *polymers* are a special kind of macromolecules: chains of simple molecules or molecular groups, the *monomers*. The chains typically consist of ten thousand up to ten million monomers, say, and may be linear (the simpler case to be treated here) or even bifurcating. *Copolymers* are built from more than one type of monomer. Polymer materials include a variety of synthetic materials such as lacquers, adhesives, PVC, and MMA. Mathematically speaking, models of *polyreaction kinetics* involve a huge set of ordinary differential equations (ODEs), usually nonlinear and stiff. The numbers of ODEs again range from ten thousand up to ten million – one per each arising polymer length. For a mathematician it is simpler to think of countably infinitely many ODEs to be called *countable* ODEs or just CODEs. Even though CODEs are usually underrated if not totally overlooked in standard mathematical textbooks on differential equations, they play an important role in several scientific fields, e.g. in environmental science (soot formation), astrophysics (cosmic dust), or medicine (pharmacokinetics). In this section we will first describe the CODE models of typical polyreaction mechanisms. A survey of the basic computational approaches will follow. In more detail, we will then present the recent concept of *adaptive discrete Galerkin methods*. This concept has been first proposed in [45] by the author and WULKOW, who has then improved the method considerably in his thesis [87]. On this mathematical and software basis he had started a spin–off firm, which – after the usual critical initial phase – meanwhile consults in chemical industry and research labs all over the world.

3.1 Polyreaction Kinetics

In order to convey an impression of the CODE problem class, we begin with a short list of polyreaction mechanisms that arise in industrial applications. Only under unrealistic simplifications few of these CODEs can be solved in closed analytic form. In industrial applications, however, the mechanisms arise within combinations, which makes an analytical treatment anyway hopeless. That is why two realistic problems are also included here in some detail. As for the notation, let $P_s(t)$ be the concentration of polymers of *chain length* or *polymer degree* s at time t. For ease of writing, we will not distinguish between the chemical species P_s, its concentration $P_s(t)$, and the *chain length distribution* $\{P_s(t)\}_{s=1,2...}$, but just rely on the context.

Chain addition polymerization or free radical polymerization.
With $M(t)$ denoting some monomer concentration, this special reaction mechanism is

$$P_s + M \xrightarrow{k_p} P_{s+1}, \quad s = 1, 2, \ldots, \tag{3.25}$$

where $k_p > 0$ is the reaction rate coefficient. The kinetics of this reaction is modeled by

$$
\begin{aligned}
P_1' &= k_p M P_1 \\
P_s' &= k_p M (P_s - P_{s-1}), \quad s = 2, 3, \ldots \\
M' &= k_p M \sum_{s=1}^{\infty} P_s
\end{aligned}
\tag{3.26}
$$

with the given initial values

$$
P_1(0) = P_{10}, \quad P_s(0) = 0, \quad s = 2, 3, \ldots, \quad M(0) = M_0 .
\tag{3.27}
$$

Coagulation and irreversible polycondensation. This mechanism can be described in chemical terms as

$$
P_s + P_r \longrightarrow P_{s+r}
\tag{3.28}
$$

and modeled mathematically by the *nonlinear* CODE

$$
P_s' = \frac{1}{2} \sum_{r=1}^{s-1} k_{r,s-r} P_r P_{s-r} - P_s \sum_{r=1}^{\infty} k_{sr} P_{rs} \quad s = 1, 2, \ldots .
\tag{3.29}
$$

Once again, the initial distribution $P_s(0)$ is usually given.

Example: Biopolymerization [22]. This problem deals with an attempt to recycle waste of synthetic materials in an ecologically satisfactory way – which is certainly an important problem of modern industrial societies. An attractive idea in this context is to look out for synthetic materials that are both produced and eaten by bacteria – under different environmental conditions, of course. A schematic illustration of the production process within such a bacterial recycling is given in Fig. 3.1: there certain bacteria use fructose as a chemical input to produce polyester (PHB) as chemical output. The macromolecular reaction steps of production and degradation of PHB can be summarized in the *chemical model*

$$
\begin{aligned}
E & & & \xrightarrow{k_a} & A & \\
A & + & M & \xrightarrow{k_i} & P_1 & \\
P_s & + & M & \xrightarrow{k_p} & P_{s+1} & \\
P_s & & & \xrightarrow{k_t} & D_s & + & E \\
D_{s+r} & & & \xrightarrow{k_d} & D_s & + & D_r
\end{aligned}
$$

with $s, r = 1, 2, \ldots$. Herein M denotes the monomer fructose, E an enzyme, A the activated enzyme, P_s the so-called "living" and D_s the so-called "dead" PHB–polymer. The *mathematical model* for the above process

Figure 3.1: Biopolymerization: bacteria eat sugar and produce polyester. White areas: polyester granules within bacteria cells.

comprises the CODE system

$$
\begin{aligned}
E' &= -k_a E + k_t \sum_{r=1}^{s_{max}} P_r \\
A' &= +k_a E - k_i A M \\
M' &= -k_p M \sum_{r=1}^{s_{max}} P_r - k_i A M \\
P_1' &= -k_p M P_1 + k_i A M - k_t P_1 \\
P_s' &= -k_p M (P_s - P_{s-1}) - k_d P_s \qquad\qquad , \quad s = 2,3,\ldots,s_{max} \\
D_s' &= +k_t P_s - k_d(s-1) D_s + 2k_d \sum_{r=s+1}^{s_{max}} D_r \quad , \quad s = 1,2,\ldots,s_{max}.
\end{aligned}
$$

Herein the truncation index s_{max} is not known a priori, practical considerations lead to roughly $s_{max} = 50.000$ – which means that the above system consists of 100.000 ODEs, each of which has roughly the same number of terms in the right side.

Copolymerization. Most industrial synthetic materials are copolymers, typically consisting of three up to seven different sorts of monomers. The mathematical modelling of such systems is often performed in terms of some multidimensionsal ansatz $P_{sr...}$ with s monomers of type A, r monomers of tape B etc. This ansatz, however, leads to an enormous blowup in terms of both computing time and storage. An alternative model has been suggested in the chemical literature (see [46] for reference). In this approach polymers are characterized by their chemically active site at one end of the chain. In the process of numerical solution of the polyreaction CODE enough information about these active sites comes up anyway. In this much simpler framework the following questions of chemical and economical relevance can still be answered: *Which portion of the monomer is consumed in the course of the reaction? What are the time dependent relative distributions of the different polymers?* A typical example with three monomers will be given at the end of Section 3.3 below.

3.2 Basic Computational Approaches

As exemplified in the previous Section 3.1, CODE initial value problems are in general infinite sets of nonlinear ODEs like

$$P_s'(t) = f_s(P_1(t), \dots, P_s(t), P_{s+1}(t), \dots), \quad s = 1, \dots \quad (3.30)$$

given together with initial values $P_s(0)$. Whenever the above right hand side f_s contains only arguments up to P_s, i.e. it has the form $f_s(P_1(t), \dots, P_s(t))$, then the system can, in principle, be solved one by one – a property called *self-closing* in the literature. If f_s has the general form as above, then it is said to be *open*. The latter case is the typical one in real life applications. In what follows we will survey and assess the basic algorithmic concepts for the numerical solution of open CODEs.

Direct numerical integration. In not too complicated cases direct stiff integration of the reaction kinetics ODEs is still a popular approach. In fact, the author and former co–workers have developed the efficient software package LARKIN (for LARge chemical KINetics) to tackle such systems, see [6]. Typically, with any such package, a sequential process is performed starting from a small number of ODEs and successively running up to larger and larger numbers. However, compared to direct integration even in this restricted application, the method to be presented in Section 3.3 has proved computational speed–up factors of more than 10.000 together with better accuracies for quantities of industrial relevance – compare [46]. Finally, since all stiff integrators require the Jacobian matrix of the right hand side, this approach suffers from a rather narrow domain of applicability, just think of the 100.000 by 100.000 nonzero Jacobian elements in the above biopolymerization example.

Lumping technique. In this kind of technique linear combinations of components are collected to certain supercomponents, for which then ODEs are derived and solved numerically. A proper collection of components requires a lot of a–priori insight into the process under consideration, sometimes just a $\log(s)$–equilibration is imposed. However, even though this technique is reported to work satisfactorily in some *linear* ODE cases, it is certainly totally unreliable for *nonlinear* ODEs, which represent the bulk of industrially relevant models.

Method of statistical moments. The canonical formulation for distributions – like $P_s(t)$ here – is in terms of statistical moments

$$\mu_k(t)[P] := \sum_{s \geq 1} s^k P_s(t), \qquad k = 0, 1, \dots, .$$

Note that *mass conservation* shows up as $\mu_0(t) = \text{const.}$ Insertion of the above definition into a polyreaction CODE (see Section 3.1 above) generates an infinite set of ODEs for these moments, a CODE again. It is easy to show that the structural property *open/selfclosing* of the original polyreaction CODE is passed on to the moment CODE. The whole approach is based on the theorem of STIELTJES, which states that the knowledge of *all* moments (if they are bounded) is equivalent to the knowledge of the distribution. If, however, only a finite number N of moments is known, in practice mostly only *few* moments, then there exists a full range of approximations $P_s^{(N)}$ with unclear representation and approximation quality.

One popular method to deal with this lack of information is to specify the distribution in advance. For example, assume that – on the grounds of scientific insight into the given problem – the unknown distribution is expected to be a POISSON distribution

$$P_s = C\, e^{-\lambda s}\, \frac{\lambda^{s-1}}{(s-1)!}\,,$$

then the unknown parameters λ, C can be determined from just the two moments

$$C = \mu_0, \qquad \lambda = \frac{\mu_1}{\mu_0}\,.$$

Once the two moments are computed, the distribution as a whole seems to be known. Sometimes this kind of specification is also hidden behind so-called *closure relations*. However, strictly speaking, none of these approaches can assure that the stated problem is really solved – without further check of whether the assumptions made are appropriate. An insidious feature of any such approach is that all the thus computed approximations look optically smooth and therefore "plausible" – even when they are totally wrong! That is why these two approaches are only recommended for situations wherein the essential features of the solution are well–studied.

Last but not least, it is not clear how many terms need to be kept in the *truncated* moment CODE. Of course, the approximate moments $\mu_k^{(N)}$ corresponding to truncation index N should be accurate enough within some prescribed tolerance compared to the exact moments μ_k. The choice of truncation index N becomes even hopeless when the reaction rate coefficients depend on the chain length – as e.g. in soot formation, cf. [45, 88].

Monte Carlo method. Markov chains play some role in the computation of stationary solutions of copolymerization problems. In the application studied here, however, Monte Carlo methods require too much computing time in comparison with the method to be presented in Section 3.3. They will play an important role in Section 4.2 below in some different context.

Galerkin methods for continuous PDE models. A rather popular approach to condense the infinite number of ODEs is to model the degree s by some *real* variable, interpreting sums as infinite integrals and thus arriving at integro–partial differential equations, usually with lag terms (like e.g. the neutron transport equation). Following this line, an unknown modelling error for low s is introduced – which is the very part of the models where measurements are typically available. Moreover, depending on the polyreaction mechanism, the arising problem may turn out to be *ill–posed*, which in turn requires some regularization to be carefully studied. In [49], GAJEWSKI/ZACHARIAS suggested *Galerkin methods* in Hilbert space based on modified LAGUERRE polynomials L_k^α for the weight function

$$\Psi(s) = \sigma^\alpha e^{-\sigma} \quad \text{with} \quad \sigma = \beta(t)s, \quad \beta = \frac{\mu_0(t)}{\mu_1(t)} \ . \tag{3.31}$$

The specification of β assures *scaling invariance* in the (continuous) variable s.

Discrete Galerkin methods for CODEs. In [45] the author had suggested to use the above PDE approach only in principle, but to avoid turning the discrete variable s artificially into a continuous one. Upon interpreting discrete inner products as infinite sums, the discrete nature of the problem can be preserved – thus keeping the proper regularization. As a first attempt on this mathematical basis, discrete GALERKIN methods based on *discrete* LAGUERRE polynomials l_k for the weight function

$$\Psi(s) = \rho^s, \qquad \rho < 1 \ . \tag{3.32}$$

were constructed. The specification of ρ via "scaling" of the argument as in (3.31) is not directly possible here, since for a discrete variable s scaling is not a proper concept (see, however, the "moving weight function" concept below). This discrete Galerkin approach turned out to be the starting point for the construction of a new class of rather efficient algorithms to be discussed in detail in the next section.

3.3 Adaptive Discrete Galerkin Methods

For ease of presentation we replace the above nonlinear CODE (3.30) by the linear CODE

$$P_s'(t) = (AP(t))_s, \qquad P_s(0) \quad \text{given} \ . \tag{3.33}$$

Herein the discrete operator $\mathcal{A}$ describing the polyreaction mechanisms may be bounded (rare) or, unbounded (typical). The key to the construction of discrete Galerkin methods is the introduction of a *discrete inner product*

$$(f, g) := \sum_{s=1}^{\infty} f(s)g(s)\Psi(s) \tag{3.34}$$

in terms of some prescribed componentwise positive *weighting function* Ψ. This product induces a set of *orthogonal polynomials* $\{l_j\}, j = 1, 2, \ldots$ satisfying the relations

$$(l_j, l_k) = \gamma_j \delta_{jk} \quad , \quad \gamma_j > 0 \quad j, k = 0, 1, 2, \ldots \tag{3.35}$$

For the solution P of the CODE we naturally try the corresponding ansatz

$$P_s(t) = \Psi(s) \sum_{k=0}^{\infty} a_k(t) l_k(s). \tag{3.36}$$

Moving weight function. There exists an interesting close connection between the statistical moments μ_k and the just introduced coefficients a_k. Upon representing the monomials s^k by the orthogonal set of polynomials, we are able to derive an infinite lower triangular system of algebraic equations of the form

$$\begin{aligned} \mu_0 &= b_{00}\gamma_0 a_0, \\ \mu_1 &= b_{10}\gamma_0 a_0 + b_{11}\gamma_1 a_1, \\ \mu_2 &= \ldots, \end{aligned} \tag{3.37}$$

where $b_{kk} \neq 0$ is guaranteed. Herein the moments μ_k arise row–wise, whereas the coefficients a_k arise column–wise. This implies that if all moments are given and bounded, then all coefficients can be computed, which is just the already mentioned STIELTJES theorem. If only N moments are given, then only N coefficients can be computed. On the side of the coefficients, however, we have a reasonable option of *truncating* the expansion (3.36), since they (unlike the moments) are known to *decrease asymptotically*, if the solution P can actually be represented by the above ansatz. In other words, the solution P must be contained in some weighted sequence space, say H_Ψ, with an associated inner product and its induced norm

$$< f, g > := \sum_{s=1}^{\infty} f(s)g(s)/\Psi(s), \qquad \|f\|^2 := < f, f > . \tag{3.38}$$

In order to enforce that $P \in H_\Psi$ throughout the whole evolution, we additionally require that $P \approx \mu_0 \Psi$ by imposing the *moving weight function* conditions

$$\nu_0[\Psi] = 1, \qquad \nu_1[\Psi] = \frac{\mu_1[P]}{\mu_0[P]} \tag{3.39}$$

wherein the ν_k denote the statistical moments of the prescribed weight function. The first condition means that Ψ is some *probability density function*, whereas the second one assures some time dependent coupling of the mean values of the unknown distribution P to the known distribution Ψ – hence the name. Insertion of (3.39) into (3.37) leads to the two equivalent conditions

$$a_0(t) = \mu_0(t), \qquad a_1(t) = 0 \tag{3.40}$$

independent of the underlying problem and of the choice of weight function. Summarizing, with these two additional conditions a much smaller number n of degrees of freedom turned out to be sufficient to characterize the dynamics of realistic polyreaction systems.

Method of lines. Upon insertion of the expansion (3.36) into the CODE (3.33), multiplication by the test function $l_j(s)$, summation over s, change of summation order, and use of the above orthogonality relations, we end up with the CODE

$$\gamma_j a_j'(t) = \sum_{k=0}^{\infty} a_k(t)(l_j, Al_k) \quad j = 0, 1, \ldots. \tag{3.41}$$

for the a_k. In the above moving weight function approach the ODE for a_1 can be dropped (recall that $a_1 = 0$) and a new ODE can be created instead, say, for the parameter ρ in the discrete LAGUERRE method based on Ψ as in (3.32). In other words, in this method of lines type approach the moving weight function induces a *moving basis* $l_k(s; \rho(t)), k = 0, 1, \ldots$ – similar to moving nodes in PDE applications with moving fronts.

For the numerical realization of (3.41), the inner products

$$(l_j, Al_k) = \sum_{s=1}^{\infty} l_j(s)A(s)l_k(s)\Psi(s)$$

must be approximated by a finite number of terms to prescribed accuracy. In rare cases analytical methods allow a closed form representation, which is cheap to evaluate. In most cases, however, numerical approximations turn out to be the only choice. We developed two efficient approximation methods: an *adaptive multigrid summation technique* (compare Section 9.7.2 in the undergraduate numerical analysis textbook [35]), a discrete variant of the MG method described in Section 1.1, and a *Gauss–Christoffel summation*, a discrete variant of Gauss–Christoffel quadrature based on the selected weight function Ψ. As for the applied time discretization, *linearly implicit* schemes in the generally nonlinear CODEs (3.30) turned out to be efficient to tackle the unbounded part of the discrete Fréchet operators of the right hand side. In addition, adaptive timestep and order control is as useful as in ordinary stiff integration.

Adaptive truncation. Truncation of this CODE by setting

$$a_j = 0, \qquad j = n, \quad n+1, \quad \ldots$$

leads to the discrete GALERKIN approximation

$$P_s^{(n)}(t) = \Psi(s) \sum_{k=0}^{n} a_k^{(n)}(t) l_k(s). \tag{3.42}$$

In the simpler case of *self-closing* CODEs, the coefficients a_k do *not* depend on the truncation index n. The key to *adaptivivity* in any (global) Galerkin method is that the *truncation error* can be estimated by

$$\left| P^{(n)} - P \right| \doteq \left| P^{(n)} - P^{(n+1)} \right| =$$

$$\left[(a_{n+1}^{(n)})(t)^2 \gamma_{n+1} + \sum_{k=0}^{n} \left(a_k^{(n)}(t) - a_k^{(n+1)}(t) \right)^2 \gamma_k \right]^{\frac{1}{2}}. \tag{3.43}$$

Note that if we want to assure a condition like

$$\left| P^{(n)} - P^{(n+1)} \right| \leq \text{TOL}$$

for some prescribed error tolerance TOL, then we might thus obtain some time dependent number $n(t)$ of terms in the Galerkin approximation.

Adaptive Rothe method. As already tacitly indicated in the above moving weight function conditions, the typical CODE initial value problem lives in some *scale* of Hilbert spaces, say $H_{\rho(t)}, t \geq t_0$ for the discrete Laguerre method, rather than just in single fixed space. This nicely shows up in a special Lipschitz condition, a nonlinear extension of a semi–continuity assumption:

$$\|f(t, u) - f(t, v)\|_{\bar{\rho}} \leq \frac{M}{(\bar{\rho} - \rho)^\gamma} \|u - v\|_\rho, \quad 0 < \gamma \leq 1,$$
$$\bar{\rho} > \rho \quad \text{and} \quad u, v \in H_\rho. \tag{3.44}$$

This condition is the essential ingredient of a uniqueness theorem for CODEs – see [87]. In analogy to the discussion for PDEs in Section 1.1 above, the appropriate order of discretization will therefore be *first time discretization, then Galerkin approximation*, which is the so–called Rothe method. Starting from some initial value $u(t) = \phi$ suppose we apply some linearly implicit Euler discretization

$$(I - \tau \mathcal{A}) \Delta u = \tau f(\varphi), \quad u_1 = \varphi + \Delta u, \tag{3.45}$$

wherein $\mathcal{A}$ is the derivative $f_u(\varphi)$. This is a linear boundary value problem in some discrete sequence space to be treated numerically by a discrete

Galerkin method. The numerical realization of the time step control requires an estimate η_1 of the error $\|u_1 - u(t + \tau)\|$ in the norm $\| \cdot \|$ induced by the corresponding inner product. For this purpose we solve the correction equation

$$(I - \tau A)\eta_1 = -\frac{1}{2}\tau^2 A f(\varphi) \,, \quad u_2 = u_1 + \eta_1 \,, \tag{3.46}$$

which obviously has the same structure as (3.45). The approximation u_2 is of order 2 in time, which implies the estimated optimal time step (with safety factor $\sigma < 1$)

$$\tau_{\text{new}} = \tau \sqrt{\frac{\sigma \tau_{\text{tol}}}{\|\eta_1\|}} \,. \tag{3.47}$$

In this realization the truncation index for (3.46) may correspond to a less stringent (absolute) error tolerance than the one for (3.45), since (3.47) requires less accuracy; in this setting the approximation u_1 is regarded as the accepted solution. If the same absolute accuracy is prescribed in both equations, then u_2 is taken as the accepted approximation.

Discrete h-p-method. For sufficiently complex problems in industry *global* Galerkin methods – such as the one given in (3.36) – have meanwhile been clearly outperformed by *local* Galerkin methods (similar as FEMs in PDEs). In order to be able to construct local multilevel bases, the weight function has to be restricted to $\Psi = 1$ on finite subintervals, which induce (discrete) Chebyshev polynomials t_k of degree k. The actually developed method combines adaptive interval refinement (h–method) with adaptive choice of the local degree (p–method) to obtain some rather sophisticated *adaptive h-p-method* on some finite interval. In order to match the asymptotic tail of the distribution, global representations still play a role.

In the course of the years the successive efficient treatment of industrially relevant polyreaction problems has led to a steady increase of modern rather sophisticated algorithmic tools that made their way into the commercial software package Predici^{TM} of WULKOW [88]. The following copolymerization example with three monomers should give some flavor of such problems.

Example: Radical terpolymerization [46]. Let P_s be the polymer with active end MMA (index a), Q_s the one with styrole (index b), and R_s the one with MSA (index c below). With this notation, the chemical reaction scheme reads
Initialization:

$$
\begin{array}{rcl}
I & \xrightarrow{k_d} & 2I* \\
I^* + MMA & \xrightarrow{k_l} & P_1^* \\
I^* + S & \xrightarrow{k_l} & Q_1^* \\
I^* + MSA & \xrightarrow{k_l} & R_1
\end{array}
$$

Chain growth:

$$P_s^* + MMA \xrightarrow{k_{paa}} P_{s+1}^*$$
$$P_s^* + S \xrightarrow{k_{pab}} Q_{s+1}^*$$
$$P_s^* + MSA \xrightarrow{k_{pac}} R_{s+1}^*$$
$$Q_s^* + MMA \xrightarrow{k_{pba}} P_{s+1}^*$$
$$Q_s^* + S \xrightarrow{k_{pbb}} Q_{s+1}^*$$
$$Q_s^* + MSA \xrightarrow{k_{pbc}} R_{s+1}^*$$
$$R_s^* + MMA \xrightarrow{k_{pca}} P_{s+1}^*$$
$$R_s^* + S \xrightarrow{k_{pcb}} Q_{s+1}^*$$

Chain termination:

$$P_s^* + P_r^* \xrightarrow{k_{caa}} D_{s+r}$$
$$P_s^* + Q_r^* \xrightarrow{k_{cab}} D_{s+r}$$
$$Q_s^* + Q_r^* \xrightarrow{k_{cbb}} D_{s+r}$$

For reasons of confidentiality some parts of the mechanism are left out. Initial values and generic reaction rate coefficients are given in [46]. Under the modelling assumption that in this specific polymerization process the living chains live only for a very short time span, the copolymer distribution can be computed as follows. Define e.g. integrated quantities like

$$\bar{P}_s(t) = \int_0^t P_s(t)dt \ .$$

which (up to normalization) counts the number of copylmers with MMA in fixed position s. Then the relation $\bar{P}_s : \bar{Q}_s : \bar{R}_s$ offers detailed insight into the composition of the copolymer.

In Fig. 3.2 such compositions are shown for times $t = 360$ min and $t = 1.080$ min. Information of this kind could not have been gained from statistical moment analysis as applied earlier in industry.

4 Hamiltonian Equations in Pharmaceutical Drug Design

The design of highly specific drugs on the computer, the so-called *rational drug design* (as opposed to irrational drug consumption), is a fairly recent dream of biochemistry and pharmaceutical industry. Typically, a lot of heuristics go with this problem, which we skip here. At first mathematical glance, drug design seems to involve the numerical integration of the Hamiltonian differential equations that describe the dynamics of the molecular

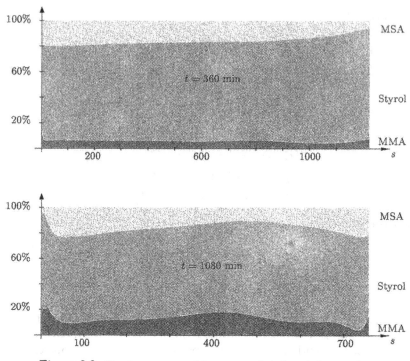

Figure 3.2: Copolymer composition versus chain length for two times t.

system under consideration. Following this idea, a huge discrepancy of time scales shows up: phenomena of interest, such as *protein folding* or *active site docking*, occur on a micro- or millisecond scale, whereas present routine computations only cover time spans of up to a few nanoseconds (at best). This gap has stimulated a lot of work in an interdisciplinary field including Numerical Analysis, Statistical Physics, Biochemistry, and Dynamical Systems. The present section reports about some recent and still ongoing collaboration of the author and his group with internationally renowned RNA biochemists including biotech firms in the Berlin region.

Our special contribution to drug design starts from the known insight that the corresponding trajectories are *chaotic*, which means that small perturbations of the initial values of such trajectories *asymptotically* lead to unbounded deviations. In terms of Numerical Analysis this means that the Hamiltonian initial value problems (IVPs) are *ill-conditioned after short time spans*. On the basis of this insight, we suggested a novel concept for the computation of *essential* features of Hamiltonian dynamical systems. The key idea presented in Section 4.1. below is to directly compute *chemical conformations* and rates of conformational changes, interpreting chemical conformations as *almost invariant sets* in the *phase space* (positions *and* momenta) of the corresponding dynamical system. In a first step, this led to an eigenproblem for eigenvalue clusters around the Perron root of the so-called FROBENIUS-PERRON operator associated with the (numerical) flux of the dynamical system.

In a second step, in Section 4.2., we interpreted chemical conformations as objects in the *position space* of the Hamiltonian dynamical system. Moreover, we abandoned the deterministic Hamiltonian systems with given initial values to turn over to ensembles of initial values in the frame of Statistical Physics. This led to the natural construction of a stochastic operator, which appears to be selfadjoint over some weighted L^2-space. Discretization of that operator by means of certain *hybrid Monte Carlo* methods (HMC) generates nearly uncoupled Markov chains that need to be computed. As it turns out, the eigenvectors associated with the Perron cluster of eigenvalues for the stochastic operator contain the desired information about the chemical conformations and their patterns of change.

The described approach is presently worked out in collaboration with biochemists that design RNA drugs in their chemical labs. Our aim is to substitute time consuming and costly experiments in the chemical RNA lab by reliable simulations in a *virtual RNA Lab*. First steps in this direction are illustrated in Section 4. 3 at some fairly complex, but moderate size RNA molecule.

4.1 Deterministic Chaos in Molecular Dynamics

In classical textbooks on Molecular Dynamics (MD), see e.g. [1], a single molecule is modelled by a Hamilton function

$$H(q,p) = \tfrac{1}{2} p^T M^{-1} p + V(q) \tag{4.48}$$

with q the 3D atomic positions, p the (generalized) momenta, M the diagonal mass matrix, and V a differentiable potential function. The Hamilton function is called *separated*, whenever the p-part and the q-part of H are separated as above. The Hamilton function defined on the phase space $\Gamma \subset \mathbb{R}^{6N}$ induces the canonical equations of motion, the *Hamiltonian equations*

$$\dot{q} = M^{-1}p, \qquad \dot{p} = -\operatorname{grad} V , \tag{4.49}$$

which describe the dynamics of the molecule in a deterministic way: For given initial state $x_0 = (q(0), p(0))$ the unique formal solution of (4.49) is usually written as $x(t) = (q(t), p(t)) = \Phi^t x_0$ in terms of the *flow* Φ^t. Numerical integration of (4.49) by any one-step method with stepsize τ leads to the discrete solution

$$x_{k+1} = \Psi^\tau x_k \quad \Rightarrow \quad x_k = (\Psi^\tau)^k x_0, \tag{4.50}$$

in terms of a *discrete flow* Ψ^τ.

Condition of the initial value problem. Given an initial perturbation δx_0 we are interested in its growth along the flow

$$\delta x(t; x_0) = \Phi^t(x_0 + \delta x_0) - \Phi^t x_0 .$$

The *condition number* $\kappa(t)$ of an initial value problem (see the textbook by DEUFLHARD AND BORNEMANN [32]) may be defined as the worst case error propagation factor in first order perturbation analysis so that (in some suitable norm $|\cdot|$)

$$|\delta x(t; x_0)| < \kappa(t)|\delta x_0| \quad \text{for all } x_0.$$

It is of utmost importance to keep in mind that $\kappa(t)$ is a quantity characterizing the analytic problem independent of any discretization. A *linear growth* result $\kappa(t) \sim t$ holds for a subclass of so–called *integrable* Hamiltonian systems such as the popular Kepler problem – see ARNOLD [3]. In real life MD problems, however, κ *increases exponentially*. In order to illustrate this behavior, Fig. 4.1 shows results for the simple Butane molecule. In order to be able to ignore any discretization error effects, unusually small time steps ($\tau = 0.005$ fs) within the Verlet scheme have been chosen. A physically negligible initial perturbation 10^{-4}Å can be seen to overgrow the nominal solution after a time span $T > 500$ fsec, which is significantly shorter than the time spans of physical interest.

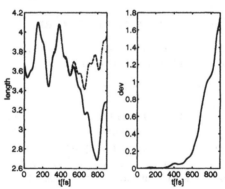

Figure 4.1: Two dynamical simulations for the Butane molecule with initial deviation 10^{-4}Å. Left: Evolutions of the total molecule length (in Å). Right: Dynamics of the deviation (in Å).

Once *long term* trajectories in MD have been identified as ill-conditioned mathematical objects, they should be avoided in actual computation. Only *short term* trajectories should be accepted as numerical input for further scientific interpretation. This seems to be in direct contradiction to the possible prediction of long term behavior of biomolecules! How can we overcome this difficulty?

Warning. There are quite a number of beautiful movies about the dynamics of biomolecules that visualize merely accidental numerical results of long term simulations – and are therefore of doubtful value for chemical interpretation.

Multiscale structure. In order to gain more insight, we proceed to a rather instructive example due to GRUBMÜLLER AND TAVAN [51]. Figure 4.2 describes the dynamics of a polymer chain of 100 CH_2 groups. Possible timesteps for numerical integration are confined to $\tau < 10$ fsec due to fast oscillations. Time scales of physical interest range from 10^3 to 10^5 psec in this problem, which is a factor 10^5 to 10^7 larger. The figure presents six different zoom levels in time, each of which scales up by a factor 10. On the smaller time scales (upper levels) the dynamical behavior is characterized by nonlinear *oscillations* around certain vague "equilibrium positions". On larger and larger time scales these oscillations become less and less important. On the largest time scale (lowest level) we observe a kind of flip–flop behavior between two "conformations".

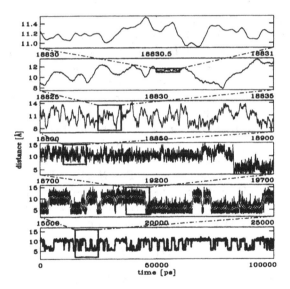

Figure 4.2: MD simulation of a polymer chain of 100 CH_2 groups due to [51]. Time scale zoom factor 10 from level to level.

This observation suggests an alternative to long term trajectory simulation: the *essential dynamical pattern* of a molecular process could as well be modelled by *probabilities* for the molecular system to stay within different *conformations*. From a chemical point of view, conformations describe clusters of *geometric* configurations associated with some specified chemical *functionality.* In a conformation, the large scale geometric structure of the molecule is understood to be conserved, whereas on smaller scales the system may well rotate, oscillate, or fluctuate. In order to understand a chemical system, conformations and their average life spans are the main objects of chemical interest. Therefore the *direct computation of conformational dynamics* seems to be the concept to pursue.

Frobenius–Perron operator. In a first approach, we recurred to the so-called FROBENIUS-PERRON operator U associated with the flow Φ^τ. This operator is defined on the set $\mathcal{M}$ of probability measures over the phase space Γ by virtue of

$$(U\mu)(G) = \mu(\Phi^{-\tau}(G)) \quad \text{for all measurable } G \subset \Gamma \text{ and arbitrary } \mu \in \mathcal{M}.$$
(4.51)

Invariant sets are the union of all those states that a dynamical system can reach, i.e. they correspond to an *infinite* duration of stay. Moreover, they are fixed points of U associated with the so-called PERRON eigenvalue $\lambda = 1$ of U. In [33], we interpreted conformations as *almost invariant subsets in phase space* corresponding to *finite, but still large* duration of stay – a phenomenon often called *metastability*. In the spirit of earlier work of DELLNITZ/JUNGE [29] we analyzed the connection of almost invariant sets with eigenmodes to real eigenvalue clusters around the Perron root, to be called *Perron clusters* hereafter – see DEUFLHARD, HUISINGA, FISCHER, AND SCHÜTTE [37].

Following HSU [57] the discretization of the Frobenius-Perron operator is done on a box covering $\{G_1, \ldots, G_n\}$ associated with characteristic functions χ_{G_i}. Since the flow Φ^τ conserves energy, these boxes would be expected to subdivide the *energy surface* $\Gamma_0(E) = \{x \in \Gamma : H(x) = E\}$. However, a typical discrete flow $(\Psi^{\tau/k})^k$ with k steps will *not* conserve energy exactly – even symplectic discretizations preserve energy only on average over long times. Therefore the boxes have to subdivide *energy cells* defined by

$$\Gamma_\delta(E) = \{x \in \Gamma, |H(x) - E| \leq \delta\}$$

in terms of some perturbation parameter δ. With $f = (\Psi^{\tau/k})^k \approx \Phi^\tau$ the discretized Frobenius-Perron operator $U_n = (u_{i,j})$ can be written componentwise as

$$u_{ij} = \frac{m(f^{-1}(G_i) \cap G_j)}{m(G_j)}, \quad i = 1, \ldots, n,$$
(4.52)

where m denotes the LEBESGUE measure; roughly speaking, this means that $m(G_j)$ is the volume of the box G_j. The geometric situation as a whole is illustrated in Fig. 4.3. The approximations of the various volumes are performed by some equidistributed Monte Carlo method. In order to speed up computations, a nested sequence of boxes is constructed for an *adaptive*

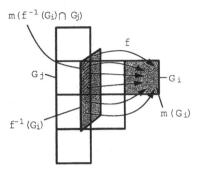

Figure 4.3: Stochastic matrix element $u_{i,j}$ computed from (4.52)

multilevel box method called subdivision algorithm by DELLNITZ/HOHMANN [28].

The realization of this first attempt, however, turned out to suffer from two important disadvantages. The first one is of a theoretical nature: since the Frobenius–Perron operator for a deterministic Hamiltonian system is *unitary* in $L^2(\Gamma)$, real eigenvalues inside the unit circle cannot exist. But they did exist and had been computed and interpreted in our method! This comes from the fact that, by subdividing energy cells rather than the energy surface, we had allowed for stochastic perturbations of the deterministic systems; in this more general setting, eigenvalues could show up also inside the unit circle and did contain the information wanted. Consequently, to model this situation correctly, a more general stochastic theory would be needed – which is still missing. Second, this approach obviously causes some *curse of dimension*, which prevents the method to be applicable to realistic molecules. To understand this perhaps unexpected and certainly undesirable effect, recall that the described subdivision method had been successfully developed in [29] for hyperbolic systems, where the dynamics is known to collapse asymptotically to some low dimensional attractor; this feature, however, does not carry over to Hamiltonian systems.

4.2 Identification of Metastable Conformations

This section describes a rather recent improved approach mainly due to SCHÜTTE [80, 81]. This approach keeps the conceptual advantages of the Dynamical Systems approach (as given in Section 4.1), but avoids the conceptual disadvantages by exploiting concepts of Statistical Physics instead. Its key new feature is the replacement of the Frobenius-Perron operator by some well-designed MARKOV operator, a *spatial transition operator* based on BOLTZMANN statistics. Most experiments in a chemical lab are performed under the conditions of constant temperature and volume, which is known to give rise to the *canonical density*

$$f_0(x) = \frac{1}{Z} \exp(-\beta H(x)), \quad \text{with} \quad Z = \int_\Gamma \exp(-\beta H(x)) dx, \quad (4.53)$$

where $\beta = 1/k_B T$, temperature T, and Boltzmann's constant k_B.

We start with some notion of *almost invariant* sets in the language of statistics. For some selected set S let χ_S denote its characteristic function, i.e. $\chi_S(x) = 1$ iff $x \in S$ and $\chi_S(x) = 0$ otherwise. Then the *transition probability* between two subsets S_1, S_2 of the phase space Γ is given by

$$w(S_1, S_2, \tau) = \frac{1}{\int_{S_1} f_0(x) dx} \int_{S_1} \chi_{S_2}(\Phi^\tau x) f_0(x) dx \quad S_1, S_2 \subset \Gamma \quad (4.54)$$

By this definition almost invariant sets will be those with $w(S, S, \tau) \approx 1$. In Section 4.1 we had understood chemical conformations as almost invariant sets in phase space. However, in reality these objects are observed in *position space*. That is why we now turn over to characterize conformations as sets B in position space $\Omega \subset \mathbb{R}^{3N}$. Upon allowing for arbitrary momenta p, we are naturally led to focus our interest on the *phase space fiber*

$$\Gamma(B) = \{(q, p) \in \Gamma, \quad q \in B\}. \quad (4.55)$$

With this notation we now call a set $B \subset \Omega$ a *conformation* whenever

$$w(\Gamma(B), \Gamma(B), \tau) \approx 1. \quad (4.56)$$

For H *separable* as in (4.48), f_0 from (4.53) splits into the product

$$f_0(x) = \underbrace{\frac{1}{Z_p} \exp\left(-\frac{\beta}{2} p^T M^{-1} p\right)}_{=\mathcal{P}(p)} \underbrace{\frac{1}{Z_q} \exp(-\beta V(q))}_{=\mathcal{Q}(q)}, \quad (4.57)$$

with $\mathcal{P}$ and $\mathcal{Q}$ normalized such that

$$\int \mathcal{P}(p) dp = \int \mathcal{Q}(q) dq = 1.$$

Upon returning to (4.55), we may specify the conditional probability for a system being in set $B \subset \Omega$ to move to set $C \subset \Omega$ after some time τ as

$$w(\Gamma(B), \Gamma(C), \tau) = \frac{1}{\int_{\Gamma(B)} f_0(x) dx} \int_{\Gamma(B)} \chi_{\Gamma(C)}(\Phi^\tau x) f_0(x) dx. \quad (4.58)$$

Construction of spatial Markov operator. With these preparations we are now ready to confine all terms to position space only. The denominator in (4.58) represents the probability for a system from the statistical ensemble *to be within* $B \subset \Omega$ which can be simplified using $dx = dqdp$ (for H separable) and the normalization of $\mathcal{P}$ to yield

$$\pi(B) = \int_{\Gamma(B)} f_0(x)\, dx = \int_B \mathcal{Q}(q)\, dq . \tag{4.59}$$

The analog treatment of the numerator in (4.58) will give rise to the definition of an operator T. Let $\xi_1(q,p) = q$ denote the projection from the variables x to the position variables and introduce an inner product in the *weighted* Hilbert space $L_{\mathcal{Q}}^2$ by

$$\langle u, v \rangle_{\mathcal{Q}} = \int_{\Omega} u^*(q)\, v(q)\, \mathcal{Q}(q)\, dq$$

together with its induced norm

$$\|u\|_{\mathcal{Q}}^2 = \langle u, u \rangle_{\mathcal{Q}} .$$

With this notation we may write

$$\int_{\Gamma(B)} \chi_{\Gamma(C)}(\Phi^\tau x)\, f_0(x)\, dx = \underbrace{\int_C \left\{ \int_{\mathbb{R}^{3N}} \chi_B(\xi_1 \Phi^\tau(q,p))\, \mathcal{P}(p)\, dp \right\} \mathcal{Q}(q)\, dq}_{=: \langle T\chi_B, \chi_C \rangle_{\mathcal{Q}}}$$
$$\tag{4.60}$$

thus defining the operator

$$Tu(q) = \int u(\xi_1 \Phi^\tau(q,p))\, \mathcal{P}(p)\, dp, \tag{4.61}$$

Upon combining (4.58) up to (4.60) we may write the conditional probability of a system being in B to move into C during time τ as

$$w(B, C, \tau) = \frac{\langle T\chi_B, \chi_C \rangle_{\mathcal{Q}}}{\pi(B)} \quad B, C \subset \Omega \tag{4.62}$$

By construction, T can be interpreted as the restriction of the Frobenius–Perron operator U, see (4.51), to position space via averaging over the momentum part of the canonical distribution. The operator T is defined on the weighted spaces

$$L_{\mathcal{Q}}^r(\Omega) = \{u : \Omega \to \mathcal{C}, \int_{\Omega} |u(q)|^r \mathcal{Q}(q)\, dq < \infty\}, \qquad r = 1, 2.$$

In terms of these spaces the important properties of T are (SCHÜTTE [80]):

1. T is bounded in $L^p_Q(\Omega)$: $\|Tu\|_Q \le \|u\|_Q$ $p = 1, 2$

2. T is a MARKOV operator on $L^1_Q(\Omega)$.

3. T is *selfadjoint* in L^2_Q, since Φ^τ is *reversible*. Hence, the spectrum $\sigma(T)$ is real-valued and bounded: $\sigma(T) \subset [-1, 1]$.

4. There exists a *Perron cluster* of discrete eigenvalues well-separated from the remaining (continuous) part of the spectrum.

Properties 1-3 hold for Hamiltonian systems in general; property 4 only holds under additional assumptions – which are, however, satisfied in the systems of interest here. Summarizing, T has just the theoretical properties needed as a basis for the computational identification of conformational subsets via the eigenmodes to Perron clusters of eigenvalues.

Discretization of Markov operator. We proceed as in the previous section by introducing a set of boxes, this time in position space rather than phase space. Let $\{G_1, \ldots, G_n\} \subset \Omega$ denote a *covering* of Ω. In view of a Galerkin approximation in L^2_Q we define the basis $V_n = \text{span}\{\chi_1, \ldots, \chi_n\}$ in terms of the characteristic functions $\chi_i = \chi_{G_i}$. Moreover, let $\pi(G_i) = \pi_i$. In this basis we may construct an (n, n)-matrix $P = (p_{ij})$ via spatial transition probabilities as

$$p_{ij} = \frac{\langle T\chi_i, \chi_j \rangle_Q}{\pi_i} = w(G_i, G_j, \tau) \qquad i, j \in \{1, \ldots, n\}. \tag{4.63}$$

By construction, this matrix P is row-wise stochastic and, since T is self-adjoint, also *reversible*. Upon observing the condition of *detailed balance* in the form

$$\pi_i\, p_{ij} = \pi_j\, p_{ji}, \qquad \forall i, j \in \{1, \ldots, n\}.$$

the matrix is *symmetric* with respect to a weighted discrete inner product, which implies that $\sigma(P) \subset [-1, 1]$, real. Moreover, there exists a Perron cluster of eigenvalues, if the boxes are well-chosen to really cover the relevant position space.

Hybrid Monte Carlo realization. The above introduced MARKOV operator T in $L^1_Q(\Omega)$ can also be interpreted as a transition operator that generates a MARKOV *chain* $\{q_k\}_{k=0,1,\ldots}$ via the *discrete stochastic dynamical system*

$$q_{k+1} = \xi_1 \Phi^\tau(q_k, p_k), \quad k = 0, 1, \ldots, \tag{4.64}$$

wherein the momenta p_k in each step are chosen randomly from the distribution $\mathcal{P}$. Obviously, upon changing from the Frobenius-Perron operator U to the transition operator T, we also change from the discrete *deterministic* dynamical system (4.50) to its stochastic counterpart (4.64). Iterations of (4.64) realize sequences $\{q_k\}$ that are asymptotically distributed according to $\mathcal{Q}$. This feature can be exploited by application of suitable *Monte Carlo* (MC) methods. After M samples $\{q_j\}$ from (4.64) the $\mathcal{Q}$-expectation value of any *spatial observable* $\mathcal{A} : \Omega \to \mathbb{R}$ is approximated by an averaged sum such that

$$\left| \frac{1}{M} \sum_{j=1}^{M} \mathcal{A}(q_j) - \int_{\Omega} \mathcal{A}(q) \mathcal{Q}(q) \, dq \right| \leq C M^{-1/2}, \qquad (4.65)$$

with a constant C not explicitly depending on $\dim(\Gamma) = 6N$. In the present context, the observable will be any of the $\frac{1}{2}n(n+1)$ integrand factors $T\chi_i\chi_j$ arising in (4.63). As the integrand contains the (short term numerical) flux, a *hybrid Monte Carlo* (HMC) method is preferably applied, which is a compromise between trajectory evaluation and MC. In more detail, the matrix elements p_{ij} are asymptotically approximated by virtue of *relative frequencies*

$$\frac{\# \left(q_k \in G_i \wedge q_{k+1} \in G_j \right)}{\# \left(q_k \in G_i \right)} \to w(G_i, G_j, \tau) = p_{ij} \qquad i, j \in \{1, \dots, n\}. \tag{4.66}$$

The theoretical result (4.65) raises the expectation that we have eventually overcome the *curse of dimension*. However, the well-known undesirable effect of *critical slowing down* of the iteration due to local trapping is still possible. For this reason, we developed an *adaptive temperature* variant called ATHMC – see FISCHER, CORDES, AND SCHÜTTE [47].

Essential degrees of freedom. In realistic RNA drug molecules we have to face around $N > 100$ atoms – see Fig. 4.5. If we subdivide each of the $3N$ state variables into m pieces, then the dimension of the stochastic matrix P would be $n = m^{3N}$ – which would be just too much even for a Krylov space iterative eigenproblem solver. Therefore, in the spirit of a suggestion due to AMADEI, LINSSEN, AND BERENDSEN [2], we start with a long term HMC series and apply some covariance analysis to it; this technique then helps to reduce the total number of state variables to a subset of $d << 3N$ *essential* variables so that only $n = m^d$ discretization boxes are needed. A realistic example will be presented in the subsequent Section 4.3.

Eigenvalue Cluster analysis. Assume that we know how to evaluate the entries of P. So we are finally left with the numerical solution of the eigenproblem

$$P\alpha = \lambda\alpha \quad \text{with} \quad \alpha = (\alpha_1, \dots, \alpha_n) .$$

for a cluster of eigenvalues around the Perron eigenvalue $\lambda = 1$. One question is that we will not know in advance how many of the eigenvalues close to 1 should be included into the Perron cluster. The basic insight from [37] is that the k eigenvalues in the Perron cluster should be interpretable as a perturbed k-fold Perron root. The decision about k is quite subtle and has to be made by careful examination of the matrix $\mathcal{W}$ defined below in (4.69). Suppose now that this decision has been made. Then the conformational sets can be computed via appropriate linear combinations of the eigenmodes corresponding to the Perron cluster. Each conformation, say B, is represented as a set of indices, say I_B, that mark the associated boxes belonging to the spatial conformation. Given such a conformation as subset $B \subset \Omega$, the *probability* for the dynamical system *to stay within* B can easily be evaluated via the relation

$$w(B, B, \tau) \;=\; \frac{1}{\displaystyle\sum_{i \in I_B} \pi_i} \sum_{i,j \in I_B} \pi_i \, p_{ij} \;. \tag{4.67}$$

This should be distinguished from the *probability* for the system *to be within* a conformation (compare also (4.59))

$$\pi(B) = \sum_{i \in I_B} \pi_i \;. \tag{4.68}$$

Finally, with k conformations $B_1, B_2, \ldots, B_k$ identified, the *transition rates between conformations* can be arranged in the (k, k)–matrix

$$\mathcal{W} = (w_{ij}) = \big(w(B_i, B_j, \tau)\big) \;, \quad i, j = 1, \ldots, k \tag{4.69}$$

which, together with the vector $\{\pi(B_1), \ldots, \pi(B_k)\}$, contains the core information of conformation dynamics. Details of this Perron cluster analysis omitted here can be found in our paper [37].

4.3   Virtual RNA Lab

In the frame of our collaboration with RNA technologists, all the above described mathematical techniques (and more) together with a variety of state of the art tools in scientific visualization are to be collected within some virtual RNA lab. Such a virtual lab is an integrated software package that permits a convenient switch between numerical code, fast visualization, and 3D graphic interaction. In order to give some flavor and at the same time to illustrate the above mathematical methods, a moderate size RNA molecule, the trinucleotide r(ACC), will be presented in some detail.

Hamilton function. The function H defined in (4.48) consists of the kinetic energy (p=momenta, M = mass tensor) and of the potential energy terms V like the covalent energy terms for bond stretching, angle bending, out-of-plane oscillations, dihedral torsions, non-bonded Lennard-Jones, and Coulomb terms. We have used the semi-emperical force field GROMOS96

[86]. The set of parameters had been adapted by the GROMOS96 group to quantum chemical calculations and experimental observations. The whole set was refined self-consistently to reproduce experimental results on chemical systems.

$$
\begin{aligned}
H(q,p) \;=\; & \tfrac{1}{2}p^T M^{-1} p \;+ \\
& \sum_{k,l} V_{\text{bond}}(q_k, q_l) \;+ \\
& \sum_{k,l,j} V_{\text{angle}}(q_k, q_l, q_j) \;+ \\
& \sum_{k,l,j,m} V_{\text{out-of-plane}}(q_k, q_l, q_j, q_m) \;+ \\
& \sum_{k,l,j,m} V_{\text{dihedral}}(q_k, q_l, q_j, q_m) \;+ \\
& \sum_{k,l} V_{\text{Lennard-Jones}}(q_k, q_l) \;+ \\
& \sum_{k,l} V_{\text{Coulomb}}(q_k, q_l)
\end{aligned}
$$

The *short term trajectories* needed in the computational model of Section 4.2 were realized by $m = 40$ steps of a Verlet discretization with local timestep $\tau/m = 2$ fsec, which means $\tau = 0.08$ psec.

Dimension reduction. The moderate size RNA molecule r(ACC) has $N = 70$ atoms, which means a phase space dimension $6N = 420$ for the Hamiltonian dynamical system and half of that for the position space. From chemical insight, a set of 37 *torsion angles* is responsible for the existence of different chemical conformations – see Fig. 4.4 left. By means of $M = 320.000$ sampling points within an ATHMC run, the covariance analysis due to [2] supplied a set of only $d = 4$ *essential* variables!

Perron cluster analysis. Upon subdividing two of the essential variables by 2, the other two by 3, we end up with only $n = 36$ boxes in position space. The approximation of the $\sim \tfrac{1}{2}n^2$ elements of the stochastic matrix P required $M = 128.000$ samplings of short term trajectories. The numerical eigenproblem solution for the $(36, 36)$-matrix P yielded the following candidates for the Perron cluster:

A first significant gap can be clearly observed after $\lambda_2 = 0.999$, which would have led to $k = 2$ conformations. A careful further analysis via the correlation matrix $\mathcal{W}$, however, led to the decision $k = 8$, which also shows

k	1	2	3	4	5	6	7	8	9	...
λ_k	1.000	0.999	0.989	0.974	0.963	0.946	0.933	0.904	0.805	...

Table 4.4: Perron eigenvalue cluster for r(ACC).

a remarkable gap to the rest of the spectrum. The eight conformations
{D1c, D1t, ... D4c, D4t} actually show significant structural differences, which
supply a lot of insight to the chemical expert. In Table 4.5 we give the additional information about the probabilities for the dynamical system *to be
within a conformation* (first row) and the probabilities *to stay within a conformation* (second row). In Fig. 4.4 two of the computed conformations are
presented. Observe that D2c and D3c differ in a turn of the torsion angle
χ and a flip–flop in the pseudo–angle P, which has appeared as an essential
variable from our computation.

conformations	D1c	D1t	D2c	D2t	D3c	D3t	D4c	D4t
prob. to be within	0.107	0.011	0.116	0.028	0.320	0.038	0.285	0.095
prob. to stay within	0.986	0.938	0.961	0.888	0.991	0.949	0.981	0.962

Table 4.5: Probabilities $\pi(B_i)$ due to (4.68) and w_{ii} due to (4.67), $i = 1, \ldots, 8$.

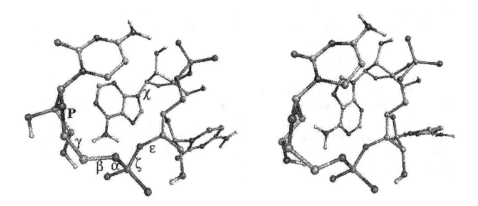

Figure 4.4: Conformations D3c (left) and D2c (right) of the r(ACC) molecule. Left: 6
out of 37 torsion angles. Look at the torsion angle χ and the pseudo-angle P to compare.

Conformational transition rates. The transition probabilities *between* the conformations are collected in the $(8,8)$-matrix $\mathcal{W}$ from (4.69) which here reads

$$
\begin{pmatrix}
0.986 & 5.3_{10}-3 & 1.4_{10}-5 & 8.1_{10}-3 & 5.3_{10}-4 & 0 & 0 & 0 \\
5.1_{10}-2 & 0.938 & 1.0_{10}-2 & 4.6_{10}-6 & 2.6_{10}-4 & 6.3_{10}-4 & 0 & 1.1_{10}-7 \\
4.0_{10}-5 & 3.1_{10}-3 & 0.949 & 4.8_{10}-2 & 0 & 3.6_{10}-7 & 3.6_{10}-5 & 2.6_{10}-5 \\
2.7_{10}-3 & 1.6_{10}-7 & 5.7_{10}-3 & 0.991 & 7.7_{10}-8 & 0 & 8.0_{10}-4 & 1.2_{10}-9 \\
4.9_{10}-4 & 2.5_{10}-5 & 0 & 2.1_{10}-7 & 0.961 & 1.3_{10}-2 & 2.5_{10}-2 & 7.3_{10}-4 \\
0 & 2.6_{10}-4 & 5.0_{10}-7 & 0 & 5.6_{10}-2 & 0.888 & 3.5_{10}-3 & 5.2_{10}-2 \\
0 & 0 & 4.7_{10}-6 & 9.0_{10}-4 & 1.0_{10}-2 & 3.4_{10}-4 & 0.981 & 7.4_{10}-3 \\
0 & 1.4_{10}-8 & 1.0_{10}-5 & 4.0_{10}-9 & 9.0_{10}-4 & 1.5_{10}-2 & 2.2_{10}-2 & 0.962
\end{pmatrix}
$$

Recall that the diagonal elements of $\mathcal{W}$ are the same as the second row in Table 4.5, whereas the first row in Table 4.5 does not show up in $\mathcal{W}$.

Extension to larger molecules. The above moderate size molecule is a quite good candidate to sharpen the mathematician's knife in this kind of collaboration. Biomolecules of real interest to our partners are larger – see e.g. the *hammerhead molecule* as represented in Fig. 4.5.

This type of biomolecule consists of about $150-200$ nucleotides; each nucleotide gives rise to roughly 12 torsion angles, which sums up to about 2.000 torsion angles that might generate conformations and should therefore be subdivided! Fortunately, chemists report that usually *the larger the molecules, the stiffer their structure*. If this is reliable, then covariance analysis for the associated HMC data will bring up "not too many" essential degrees of freedom that should still be tractable within our algorithmic framework. Future developments will include parallelization of the ATHMC part, a hierarchical framework to sample all relevant spatial configuration data, and a telescoping of the method with some kind of subspace multigrid eigenproblem solver similar to the one presented in [48].

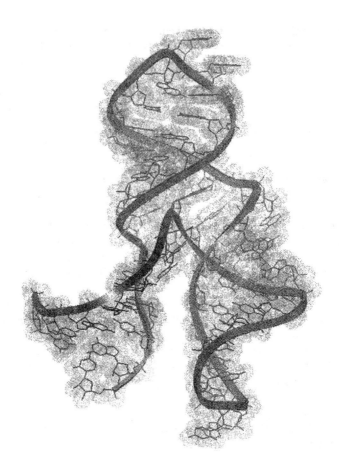

Figure 4.5: Hammerhead molecule, a flexible candidate for RNA drug design.

References

[1] M. P. Allen and D. J. Tildesley: *Computer Simulations of Liquids.* Clarendon, Oxford (1990).

[2] A. Amadei, A. Linssen, and H. Berendsen: *Essential dynamics of proteins.* Proteins, Vol. 17 (1993).

[3] V.I. Arnold: *Mathematical Methods of Classical Mechanics.* Second Edition. Springer (1989).

[4] I. Babuška, A. Miller: *A feedback finite element method with a–posteriori error estimation: part I.* Comput. Meth. Appl. Mech. Engrg. **61** pp. 1–40 (1987).

[5] I. Babuška, W.C. Rheinboldt: *Estimates for adaptive finite element computations.* SIAM J. Numer. Anal. **15** pp. 736–754 (1978).

[6] G. Bader, U. Nowak, P. Deuflhard: *An advanced simulation package for large chemical reaction systems.* In: R. Aiken (ed.), Stiff Computation, Oxford University Press, pp. 255–264 (1985).

[7] A. Bachem, M. Jünger, R. Schrader (eds.): *Mathematik in der Praxis.* Springer–Verlag (1995).

[8] R.E. Bank: *PLTMG: A Software Package for Solving Elliptic Partial Differential Equations. Users' Guide 8.0* SIAM (1998).

[9] R.E. Bank, A.H. Sherman, A. Weiser: *Refinement algorithms and data structures for regular local mesh refinement.* In: R. Stapleman et al (eds.), Scientific Computing, North–Holland, pp. 3–17 (1983).

[10] R.E. Bank, A. Weiser: *Some a posteriori error estimates for elliptic partial differental equations.* Math. Comp. **44**, pp. 283–301 (1985).

[11] P. Bastian, K. Birken, K. Johannsen, S. Lang, N. Neuss, H. Rentz-Reichert, C. Wieners: *UG – A flexible software toolbox for solving partial differential equations.* Computing and Visualization in Science, Vol. 1, pp. 27–40 (1997).

[12] P. Bastian, G. Wittum: *Adaptive multigrid methods: The UG concept.* In: W. Hackbusch and G. Wittum (eds.), Adaptive Methods – Algorithms, Theory and Applications, Series Notes on Numerical Fluid Mechanics, Vol. 46, pp. 17–37, Vieweg, Braunschweig (1994).

[13] R. Beck, P. Deuflhard, R. Hiptmair, B. Wohlmuth, R.H.W. Hoppe: *Adaptive Multilevel Methods for Edge Element Discretizations of Maxwell's Equations.* Surveys of Mathematics for Industry, accepted for publication (1999).

[14] R. Beck, B. Erdmann, R. Roitzsch: *KASKADE 3.0 – User's Guide.* ftp://ftp.zib.de/pub/kaskade (1996).

[15] J. Bey: *Tetrahedral Grid Refinement.* Computing, Vol. 55, No. 4, pp. 355–378 (1995).

[16] F.A. Bornemann, B. Erdmann, R. Kornhuber: *Adaptive multilevel methods in three space dimensions.* Int. J. Num. Meth. in Eng. **36**, pp. 3187–3203 (1993).

[17] F.A. Bornemann, B. Erdmann, R. Kornhuber: *A posteriori error estimates for elliptic problems in two and three space dimensions.* SIAM J. Numer. Anal. **33**, pp. 1188–1204 (1996).

[18] F. Bornemann: *An adaptive multilevel approach to parabolic equations I. General theory and implementation.* IMPACT Comput. Sci. Engrg. **2**, pp. 279–317 (1990).

[19] F. Bornemann: *An adaptive multilevel approach to parabolic equations II. Variable–order time discretization based on a multilevel error correction.* IMPACT Comput. Sci. Engrg. **3**, pp. 93–122 (1991).

[20] F. Bornemann, P. Deuflhard: *The Cascadic Multigrid Method for Elliptic Problems.* Numer. Math. 75, Springer International, pp. 135–152 (1996).

[21] A. Bossavit: *Solving Maxwell's equations in a closed cavity and the question of spurious modes.* IEEE Trans. Mag., 26(2), pp. 702–705 (1990).

[22] R. Bradel, A. Kleinke, K.-H. Reichert: *Molar Mass Distribution of Microbial Poly (D-3-Hydroxybutyrate) in the Course of Intracellular Synthesis and Degradation.* Makromol. Chem., Rapid Commun. **12**, p. 583 (1991).

[23] D. Braess, W. Hackbusch: *A new convergence proof for the multigrid method including the V-cycle.* SIAM. J. Numer. Anal., 20, pp. 967–975 (1983).

[24] J. Bramble, J. Pasciak, J. Xu: *Parallel multilevel preconditioners.* Math. Comp. **55**, pp. 1–22 (1990).

[25] J. Bramble, J. Pasciak, J. Wang, J. Xu: *Convergence estimates for multigrid algorithms without regularity assumptions.* Math. Comp. **57**, pp. 23–45 (1991).

[26] A. Brandt: *Multi–level adaptive solutions to boundary–value problems.* Math. Comp. **31**, pp. 333–390 (1977).

[27] M.-O. Bristeau, G. Etgen, W. Fitzigibbon, J.-L. Lion, J. Periaux, M. Wheeler (eds.): *Computational Science for the 21st Century.* Tours, France. Wiley–Interscience–Europe (1997).

[28] M. Dellnitz, A. Hohmann: *A subdivision algorithm for the computation of unstable manifolds and global attractors.* Numer. Math. 75, pp. 293–317 (1997).

[29] M. Dellnitz, O. Junge: *On the approximation of complicated dynamical behavior.* SIAM J. Num. Anal. 36(2), pp. 491–515 (1999).

[30] P. Deuflhard: *Cascadic Conjugate Gradient Methods for Elliptic Partial Differential Equations. Algorithm and Numerical Results.* In [61], pp. 29–42 (1994).

[31] P. Deuflhard: *Uniqueness Theorems for Stiff ODE Initial Value Problems.* In: D.F. Griffiths and G.A. Watson (eds.): *Numerical Analysis 1989*, Longman Scientific & Technical, Harlow, Essex, UK, pp. 74–205 (1990).

[32] P. Deuflhard and F. Bornemann: *Numerische Mathematik II — Integration gewöhnlicher Differentialgleichungen.* Walter de Gruyter, Berlin, New York (1994).

[33] P. Deuflhard, M. Dellnitz, O. Junge, Ch. Schütte: *Computation of Essential Molecular Dynamics by Subdivision Techniques.* In [36], pp. 98–115 (1998).

[34] P. Deuflhard, T. Friese, F. Schmidt, R. März, H.-P. Nolting: *Effiziente Eigenmodenberechnung für den Entwurf integriert-optischer Chips.* In [56], pp. 267–279 (1996).

[35] P. Deuflhard, J. Hohmann: *Numerical analysis. A First Course in Scientific Computation.* Berlin, New York: de Gruyter (1995).

[36] P. Deuflhard, A. Hermans, B. Leimkuhler, A.E. Mark, S. Reich, and R.D. Skeel (eds.): *Computational Molecular Dynamics: Challenges, Methods, Ideas.* Lecture Notes in Computational Science and Engineering, Vol. 4, Springer-Verlag (1998).

[37] P. Deuflhard, W. Huisinga, A. Fischer, and Ch. Schütte: *Identification of almost invariant aggregates in reversible nearly uncoupled Markov chains.* Lin. Alg. Appl., accepted.

[38] P. Deuflhard, J. Lang, U. Nowak: *Recent Progress in Dynamical Process Simulation.* In: H. Neunzert (ed.): *Topics in Industrial Mathematics*, Wiley & Teubner Publishers, pp. 122–137 (1996).

[39] P. Deuflhard, P. Leinen, H. Yserentant: *Concepts of an Adaptive Hierarchical Finite Element Code.* IMPACT Comp. Sci. Eng. 1, pp. 3–35 (1989).

[40] P. Deuflhard, M. Seebass: *Adaptive Multilevel FEM as Decisive Tools in the Clinical Cancer Therapy Hyperthermia.* In: Choi-Hong Lai, Peter Bjørstad, Mark Cross, O. Widlund (eds.), Procs. 11th International Conference on Domain Decomposition Methods (DD11), UK, 1998 (to appear 1999).

[41] P. Deuflhard, M. Seebass, D. Stalling, R. Beck, H.C. Hege: *Hyperthermia Treatment Planning in Clinical Cancer Therapy: Modelling, Simulation, and Visualization.* Plenary Keynote talk, 15th IMACS World Congress 1997. In: Achim Sydow (ed.), Vol. 3, *Computational Physics, Chemistry and Biology.* Wissenschaft and Technik Verlag, pp. 9–17 (1997).

[42] P. Deuflhard, M. Weiser: *Local Inexact Newton Multilevel FEM for Non-linear Elliptic Problems.* In [27], pp. 129–138 (1997).

[43] P. Deuflhard, M. Weiser, M. Seebass: *A New Nonlinear Elliptic Multilevel FEM Appied to Regional Hyperthermia.* Konrad Zuse Zentrum, Preprint SC 98-35 (1998).

[44] P. Deuflhard, M. Weiser: *Global Inexact Newton Multilevel FEM for Nonlinear Elliptic Problems.* In [53], pp. 71–89 (1998).

[45] P. Deuflhard, M. Wulkow: *Computational Treatment of Polyreaction Kinetics by Orthogonal Polynomials of a Discrete Variable.* IMPACT Comp. Sci. Eng.1, pp. 269–301 (1989).

[46] P. Deuflhard, M. Wulkow: *Simulationsverfahren für die Polymerchemie.* In [7], pp. 117–136 (1995).

[47] A. Fischer, F. Cordes, and Ch. Schütte: *Hybrid Monte Carlo with adaptive temperature in mixed canonical ensemble: Efficient conformational analysis of RNA.* J. Comput. Chem., 19(15), pp. 1689-1697 (1998).

[48] T. Friese, P. Deuflhard, F. Schmidt: *A Multigrid Method for the Complex Helmholtz Eigenvalue Problem.* In: Procs. 11th International Conference on Domain Decomposition Methods (DD11), UK, 1998 (to appear 1999).

[49] H. Gajewski, K. Zacharias: *On an Initial Value Problem for a Coagulation Equation with Growth Term.* Math. Nachr. 109, pp. 135–156 (1982).

[50] G.H. Golub, C.F. Van Loan: *Matrix Computations.* The Johns Hopkins University Press, Baltimore, London, 2nd Edition (1989).

[51] H. Grubmüller, P. Tavan. *Molecular dynamics of conformational substates for a simplified protein model.* J. Chem. Phys. 101 (1994).

[52] W. Hackbusch: *Multi-Grid Methods and Applications.* Springer Verlag, Berlin, Heidelberg, New York (1995).

[53] W. Hackbusch, G. Wittum (eds.): *Multigrid Methods, Lecture Notes in Computational Science and Engineering.* Vol. 3, Springer–Verlag (1998).

[54] H.-C. Hege, M. Seebaß, D. Stalling, M. Zöckler: *A Generalized Marching Cubes Algorithm Based On Non-Binary Classifications.* ZIB Preprint SC 97–05 (1997).

[55] R. Hiptmair: *Multilevel Preconditioning for Mixed Problems in Three Dimensions.* Ph.D. Thesis, Wissner, Augsburg (1996).

[56] K.-H. Hoffmann, W. Jäger, T. Lohmann, H. Schunk (eds.): *MATHE-MATIK Schlüsseltechnologie für die Zukunft.* Springer (1997).

[57] C.S. Hsu: *Cell-to-Cell Mapping. A Method of Global Analysis for Nonlinear Systems.* In: F. John, J.E. Marsden, L. Sirovich (eds.): *Applied Mathematical Sciences*, Vol. 64, Springer–Verlag.

[58] W. Huisinga, Ch. Best, R. Roitzsch, Ch. Schütte, and F. Cordes: *From Simulation Data to Conformational Ensembles: Structure and Dynamics Based Methods.* Konrad-Zuse-Zentrum Berlin. Preprint SC 98–36 (1998).

[59] P. Iedema, M. Wulkow, H. Hoefsloot: *Modeling Molecular Weight and Degree of Branching Distribution of Low Density Polyethylene.* Submitted to Macromolecules (1999).

[60] V.E. Katsnelson: *Conditions under which systems of eigenvectors of some classes of operators form a basis.* Funkt Anal. Appl.1, pp. 122–132 (1967).

[61] D.E. Keyes, J. Xu (eds.): *Domain Decomposition Methods in Scientific and Engineering Computing.* AMS Series Contemporary Mathematics, Vol. 180, Providence (1994).

[62] R. Kornhuber, R. Roitzsch: *On adaptive grid refinement in the presence of internal or boundary layers.* IMPACT Comput. Sci. Engrg. 2, pp. 40–72 (1990).

[63] J. Lang: *Adaptive FEM for Reaction–Diffusion Equations.* Appl. Numer. Math. 26, pp. 105–116 (1998).

[64] J. Lang: *Adaptive Multilevel Solution of Nonlinear Parabolic PDE Systems. Theory, Algorithm, and Applications.* Habilitation Thesis, Free University of Berlin, 1999. Konrad–Zuse-Zentrum Berlin, Preprint SC 99–20 (1999).

[65] J. Lang, B. Erdmann, M. Seebass: *Impact of Nonlinear Heat Transfer on Temperature Distribution in Regional Hyperthermia.* Accepted for publication in IEEE Transaction on Biomedical Engineering (1999).

[66] W. E. Lorensen, H. E. Cline: *Marching Cubes: A high resolution 3D surface construction algorithm.* Computer Graphics 21:4, pp. 163-169 (1987).

[67] J. Mandel, S.F. McCormick: *A multilevel variational method for $Au = \Lambda Bu$ on composite grids.* J. Comp. Phys. 80, pp. 442-452 (1989).

[68] B. Mayfield: *Nonlocal Boundary Conditions for the Schrödinger Equation.* PhD thesis, University of Rhode Island, Providence, RI, (1989).

[69] A.H.E. Müller, D. Yan, M. Wulkow: *Molecular Parameters of Hyperbranched Polymers Made by Self-condensing Vinyl Polymerization. I. Molecular weight distribution.* Macromolecules 30, pp. 7015 (19997).

[70] J.C. Nédeléc: *Mixed finite elements in* $\mathbb{R}^3$. Numer. Math. 35, pp. 315–341 (1980).

[71] M.E. Go Ong: *Hierarchical Basis Preconditioners for Second Order Elliptic Problems in Three Ddimensions.* PhD Thesis, University of California, Los Angeles (1989).

[72] H.H. Pennes: *Analysis of tissue and arterial blood temperatures in the resting human forearm.* J. Appl. Phys. 1, pp. 93-122 (1948).

[73] M.C. Rivara: *Algorithms for refining triangular grids suitable for adaptive and multigrid techniques.* Int. J. Numer. Meth. Engrg. 20, pp. 745-756 (1984).

[74] Y. Saad: *Numerical Methods for Large Eigenvalue Problems.* Manchester University Press (1992).

[75] J.M. Sanz-Serna, M.P. Calvo: *Numerical Hamiltonian Problems.* Chapman & Hall (1994).

[76] F. Schmidt: *An adaptive approach to the numerical solution of Fresnel's wave equation.* IEE Journal of Lightwave Technology, 11(9), pp. 1425–1435 (1993).

[77] F. Schmidt: *Computation of discrete transparent boundary conditions for the 2D Helmholtz equation.* Konrad–Zuse–Zentrum Berlin. Preprint SC 97-64 (1997).

[78] F. Schmidt, P. Deuflhard: *Discrete Transparent Boundary Conditions for the Numerical Solution of Fresnel's equation.* Computers Math. Applic. **29**, No. 9, pp. 53–76 (1995).

[79] F. Schmidt, D. Yevick: *Discrete transparent boundary conditions for Schrödinger-type equations.* J. Comput. Phys., Vol. 134, pp. 96–107 (1997).

[80] Ch. Schütte: *Conformational Dynamics: Modelling, Theory, Algorithm, and Application to Biomolecules.* Habilitation Thesis. Konrad–Zuse–Zentrum Berlin, Preprint SC 99–18 (1999).

[81] Ch. Schütte, A. Fischer, W. Huisinga, and P. Deuflhard: *A Direct Approach to Conformational Dynamics based on Hybrid Monte Carlo.* J. Comput. Phys. **151**, Special Issue on Computational Molecular Biophysics, pp. 146–168 (1999).

[82] V.V. Shaidurov: *Some estimates of the rate of convergence for the cascadic conjugate-gradient method.* Preprint, Otto–von–Guericke–Universität. Magdeburg (1994).

[83] M. Seebass, D. Stalling, M. Zöckler, H.-C. Hege, P. Wust, R. Felix, P. Deuflhard: *Surface Mesh Generation for Numerical Simulations of Hyperthermia Treatments.* In: Procs. 16th Annual Meeting of the European Society for Hyperthermic Oncology, p. 146, Berlin (1997).

[84] M. Seebass, D. Stalling, J. Nadobny, P. Wust, R. Felix, P. Deuflhard: *Three-Dimensional Finite Element Mesh Generation for Numerical Simulations of Hyperthermia Treatments,* Proc. 7th Int. Congress on Hyperthermic Oncology, Rome, Vol. 2, pp. 547-548 (1996).

[85] C.W. Song, A. Lokshina, J.G. Rhee, M. Patten, and S.H. Levitt: *Implication of blood flow in hyperthermic treatment of tumors.* IEEE Trans. Biomed. Engrg. 31, pp. 9-16 (1984).

[86] W.F. van Gunsteren, S.R. Billeter, A.A. Eising, P.H. Hünenberger, P. Krüger, A.E. Mark, W.R.P. Scott, and I.G. Tironi: *Biomolecular Simulation: The GROMOS96 Manual and User Guide.* vdf Hochschulverlag AG, ETH Zürich (1996).

[87] M. Wulkow: *Numerical Treatment of Countable Systems of Ordinary Differential Equations.* Dissertation, Konrad–Zuse–Zentrum Berlin. Technical Report TR 90–08 (1990).

[88] M. Wulkow: *Feature article – The Simulation of Molecular Weight Distributions in Polyreaction Kinetics by Discrete Galerkin Methods.* Macromol. Theory Simul. 5, pp. 393–416 (1996).

[89] J. Xu: *Theory of Multilevel Methods.* PhD Thesis, Report No. AM 48, Department of Mathematics, Pennsylvania State University (1989).

[90] J. Xu: *Iterative methods by space decomposition and subspace correction.* SIAM Review 34 pp. 581–613 (1992).

[91] H. Yserentant: *On the multilevel splitting of finite element spaces.* Numer. Math. 58 pp. 163–184 (1986).

Inverse Problems and Their Regularization

Heinz W. Engl

Industrial Mathematics Institute, Johannes Kepler Universität,
Altenbergerstr. 69, A-4040 Linz, Austria
e-mail: engl@indmath.uni-linz.ac.at
http://www.indmath.uni-linz.ac.at

Contents

1   Introduction

Inverse Problems are concerned with determining *causes* for *desired* or *observed* effects; from this formulation, it is clear that inverse problems appear frequently in all kinds of applied fields. Examples of important classes of inverse problems are

- *tomography* (cf. [57]), which is important both in medical applications and in nondestructive testing [28]

- *inverse scattering* (cf. [16], [61])

- *inverse heat conduction problems* like solving a heat equation backwards in time or "sideways" (i.e., with Cauchy data on part of the boundary) (cf. [29], [4])

- *geophysical inverse problems* like determining a spatially varying density distribution in the earth from gravity measurements (cf. [27])

- inverse problems in *imaging* like deblurring and denoising (cf. [5])

- *identification of parameters* in (partial) differential equations from interior or boundary measurements of the solution (cf. [3], [46]), the latter case appearing e.g. in *impedance tomography* ([45]).

Detailed references for these and many more classes of inverse problems can be found e.g. in [30], [21], [24],[35],[53], [49],[42],[15].

We now give one example each for inverse problems concerned with determining causes for an *observed* and a *desired* effect, respectively. Usually, in the first case, one is interested in uniqueness, since one needs a unique cause for the observation, while in the second case, where not the cause but the desired effect is important, uniqueness is not even desirable: it can be advantageous to have several possible strategies for obtaining a desired effect available. Both our examples stem from different aspects of iron and steel production:

Example 1 (cf. [59]): The blast furnace process is still the most important technology in steel production to get iron from ore. Due to the high temperatures (up to $1500°C$), measurements of any kind are impossible inside the furnace. Therefore, the development of mathematical models allowing the numerical simulation of the process is important (see [20] for a kinetic model of the blast furnace process). The problem treated in [59] has to do with operational safety: The inside lining of the furnace consists of bricks of different materials (with different thermal conductivities) and is subject to physical and chemical wear, leading to the erosion of the bricks, i.e., the thickness of the lining is decreasing. At the same time, due to variations in the working conditions of the furnace, slag may deposit; this happens mainly on the sides, but also on the bottom of the furnace, increasing the thickness of the lining.

If the lining becomes too thin, the process must be stopped to avoid a breakthrough which not only would cause tremendous damage but is also very dangerous to the staff. On the other hand, since it is problematic to reactivate a stopped furnace, to halt production entrains almost inevitably the relining of the furnace. During that time, the steel plant has to deal with an enormous loss of production of pig-iron, which may trouble even subsequent production steps, besides high expenses for the relining. Moreover, at the beginning / restart of a campaign it takes some time to get the whole process running, and the produced pig-iron is of lower quality. Altogether, long campaigns are desired, so that the process should not be stopped too early. This given, the problem is to reliably calculate the thickness of the wall in the hearth in order to stop the furnace right in time. To this purpose, temperature measurement devices are placed inside the wall when the lining is bricked. One wants to calculate the shape of the wall on the inner, inaccessible side of the furnace, based on these measurements.

Since the furnace is (essentially) rotationally symmetric, the wall of the furnace in the region considered can be modelled in cylindrical coordinates. The (reasonable) assumption of stationarity and rotational symmetry lead to the following nonlinear elliptic equation in the radial and height variables (since the temperature u does not depend on the angular variable):

$$\frac{\partial}{\partial r}(\lambda \frac{\partial u}{\partial r}) + \frac{\lambda}{r} \cdot \frac{\partial u}{\partial r} + \frac{\partial}{\partial z}(\lambda \frac{\partial u}{\partial z}) = 0 \quad \text{in } \Omega, \tag{1}$$

where Ω denotes a radial cross-section of the wall (see Figure 1) with the boundary $\Gamma = \Gamma_1 \cup \Gamma_2 \cup \Gamma_3 \cup \Gamma_4$.

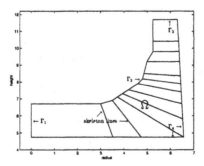

Figure 1: Sketch of the furnace: two-dimensional, rotationally symmetric

The heat conductivity depends on the material, which changes over Ω (since Ω is covered by different types of brick and possibly slag), and hence on position, and also on the temperature:

$$\lambda = \lambda(r, z, u). \tag{2}$$

Due to rotational symmetry and the fact that the lining continues past Γ_3, we have the boundary conditions

$$\frac{\partial u}{\partial n} = 0 \quad \text{on} \quad \Gamma_1 \quad \text{and on} \quad \Gamma_3. \tag{3}$$

The outer surface is cooled by water with temperature T_w, which leads to the boundary condition

$$-\lambda \frac{\partial u}{\partial n} = \alpha_0 \cdot (u - T_w) \quad \text{on} \quad \Gamma_4 \tag{4}$$

with a (measurable) heat transfer coefficient α_0. At the inner surface of the wall, an analogous condition

$$-\lambda \frac{\partial u}{\partial n} = \alpha \cdot (u - T_i) \quad \text{on} \quad \Gamma_2 \tag{5}$$

holds, where the heat transfer coefficient α depends on the actual material present at Γ_2 and T_i is the interior temperature of the furnace (i.e., of molten iron).

If Ω, i.e., the shape of the inside lining, were known, we could solve (1)-(5), which is called the *direct problem*, and compute the temperature field u in Ω. Especially, we could predict the temperatures

$$\tilde{u}_j := u(x_j), j \in \{1, \cdots, m\} \tag{6}$$

measured by thermocouples located at finitely many points $x_j \in \Omega$. Our *inverse problem* now is to determine the inner contour Γ_2 and hence Ω from measurements of $\tilde{u}_j$. Note that this problem could also be viewed as a shape optimization or as a free boundary problem.

Of course, one cannot expect uniqueness in this setup, since one has only finitely many data $\tilde{u}_1, \cdots, \tilde{u}_m$ to determine the curve Γ_2. One could consider the corresponding infinite-dimensional problem of determining Γ_2 from temperature values at a whole curve inside Ω or, which is more practical, describe Γ_2 by finitely many parameters. In [59], Γ_2 has been described by the distances p_i from the outer contour Γ_4 along the skeleton lines (see Figure 1). If we denote, for $p = (p_1, \cdots, p_n)$, by u_p the temperature field determined by (1)-(5) with Γ_2 (and hence Ω) determined by p, then our inverse problem can be reformulated as the least squares problem

$$\Phi(p) := \sum_{j=1}^{m} (u_p(x_j) - \tilde{u}_j)^2 \rightarrow \min., \quad p \in C, \tag{7}$$

where C symbolizes constraints on p (see [59]). It turns out, however, that this least squares problem is quite unstable: Figure 2 shows two domains Ω, which are markedly different, but lead to about the same value for Φ. This can be remedied by *regularization*, which will be explained below.

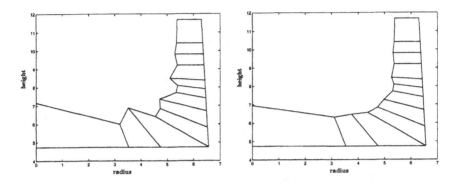

Figure 2: Two possible solutions, the first one being not very realistic, while the second one is physically reasonable

Note that uniqueness is of importance here, since one wants to find a unique inner lining which gives information about the current status of the furnace, and not several (or even infinitely many) possible inner contours. Of course, with finitely many data, one can at best obtain uniqueness for finitely many parameters describing Γ_2, i.e., one needs some a-priori assumptions about Γ_2.

Example 2 (cf. [33]): At the beginning of the continuous casting process the liquid steel is cooled in a water–cooled mould to form a solidified shell which can support the liquid pool at the mould exit. Typical temperatures at the end of the mould are 1100 °C at the strand surface and 1550 °C in the center of the strand. Since steel does not solidify at a fixed temperature, but over a temperature interval, there is a mushy region where the steel is neither completely solid nor completely liquid. Steel grades used in continuous casting are completely solid below a temperature T_{SOLID} of 1350 to 1500 °C, and the solidification starts at a temperature T_{LIQUID} between 1450 and 1525 °C (strongly depending on the content of alloying elements). Liquid steel is provided with a temperature which is typically 30 °C higher than the liquidus temperature. In the mushy region segregation occurs, depending on parameters like content of alloying elements, solidification rate and superheat temperature.

After the mould, the cooling process is continued in secondary cooling zones by spraying water onto the strand. Near the point of final solidification (*crater end*) the segregation effects are pronounced. In this region the temperature gradient in the center of the strand is much larger than the temperature gradient at the strand surface. The increased shrinkage of the strand center causes rest melt enriched with impurifications to flow towards the crater end which decreases the product quality. One possibility to reduce center segregation is to compensate shrinkage of the strand center by slightly reducing the strand thickness at the crater end. This process is called *soft-reduction*.

In order to make soft–reduction work efficiently, one has to be able to find a cooling strategy which, for a known but variable casting speed, makes sure that the crater end always remains in the soft–reduction zone.

For a constant casting speed, the more general inverse problem of controlling the whole solidification front by secondary cooling was considered e.g. in [26]. The problem for a variable casting speed, even if one only needs to control the location of the crater end, is much more complicated and complex since no cross–section of the steel strand can now be considered independently from the others. In practice, a change of the casting speed is e.g. necessary if there are problems in the steel making or in the hot rolling process. It also could be that the casting process has to stop for a short time and then start again to make it possible to change the setup of the casting machine, e.g., the width.

The control parameter for this problem is the amount of water sprayed onto the strand in the secondary cooling zones. For technical reasons the amount of water sprayed onto the strand has to be constant in each cooling zone (there are typically six cooling zones), but can vary in time; thus one has to admit piecewise constant, i.e., non–smooth heat transfer functions in the mathematical model. Furthermore, there are upper and lower bounds for the amount of water.

In [33], the problem is described in Lagrangian coordinates. The strand moves in the casting direction z with the casting speed $v(t)$, y is the direction of width and x the direction of thickness of the strand in which the spray cooling takes place. It is justified to neglect the heat conduction in the casting direction since it is much smaller than the heat conduction in the directions perpendicular to it. If the strand width is much larger than the strand thickness, which is fulfilled for slabs whose cross–section is around 120×25 cm, it is also justified to assume that for a fixed cross–section the temperature in the area near the center of the strand width–direction only depends on the x–coordinate and time. If d is the strand thickness and if the cooling is done in a symmetric way, the temperature is symmetric with respect to the plane $x = d/2$, provided the initial temperature $f(x, t)$ fulfills this condition. Therefore, the partial derivative u_x of the temperature u vanishes for $x = d/2$.

The amount of cooling water with temperature U_w sprayed onto the strand in the secondary cooling region enters into the boundary conditions via a heat transfer function $g(z, t)$. The model also includes cooling due to radiation (U_a being the temperature of the surrounding air).

The material parameters of steel primarily relevant for the problem are the thermal conductivity $k(u)$, the density $\rho(u)$ and the specific heat $c(u)$, which all depend on the temperature $u = u(x, z, t)$. σ is the Stefan–Boltzmann constant and ϵ a further material parameter that depends on the steel grade.

With these notations, the temperature field is described by the following mathematical model:

$$[k(u)u_x]_x = \rho(u)c(u)u_t \tag{8}$$
$$u(x, z = -\textstyle\int_0^t v(\tau)\, d\tau, t) = f(x, t) \tag{9}$$
$$u_x(d/2, z, t) = 0 \tag{10}$$
$$k(u(0, z, t))u_x(0, z, t) = g(z + \textstyle\int_0^t v(\tau)\, d\tau, t)(u(0, z, t) - U_w) \tag{11}$$
$$+ \sigma\epsilon(u^4(0, z, t) - U_a^4).$$

The Eulerial coordinate 0 describes the lower end of the mould. At time t the cross–section with Lagrangian coordinate $z = -\int_0^t v(\tau)\, d\tau$ passes this end of the mould. The temperature $f(x, t)$ at the end of the mould in (9) for this cross–section can be computed by calculating the temperature in the mould. It is assumed that at the beginning of the mould there is a constant initial temperature. The function f depends on time since the casting speed is not constant, so that different cross–sections remain in the mould for different amounts of time and thus cool down in a different way. $z + \int_0^t v(\tau)\, d\tau$ is the Eulerian coordinate of the cross–section with Lagrangian coordinate z at time t. This is used to compute the cooling for the cross–section z in (11).

If the material parameters k, ρ and c, the casting speed v as a function of time, the initial temperature f and the heat transfer function g are known, we call the problem of solving (8)–(11) for the temperature u the *direct problem*. The neglection of the heat conduction in the casting direction implies that one can look at each cross–section of the strand separately for solving the direct problem: The temperatures of different cross–sections evolve independently from each other. From the solution of the direct problem, the point of complete solidification of each cross–section can be obtained. Note that mathematical questions like existence and uniqueness of a solution of (8)–(11) are non–trivial due to the fact that g is piecewise constant (see [34]).

The problem to find a cooling strategy that makes sure that the crater end always remains within the soft–reduction zone amounts to solving the following *inverse problem*: If the material parameters, the casting speed, the initial temperature are known and the position of the soft–reduction zone is prescribed, a heat transfer function g has to be computed in such a way that with this cooling, the resulting solution of the direct problem is such that the crater end of each cross–section of the strand is in the given region defined by the soft–reduction zone.

In the direct problem one can look at each cross–section separately to compute the temperature if one neglects the heat conduction in the casting direction. This decoupling could also be done for the inverse problem if the casting speed would be constant: In this case, one could compute a time–independent heat transfer function *g* for one cross–section and use it also for the other cross–sections. In the present case, however, one cannot use a time–independent heat transfer function since the casting speed varies. But then, since the heat transfer function has to be piecewise constant in space and time, one has to look at all cross–sections simultaneously. This increases the complexity of the problem considerably and requires to look for an efficient algorithm to solve the inverse problem and hence also the direct problem.

Note that for this inverse problem, uniqueness is not important: it is fine, maybe even advantageous, if these are several cooling strategies leading to the *desired effect* to keep the crater and within the soft reduction zone.

The mathematical models for inverse problems usually are not *well-posed* in the sense of Hadamard, i.e., they do not fulfill the postulates that for all admissible data,

- a solution *exists*,

- the solution is *unique*,

- the solution *depends continuously* on the data.

Thus, inverse problems are the source for *ill-posed* problems, which is the reason that the analysis and numerics of such problems has been extensively studied in recent years. Of special importance is the instability (discontinuous dependence on the data) usually connected with ill–posedness, since it obviously creates severe numerical problems: algorithms one uses for well-posed problems usually do not work for ill–posed problems unless special features are added which remedy the instability; these lead to *regularization methods* which we will discuss in this paper. Numerical examples which clearly show the instability of traditional methods (using no regularization) applied to inverse problems can be found for a deconvolution problem arising in physical chemistry in [51], for inverse scattering in [18] (see also [17]).

This instability in the inverse problem is usually connected with a smoothing effect in the *direct (forward) problem*: Convolution smoothes, so does heat conduction forward in time, which means that rough details are smoothed out in the forward problem; this in turn has the effect that in the inverse problem, small errors may be amplified into large artefacts.

Both examples treated above were *nonlinear* inverse problems, as are many inverse problems arising in practice. Also, note that even if the corresponding direct problem is linear, an inverse problem connected with it may be nonlinear, as is the case e.g. in the problem of identifying a parameter

$a = a(x)$ from the values of the solution of the problem

$$(au_x)_x = f \quad \text{in} \quad [0,1] \tag{12}$$
$$u_x(0) = u(1) = 0:$$

Its solution is

$$a(x) = \frac{1}{u_x(x)} \int_0^x f(s)ds$$

and hence depends in a nonlinear way on u. Note also that it involves differentiating the data u, which is unstable.

Nevertheless, we first present aspects of the theory of *linear* inverse problems, since it is more complete and, of course, easier. A prototype of linear inverse problems are *integral equations of the first kind*

$$\int_G k(s,t)x(t)dt = f(s) \tag{13}$$

for an unknown function x (cf. [30]).

2 Regularization of Linear Ill–Posed Problems

We phrase linear ill-posed problems in the abstract setting of linear operator equations

$$Tx = y \tag{14}$$

with a bounded linear operator T between Hilbert spaces X and Y, which especially includes first-kind integral equations like (13). As concept of solution we use that of a *best–approximate solution*, which is a minimizer of the residual $\|Tx - y\|$ and minimizes $\|x\|$ among all minimizers of the residual. We thus concentrate on the stability issue, doing away with non–uniqueness. The operator $T^\dagger$ which maps $y \in D(T^\dagger)$ to the best–approximate solution of (14) is known as the *Moore-Penrose (generalized) inverse* of T (see e.g. [56], [36]). Its domain is

$$D(T^\dagger) = R(T) \dotplus R(T)^\perp, \tag{15}$$

and $T^\dagger$ is bounded if and only if $R(T)$ is closed. Hence, the problem of determining the best–approximate solution of (14) is well–posed if and only if $R(T)$ is closed, which is not the case if T is compact and $\dim R(T) = \infty$. Hence, *integral equations of the first kind* with a non–degenerate kernel such that the integral operator is compact are *always ill–posed*.

If T is compact with singular system $\{\sigma_n; u_n, v_n\}$, then $y \in D(T^\dagger)$ if and only if

$$\sum_{n=1}^{\infty} \frac{|\langle y, v_n \rangle|^2}{\sigma_n^2} < \infty, \tag{16}$$

which is called the *Picard criterion*; then, the best–approximate solution of (14) has the series representation

$$T^\dagger y = \sum_{n=1}^{\infty} \frac{\langle y, v_n \rangle}{\sigma_n} u_n. \tag{17}$$

This formula clearly shows the instability: errors in Fourier components of y with respect to v_n, i.e., in $\langle y, v_n \rangle$, are amplified by $\frac{1}{\sigma_n}$, which grows without bound since $\lim_{n \to \infty} \sigma_n = 0$ (if dim $R(T) = \infty$). This error amplification is the worse the faster the σ_n decay to 0, i.e., for the case of integral operators, the smoother the kernel is! Thus, integral equations of the first kind with a smooth kernel are *severely ill–posed*, the decay rate of the singular values can be used to quantify ill-posedness.

The formula (17) also indicates how one can stabilize (*regularize*) the problem: using that $T u_n = \sigma_n v_n$, (17) can also be written in symmetric form as

$$T^\dagger y = (T^*T)^\dagger T^* y = \sum_{n=1}^{\infty} \frac{\langle T^* y, u_n \rangle}{\sigma_n^2} u_n. \tag{18}$$

We now replace the amplification factors $\frac{1}{\sigma_n^2}$ by a *filtered* version $U_\alpha(\sigma_n^2)$, where the filter function U_α is such that for $\alpha > 0, U_\alpha$ is piecewise continuous on $[0, +\infty[$ (especially, does not have a pole at 0) and that

$$\lim_{\alpha \to 0} U_\alpha(\lambda) = \frac{1}{\lambda} \quad \text{for} \quad \lambda > 0; \tag{19}$$

this convergence is pointwise only, and $|\lambda \cdot U_\alpha(\lambda)|$ is required to stay bounded.

As *regularized solution* (with noisy data y^δ fulfilling $\|y - y^\delta\| \le \delta$) we take:

$$x_\alpha^\delta := \sum_{n=1}^{\infty} U_\alpha(\sigma_n^2) \cdot \langle T^* y, u_n \rangle u_n = \sum_{n=1}^{\infty} \sigma_n \cdot U_\alpha(\sigma_n^2) \cdot \langle y, v_n \rangle u_n. \tag{20}$$

The constant α is called the *regularization parameter*.

Different choices of the filter function U_α now lead to different regularization methods: E.g., for

$$U_\alpha(\lambda) := \frac{1}{\alpha + \lambda} \tag{21}$$

we obtain *Tikhonov regularization*

$$x_\alpha^\delta = \sum_{n=1}^{\infty} \frac{\sigma_n}{\alpha + \sigma_n^2} \langle y^\delta, v_n \rangle u_n = (\alpha I + T^*T)^{-1} T^* y^\delta, \tag{22}$$

which can also be characterized in variational form as minimizer of the functional

$$x \to \|Tx - y^\delta\|^2 + \alpha \|x\|^2. \tag{23}$$

See [30] for other choices of filter functions U_α leading to different regularization methods.

In any regularization method, the choice of the *regularization parameter* α is crucial. As can be seen from (23), it represents a compromise between accuracy and stability: if α is small, the residual term $\|Tx - y^\delta\|^2$ dominates, so that minimizers tend to lead to a small residual, but may have a large norm; if the norm contains derivatives, e.g., is the H^2-norm, this means that solutions tend to be heavily oscillating. Such oscillations can be seen (in a nonlinear inverse problem) the left picture in Figure 2 above; note that this picture comes from minimizing the least-squares functional (7) without a regularization term.

One distinguishes between two classes of parameter choice rules: In *a-priori rules*, one chooses the regularization parameter as a function of the noise level only, i.e., $\alpha = \alpha(\delta)$, while in *a-posteriori rules*, one uses the noise level and the actual data, i.e., $\alpha = \alpha(\delta, y^\delta)$. A classical a-posteriori rule is the so-called *discrepancy principle*, where α is chosen such that

$$\|Tx_\alpha^\delta - y^\delta\| = C\delta \tag{24}$$

holds (with some $C > 1$), which is a nonlinear equation in α.

One can show (see [30]) that *error-free* strategies, where $\alpha = \alpha(y^\delta)$ does not depend on δ, cannot lead to convergence as $\delta \to 0$ in the sense that $\lim_{\delta \to 0} x_\alpha^\delta = T^\dagger y$ for all y^δ with $\|y - y^\delta\| \leq \delta$. These error-free strategies include the popular methods of generalized cross-validation ([71]) and the *L-curve method* ([40]); for its non–convergence, see [23] and [70].

Crucial questions in applying regularization methods are convergence *rates* and how to choose regularization parameters to obtain optimal convergence rates. By convergence rates we mean rates for the worst-case error

$$\sup\{\|x_\alpha^\delta - T^\dagger y\| / \|y - y^\delta\| \leq \delta\} \tag{25}$$

for $\delta \to 0$ and $\alpha = \alpha(\delta)$ or $\alpha = \alpha(\delta, y^\delta)$ chosen appropriately. For an ill–posed problem, no uniform rate valid for all $y \in Y$ can be given ([67]), such rates can only be given on compact subsets of X (cf. [52]). An example of this are convergence rates valid under *source conditions* (with a $\nu > 0$)

$$T^\dagger y \in R((T^*T)^\nu), \tag{26}$$

which can, using a singular system, also be written as

$$\sum_{n=1}^\infty \frac{|\langle y, v_n \rangle|^2}{\sigma_n^{2+4\nu}} < \infty, \tag{27}$$

which puts the Picard criterion into a scale of such conditions (for $\nu = 0$): under the condition (26), which can be (due to the fact that usually T is smoothing) thought of as (abstract) a–priori smoothness condition, Tikhonov regularization converges with the rate

$$\|x_\alpha^\delta - T^\dagger y\| = O(\delta^{\frac{2\nu}{1+2\nu}}) \tag{28}$$

for the a–priori choice

$$\alpha \sim \delta^{\frac{2}{1+2\nu}} \tag{29}$$

and $\nu \leq 1$.

This (as it turns out, optimal under (26)) rate is also achieved with the a–posteriori parameter choice (24), but only for $\nu \leq \frac{1}{2}$. For a–posteriori parameter choice rules that always lead to optimal convergence rates see [22] and [63].

The total error of a regularization method has the form outlined in Fig. 3: the regularization error $\|x_\alpha^0 - T^\dagger y\|$ goes to 0 as $\alpha \to 0$, while the propagated data error $\|x_\alpha^0 - x_\alpha^\delta\|$ grows without bound as $\alpha \to 0$; the method of [22] for optimally choosing the regularization parameter is based on minimizing the total error, which is not as easy as it sounds since the curves in Figure 3 are not computable unless one knows the exact solution of the problem.

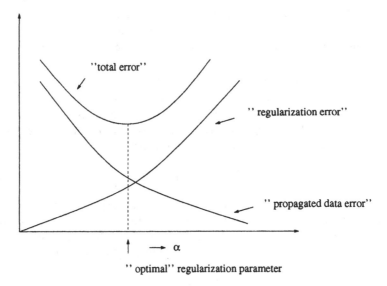

Figure 3: Typical error behaviour

Methods which are closely related to regularization methods are *mollifier methods*, where one looks for a smoothed version of the solution, cf. [55]. While these are still linear methods, there are also *nonlinear methods* for solving linear ill–posed problems, e.g., the *Backus–Gilbert method* ([1], [50]) and the *conjugate gradient method* ([9], [54], [37]). For details about these methods and other aspects we do not touch here (like the non–compact case and numerical aspects in the framework of combining regularization in Hilbert space with projection into finite-dimensional spaces) see [30]. We mention in closing that there is a close connection between regularization and approximation by neural networks (cf. [12] and the references quoted there).

3 Nonlinear Ill–Posed Problems

We describe nonlinear ill–posed problems as nonlinear operator equations

$$F(x) = y, \tag{30}$$

where F maps a domain $D(F)$ in a Hilbert space X into a Hilbert space Y. The basic assumptions for a reasonable theory are that F is continuous and weakly sequentially closed (cf. [25]). In contrast to linear problems, F is usually not explicitly given in models on nonlinear inverse problems; it symbolizes the operator describing the direct problem, the *forward map*. E.g., in the situation of Example 1, F is the operator mapping the inner contour Γ_2 to the observed temperatures u on a curve in Ω (if we consider the infinite-dimensional problem) or the operator mapping the parameter vector p (which in turn determines Γ_2) to the observation vector $(\tilde{u}_1, \cdots, \tilde{u}_m)$. Although, in the latter situation, F maps a finite-dimensional vector to another finite-dimensional vector, it involves solving the infinite-dimensional problem (1)-(5).

In the situation of Example 2, F could be the operator mapping a given cooling strategy to the distance (as a function of time) from the crater end to the soft-reduction zone, $y \equiv 0$.

In the simple parameter identification (12), F is the *parameter-to-solution-map* mapping a to the solution u of the boundary value problem (12).

Like in the linear case, we introduce a generalized solution concept (for simplicity, for the case that y is *attainable*, i.e., (30) has a solution; see [8] for the non-attainable case): For $x^* \in X$, we call a solution of (30) which minimizes $\|x - x^*\|$ among all solutions an x^*-*minimum–norm solution* ($x^* - MNS$) of (30); neither existence nor uniqueness are guaranteed in the nonlinear case. The choice of x^*, which will also appear in algorithms, is crucial; it should include available a–priori information like positions of singularities in x if they happen to be available.

For the question when a nonlinear equation (30) is ill–posed, we refer to [25]; like in the linear case, compactness (and some kind of non–degeneracy) of F implies ill–posedness.

The regularization methods for linear problems based on spectral theory described in Section 2 cannot be directly carried over to the nonlinear setting; however, Tikhonov regularization in its variational form of minimizing the functional (23) makes also sense for a nonlinear problem: For noisy data y^δ

with $\|y - y^\delta\| \leq \delta$, we denote by x_α^δ *any* global solution of the minimization problem

$$\|F(x) - y^\delta\|^2 + \alpha\|x - x^*\|^2 \to \min. \text{ over } D(F). \tag{31}$$

Minimizers always exist for $\alpha > 0$, but need not be unique. Hence, convergence has to be understood in an appropriate set–valued sense (cf.[68]).

The basic result about convergence rates for solutions of (31) says that, under a Lipschitz condition on F' (assumed to exists as Fréchet derivative), x_α^δ converges to an x^*-MNS $x^\dagger$ of (30) with rate $O(\sqrt{\delta})$ if $\alpha \sim \delta$ and a source condition

$$x^\dagger - x^* = F'(x^\dagger)^* w \tag{32}$$

holds with a sufficiently small w (see [25]). Formally, this condition and the obtained convergence rate correspond to (26) and (28) for $\nu = \frac{1}{2}$, respectively. Again, (32) is an abstract smoothness condition for the *difference* between the true solution $x^\dagger$ and the a–priori guess x^* used in the algorithm; this underpins the remark made above that all known rough details of $x^\dagger$ should be put into the a–priori guess x^*.

Example 1, continued: In [59], Tikhonov regularization has been applied to stably solving the inverse problem. The least squares minimization problem (7) is replaced by

$$\sum_{j=1}^{m}(u_p(x_j) - \tilde{u}_j)^2 + \alpha\sum_{j\equiv 1}^{n+1}(\psi_j(p) - \psi_{j-1}(p))^2 \to \min., p \in C, \tag{33}$$

with $\psi_0(p) = \psi_{n+1}(p) = \frac{\pi}{2}$ and $\psi_j(p)(j = 1, \cdots, n)$ being the angle between the j-tn skeleton line and Γ_2 (see Figure 1). This regularization term penalizes oscillations in Γ_2 as those in the left picture of Figure 2. We quote a result obtained in [59] with this regularization method:

In each picture in Figure 4, the dashed line is the starting contour, the continuous line is the solution calculated by the regularization algorithm, and the third line (---)is the "true solution" which was used to generate the data, which were then corrupted by 10 % noise. The boxes show the measurement points. The pictures show that without regularization, no useful solution can be obtained, while regularization yields quite good results (considering the high noise level). The unusual values for α result from the scaling of the problem.

We finally remark that the same practical problem was also treated in [69] for the simple case of a constant heat conductivity; instead of Tikhonov regularization, Kalman filtering was used there.

A basic disadvantage of Tikhonov regularization for nonlinear problems is that as opposed to the linear case, the functional in (31) need no longer be convex, so that it is difficult to minimize it (see [14] for a case where this is still possible).This fact makes iterative methods an attractive alternative:

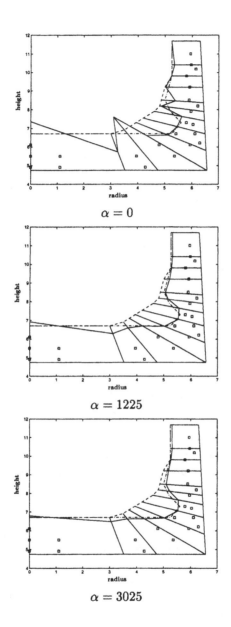

$\alpha = 0$

$\alpha = 1225$

$\alpha = 3025$

Figure 4: Solutions with simulated data and different regularization parameters, 10% noise

A natural candidate for an iterative method for solving (30) would be Newton's method, where the $(n + 1)$-st iterate x_{n+1} is determined by the linear equation

$$F'(x_n)(x_{n+1} - x_n) = -(F(x_n) - y) \qquad (34)$$

or (to avoid existence problems as far as possible) by the corresponding normal equation

$$F'(x_n)^* F'(x_n)(x_{n+1} - x_n) = -F'(x_n)^*(F(x_n) - y). \qquad (35)$$

However, if F is compact, so is $F'(x_n)$, so that the linear problems (34) and (35) are ill–posed. In order to make these linear problems well–posed, the operator $F'(x_n)^* F'(x_n)$ has to be replaced by an operator which has a bounded inverse. If this replacement is just $\omega \cdot I$, we obtain the (nonlinear damped) *Landweber method*

$$x_{n+1}^\delta = x_n^\delta - \omega F'(x_n^\delta)^*(F(x_n^\delta) - y^\delta), \qquad (36)$$

if this replacement is defined via (linear) Tikhonov regularization, we obtain the *Levenberg–Marquardt method*

$$x_{n+1}^\delta = x_n^\delta - [F'(x_n^\delta)^* F'(x_n^\delta) + \alpha_n I]^{-1} F'(x_n^\delta)^*(F(x_n^\delta) - y^\delta); \qquad (37)$$

a variant of this method which has slightly better stability properties for ill–posed problems is the *iteratively regularized Gauß–Newton method (IRGN)*

$$\begin{aligned} x_{n+1}^\delta &= x_n^\delta - [F'(x_n^\delta)^* F'(x_n^\delta) + \alpha_n I]^{-1} \\ &\quad [F'(x_n^\delta)^*(F(x_n^\delta) - y^\delta) + \alpha_n(x_n^\delta - \zeta)] \end{aligned} \qquad (38)$$

with an appropriately chosen a–priori guess ζ.

In [33], a Quasi-Newton method was used for solving the inverse problem numerically. A crucial point there (as often with iterative methods involving F') was the efficient computation of the derivative, which was done using an adjoint problem; see [33] for details and numerical results.

Crucial questions for iterative methods are the choice of the stopping index $n = n(\delta, y^\delta)$ and convergence rates. These have been studied e.g. in [2], [39], [38], [7], [6], [47], [48], [58], [65], [60], [41]. Multilevel iterative methods habe been considered in [64], [66], [32]. In [32], a general approach for analyzing convergence of iterative methods for solving nonlinear ill–posed problems has been developed (see also [31]); this approach is based on invariance principles as they had been used for well–posed problems in [19]. This approach also allows to prove convergence rates under logarithmic source conditions of the type

$$x^\dagger - x_0 = f_p(F'(x^\dagger)^* F'(x^\dagger))w \qquad (39)$$

instead of (32) (with x^* replaced by the initial iterate x_0) with

$$f_p(\lambda) = (1 - \ln \lambda)^{-p}, 0 < \lambda \le 1. \qquad (40)$$

Such conditions turn out to be appropriate for severely ill–posed problems for which the forward operator is exponentially smoothing like inverse scattering (see [43], [44]). The resulting convergence rate is logarithmic; e.g.,

$$\|x_{n_*}^\delta - x^\dagger\| = O((-\ln \delta)^{-p}) \tag{41}$$

holds under conditions (39), (40) (and some technical conditions, see [43], [32]) for Landweber iteration and IRGN, if the stopping index n_* is determined via a discrepancy principle

$$\|F(x_{n_*}^\delta) - y^\delta\| \le C\delta < \|F(x_n^\delta) - y^\delta\|, 0 \le n < n_* \tag{42}$$

with a suitable $C > 2$.

We close by reporting about the application of Landweber iteration to a parameter identification problem arising in modelling the crystallization of polymers; this provides a link to the survey paper [13] in this volume.

A model for one-dimensional crystallization of polymers leads to initial-boundary value problems of the form

$$\frac{\partial T}{\partial t} = D\frac{\partial^2 T}{\partial x^2} + L\frac{\partial \xi}{\partial t} \tag{43}$$

$$\frac{\partial}{\partial t}\left(\frac{1}{\widetilde{G}(T)(1-\xi)}\frac{\partial \xi}{\partial t}\right) = \frac{\partial}{\partial x}\left(\frac{\widetilde{G}(T)}{1-\xi}\frac{\partial \xi}{\partial x}\right) + 2\frac{\partial}{\partial t}(\widetilde{N}(T)) \tag{44}$$

with initial conditions

$$T(x,0) = T^0(x) \tag{45}$$
$$\xi(x,0) = 0 \tag{46}$$
$$\frac{\partial \xi}{\partial t}(x,0) = 0 \tag{47}$$

and boundary conditions

$$\frac{\partial T}{\partial n}(x,t) = \alpha(T(x,t) - T^1(x,t)) \qquad \text{for } x \in \partial\Omega \tag{48}$$

$$\frac{\partial \xi}{\partial t}(x,t) + \widetilde{G}(T)\frac{\partial \xi}{\partial n}(x,t) = 0 \qquad \text{for } x \in \partial\Omega, \tag{49}$$

where T^1 in (48) represents an exterior temperature in the cooling process at the boundary (see[11]).

In these equations, T denotes the temperature, ξ the degree of crystallinity. The parameters arising in the heat transfer problem, i.e., the diffusion coefficient D, the latent heat L and the heat transition coefficient α, can be determined experimentally. The function $\widetilde{N}$ represents an equivalent nucleation rate per unit of length and $\widetilde{G}$ the radial growth rate of a nucleus.

Note that $\tilde{G}$ and $\tilde{N}$ depend on temperature rather strongly and cannot (at least not simultaneously) be measured as functions of temperature. Some materials admit experimental determination of the growth rate (cf.[62]), in these cases the identification of $\tilde{N}$ for known $\tilde{G}$ is of special interest.

The measurable quantities in usual experiments are the temperature T on the boundary of the crystallization domain Ω (or at least on a part of the boundary $\Gamma \subset \partial\Omega$) as well as the degree of crystallinity $\xi(x, t_*)$ at the end of the process. The temperature at the end of the experiment is not measurable, because the structure is frozen in by a final quench below the melting point.

Via various transformations (see [11]), (44) - (49) can be rewritten as

$$
\begin{aligned}
u_t &= Du_{xx} + Le^{-v}v_t \\
v_t &= a(u)w \\
w_t &= (a(u)v_x)_x + b(u)_t
\end{aligned}
\tag{50}
$$

$$
\begin{aligned}
u|_{t=0} &= u^0 \\
v|_{t=0} &= 0 \\
w|_{t=0} &= 0
\end{aligned}
\tag{51}
$$

$$
\begin{aligned}
u_x(0, t) &= -\alpha(u(0, t) - u^1(0, t)) \\
w(0, t) - v_x(0, t) &= 0
\end{aligned}
\tag{52}
$$

$$
\begin{aligned}
u_x(1, t) &= 0 \\
v_x(1, t) &= 0.
\end{aligned}
\tag{53}
$$

The quantities $a(u)$ and $b(u)$ in (50) are the equivalents to $\tilde{G}(u)$ and $\tilde{N}(u)$, transformed to the new reference variable u. The function u is a scaled temperature, i.e., $u = 1$ is above the equilibrium melting point and $u = 0$ below the glass transition temperature, hence the temperature range of interest $[u_1, u_2]$ is a subset of $[0, 1]$. In practice, the temperature at the beginning of the experiment is always above the melting point, i.e., u_2 may be chosen as the melting point. The fact that no nucleation occurs above this temperature implies that $b(u) = 0$ for $u \geq u_2$, so one can assume that $b \in H^1([u_1, u_2])$ and $b(u_2) = 0$.

In [11], it is assumed that the temperature data are given at the point $x = 0$ by $u(0, t) = u_B(t)$ in the time interval $I = (0, t_*)$. From the available observation of ξ at $t = t_*$, one can compute the value v_* for $v|_{t=t_*}$, which can therefore be assumed to be given.

Identification of the nucleation rate $\tilde{N} = \tilde{N}(T)$ in (43) - (49) is equivalent to identifying $b = b(u)$ in (50) - (53). Thus, this identification problem can formally be written as the nonlinear operator equation

$$
F(b) = (u_B^\delta, v_*^\delta),
\tag{54}
$$

where u_B^δ and v_*^δ denote noisy measurements of u_B and v_* with noise level δ. The operator F, the *parameter-to-output-map*, maps an admissible value

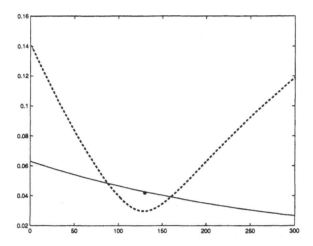

Figure 5: Residual (solid) and error (dashed) vs. iteration number. The stopping index is marked by ∗.

of the parameter b to the values u_B and v_* obtained from solving (50) - (53) with this parameter.

Equation (54) has the abstract from (30). In [11], Landweber iteration (36) is applied to this inverse problem leading to an iteration

$$b^{k+1} = b^k - \omega F'(b^k)^*(F(b^k) - (u_B^\delta, v_*^\delta)). \tag{55}$$

In [11], it is shown how $F'(b^k)^*$ can be computed efficiently by solving a system of linear partial differential equations. For results of Landweber iteration (with the stopping index determined by a discrepancy principle (42)), we refer to [11]. We just quote one Figure, which shows the residual and the error $\|b^k - b^{\text{true}}\|$ versus the iteration number for simulated data with 2 % noise:

The error in Figure 5 shows the same behaviour as predicted by the theory (cf. Figure 3), the stopping index determined by the discrepancy principle is remarkably close to the minimal error; if one iterates beyond this stopping index, the error increases quickly, which is a behaviour one should always keep in mind when solving inverse problems iteratively.

In [10], the iteratively regularized Gauß-Newton method (38) has been applied to the same inverse problem.

Acknowledgement: Partially supported by the Austrian Science Foundation FWF, projects F 1308 and P13478-INF.

References

[1] G. Backus and F. Gilbert, Numerical applications of a formalism for geophysical inverse problems, *Geophys. J. R. Astron. Soc.*, **13**(1967) 247-276

[2] A.B. Bakushinskii, The problem of the convergence of the iteratively regularized Gauß-Newton method, *Comput. Maths. Math. Phys.*, **32**(1992) 1353-1359

[3] H. Banks and K. Kunisch, Parameter Estimation Techniques for Distributed Systems, Birkhäuser, Boston, 1989

[4] J. Beck, B. Blackwell, C.S. Clair, Inverse Heat Conduction: Ill–Posed Problems; Wiley, New York 1985

[5] M. Bertero, P. Boccacci, Introduction to Inverse Problems in Imaging; Inst. of Physics Publ., Bristol 1998

[6] B. Blaschke, A. Neubauer, and O. Scherzer, On convergence rates for the iteratively regularized Gauss-Newton method, *IMA J. Numer. Anal.*, **17**(1997) 421–436

[7] A. Binder, M. Hanke, and O. Scherzer, On the Landweber iteration for nonlinear ill-posed problems, *J. Inverse Ill-Posed Probl.*, **4**(1996) 381–389

[8] A. Binder, H. W. Engl, C. W. Groetsch, A. Neubauer and O. Scherzer, Weakly closed nonlinear operators and parameter identification in parabolic equations by Tikhonov regularization, *Appl. Anal.*, **55**(1994) 215–234

[9] H. Brakhage, On ill-posed problems and the method of conjugate gradients, in [24], 165–175

[10] M. Burger, Iterative regularization of a parameter identification problem occurring in polymer crystallization, submitted

[11] M. Burger, V. Capasso, H.W. Engl, Inverse problems related to crystallization of polymers, *Inverse Problems*, **15** (1999) 155-173

[12] M. Burger, H.W. Engl, Training neural networks with noisy data as an ill–posed problem, to appear in *Advances in Comp. Mathematics*

[13] V. Capasso, Mathematical Models for Polymer Crystallization, This volume

[14] G. Chavent, K. Kunisch, On weakly nonlinear inverse problems, *SIAM J. Appl. Math.*, **56**(1996) 542–572

[15] D. Colton, R. Ewing, and W. Rundell, Inverse Problems in Partial Differential Equations, SIAM, Philadelphia, 1990

[16] D. Colton and R. Kress, Inverse Acoustic and Electromagnetic Scattering Theory, Springer, Berlin, 1992

[17] D. Colton, A. Kirsch, A simple method for solving inverse scattering problems in the resonance region, *Inverse Problems*, **12**(1996) 383–393.

[18] D. Colton, M. Piana, R. Potthast, A simple method using Morozov's discrepancy principle for solving inverse scattering problems, *Inverse Problems*, **13**(1997) 1477-1493

[19] P. Deuflhard and G. Heindl, Affine invariant convergence theorems for Newton's method and extensions to related methods, *SIAM J. Numer. Anal.*, **16**(1979) 1–10

[20] H. Druckenthaner, H. Zeisel, A. Schiefermüller, G. Kolb, A. Ferstl, H. Engl, A. Schatz, Online simulation of the blast furnace, *Advanced Steel*, (1997-98) 58-61

[21] H.W. Engl, Regularization methods for the stable solution of inverse problems, *Surveys on Mathematics for Industry*, **3**(1993) 71–143

[22] H. W. Engl and H. Gfrerer, A posteriori parameter choice for general regularization methods for solving linear ill-posed problems, *Appl. Numer. Math.* **4**(1988) 395–417

[23] H. W. Engl and W. Grever, Using the L-curve for determining optimal regularization parameters, *Numer. Math.*, **69**(1994) 25–31

[24] H. W. Engl and C. W. Groetsch, Inverse and Ill-Posed Problems, Academic Press, Orlando, 1987

[25] H. W. Engl, K. Kunisch and A. Neubauer, Convergence rates for Tikhonov regularization of nonlinear ill-posed problems,*Inverse Problems* **5**(1989) 523–540

[26] H.W. Engl, T. Langthaler, Control of the solidification front by secondary cooling in continuous casting of steel, in: H.W. Engl, H. Wacker and W. Zulehner, eds. Case Studies in Industrial Mathematics Teubner, Stuttgart, 1988, 51-77

[27] H. W. Engl, A. K. Louis and W. Rundell (eds.), Inverse Problems in Geophysics, SIAM, Philadelphia, 1996

[28] H. W. Engl, A. K. Louis and W. Rundell (eds.), Inverse Problems in Medical Imaging and Nondestructive Testing, Springer, Vienna, New York, 1996

[29] H. W. Engl and W. Rundell (eds.), Inverse Problems in Diffusion Processes, SIAM, Philadelphia, 1995

[30] H.W. Engl, M. Hanke, A. Neubauer, Regularization of Inverse Problems, Kluwer Academic Publishers, Dordrecht, 1996

[31] H.W. Engl and O. Scherzer, Convergence rates results for iterative methods for solving nonlinear ill–posed problems, to appear in: D. Colton, H.W. Engl, A.K. Louis, J. McLaughlin, W.F. Rundell (eds.), Solution Methods for Inverse Problems, Springer, Vienna/New York, 2000

[32] P. Deuflhard, H.W. Engl, and O. Scherzer, A convergence analysis of iterative methods for the solution of nonlinear ill–posed problems under affinely invariant conditions, *Inverse Probl.*, **14**(1998) 1081–1106

[33] W. Grever, A. Binder, H.W. Engl, K. Mörwald, Optimal cooling strategies in continuous casting of steel with variable casting speed, *Inverse Problems in Engineering*, **2** (1996) 289-300

[34] W. Grever A nonlinear parabolic initial boundary value problem modelling the continuous casting of steel with variable casting speed, *ZAMM*, **78** (1998) 109-119

[35] C. W. Groetsch, Inverse Problems in the Mathematical Sciences, Vieweg, Braunschweig, 1993

[36] C. W. Groetsch, Generalized Inverses of Linear Operators: Representation and Approximation, Dekker, New York, 1977

[37] M. Hanke, Conjugate Gradient Type Methods for Ill-Posed Problems, Longman Scientific & Technical, Harlow, Essex, 1995

[38] M. Hanke, A regularizing Levenberg-Marquardt scheme, with applications to inverse groundwater filtration problems. *Inverse Probl.*, **13**(1997) 79–95

[39] M. Hanke, A. Neubauer, and O. Scherzer. A convergence analysis of Landweber iteration for nonlinear ill-posed problems, *Numer. Math.*, **72**(1995) 21–37

[40] P. C. Hansen and D. P. O'Leary, The use of the L-curve in the regularization of discrete ill-posed problems, *SIAM J. Sci. Comput.*, **14** (1993) 1487–1503

[41] F. Hettlich and W. Rundell, The determination of a discontinuity in a conductivity from a single boundary measurement, *Inverse Probl.*, **14**(1998) 67–82

[42] B. Hofmann, Mathematik inverser Probleme, Teubner, Stuttgart 1999

[43] T. Hohage, Logarithmic convergence rates of the iteratively regularized Gauß–Newton method for an inverse potential and an inverse scattering problem, *Inverse Problems*, **13**(1997) 1279-1299

[44] T. Hohage, Convergence rates of a regularized Newton method in sound–hard inverse scattering, *SIAM J. Numer. Anal.*, **36**(1998) 125–142

[45] D. Isaacson and J.C. Newell, Electrical Impedance Tomography, *SIAM Review*, **41**(1999) 85–101

[46] V. Isakov, Inverse Problems in Partial Differential Equations, Springer, Berlin, New York, 1998

[47] B. Kaltenbacher, Some Newton-type methods for the regularization of nonlinear ill-posed problems, *Inverse Probl.*, **13**(1997) 729–753

[48] B. Kaltenbacher, A posteriori parameter choice strategies for some Newton type methods for the regularization of nonlinear ill–posed problems, *Numer. Math.*, **79**(1998) 501–528

[49] A. Kirsch, An Introduction to the Mathematical Theory of Inverse Problems, Springer, New York 1996

[50] A. Kirsch, B. Schomburg and G. Berendt, The Backus-Gilbert method, *Inverse Problems*, **4**(1988) 771–783

[51] G. Landl, T. Langthaler, H. W. Engl and H. F. Kauffmann, Distribution of event times in time-resolved fluorescence: the exponential series approach – algorithm, regularization, analysis, *J. Comput. Phys.*, **95**(1991) 1–28

[52] A. Leonov and A. Yagola, Special regularizing methods for ill–posed problems with sourcewise represented solutions, *Inverse Problems*, **14**(1998) 1539–1550

[53] A. K. Louis, Inverse und schlecht gestellte Probleme, Teubner, Stuttgart, 1989

[54] A. K. Louis, Convergence of the conjugate gradient method for compact operators, in [24], 177–183

[55] A. K. Louis and P. Maass, A mollifier method for linear operator equations of the first kind, *Inverse Problems*, **6**(1990) 427–440

[56] M. Z. Nashed, Generalized Inverses and Applications, Academic Press, New York, 1976

[57] F. Natterer, The Mathematics of Computerized Tomography, Teubner, Stuttgart, 1986

[58] A. Neubauer and O. Scherzer, A convergence rate result for a steepest descent method and a minimal error method for the solution of nonlinear ill-posed problems, *Z. Anal. Anwend.*, **14**(1995) 369–377

[59] E. Radmoser, R. Wincor, Determining the inner contour of a furnace from temperature measurements, Industrial Mathematics Institute, Johannes Kepler Universität Linz, *Technical Report* 12 (1998)

[60] R. Ramlau, A modified Landweber-method for inverse problems *Num. Funct. Anal. Opt.*, **20**(1999) 79–98

[61] A. Ramm, Scattering by Obstacles, Reidel, Dordrecht, 1986

[62] E.Ratajski, H.Janeschitz-Kriegl, How to determine high growth speeds in polymer crystallization, *Colloid Polym. Sci.*, **274** (1996) 938-951

[63] T. Raus, Residue principle for ill-posed problems, *Acta et Comment. Univers. Tartuensis*, **672**(1984) 16-26

[64] A. Rieder, A wavelet multilevel method for ill-posed problems stabilized by Tikhonov regularization, *Numer. Math.*, **75**(1997) 501–522

[65] O. Scherzer, A convergence analysis of a method of steepest descent and a two-step algorithm for nonlinear ill-posed problems, *Numer. Funct. Anal. Optimization*, **17**(1996) 197–214

[66] O. Scherzer, An iterative multi level algorithm for solving nonlinear ill-posed problems, *Numer. Math.*, **80**(1998) 579–600

[67] E. Schock, Approximate solution of ill-posed equations: arbitrarily slow convergence vs. superconvergence, in: G. Hämmerlin and K. H. Hoffmann, eds., Constructive Methods for the Practical Treatment of Integral Equations, Birkhäuser, Basel, 1985, 234–243

[68] T. I. Seidman and C. R. Vogel, Well-posedness and convergence of some regularization methods for nonlinear ill-posed problems, *Inverse Problems*, **5**(1989) 227–238

[69] M. Tanaka, T. Matsumoto, S. Oida, Identification of unknown boundary shape of rotationally symmetric body in steady heat conduction via BEM and filter theories, in: M. Tanaka and G.S. Dulikravich eds., Inverse Problems in Engineering Mechanics, Elsevier Science B.V., Tokyo, 1998, 121-130

[70] C. R. Vogel, Non-convergence of the L-curve regularization parameter selection method,*Inverse Problems*, **12**(1996) 535–547

[71] G. Wahba, Spline Models for Observational Data, SIAM, Philadelphia, 1990

Aerodynamic Shape Optimization Techniques Based On Control Theory

Antony Jameson [+]   Luigi Martinelli [*]

[+] Department of Aeronautics & Astronautics
Stanford University
Stanford, California 94305 USA
[*]Department of Mechanical & Aerospace Engineering
Princeton University
Princeton, NJ 08544, USA

Abstract

These Lecture Notes review the formulation and application of optimization techniques based on control theory for aerodynamic shape design in both inviscid and viscous compressible flow. The theory is applied to a system defined by the partial differential equations of the flow, with the boundary shape acting as the control. The Frechet derivative of the cost function is determined via the solution of an adjoint partial differential equation, and the boundary shape is then modified in a direction of descent. This process is repeated until an optimum solution is approached. Each design cycle requires the numerical solution of both the flow and the adjoint equations, leading to a computational cost roughly equal to the cost of two flow solutions. Representative results are presented for viscous optimization of transonic wing-body combinations and inviscid optimization of complex configurations.

Contents

1   Introduction: Aerodynamic Design

The definition of the aerodynamic shapes of modern aircraft relies heavily on computational simulation to enable the rapid evaluation of many alternative designs. Wind tunnel testing is then used to confirm the performance of designs that have been identified by simulation as promising to meet the performance goals. In the case of wing design and propulsion system integration, several complete cycles of computational analysis followed by testing of a preferred design may be used in the evolution of the final configuration. Wind tunnel testing also plays a crucial role in the development of the detailed loads needed to complete the structural design, and in gathering data throughout the flight envelope for the design and verification of the stability and control system. The use of computational simulation to scan many alternative designs has proved extremely valuable in practice, but it still suffers the limitation that it does not guarantee the identification of the best possible design. Generally one has to accept the best so far by a given cutoff date in the program schedule. To ensure the realization of the true best design, the ultimate goal of computational simulation methods should not just be the analysis of prescribed shapes, but the automatic determination of the true optimum shape for the intended application.

This is the underlying motivation for the combination of computational fluid dynamics with numerical optimization methods. Some of the earliest studies of such an approach were made by Hicks and Henne [1, 2]. The principal obstacle was the large computational cost of determining the sensitivity of the cost function to variations of the design parameters by repeated calculation of the flow. Another way to approach the problem is to formulate aerodynamic shape design within the framework of the mathematical theory for the control of systems governed by partial differential equations [3]. In this view the wing is regarded as a device to produce lift by controlling the flow, and its design is regarded as a problem in the optimal control of the flow equations by changing the shape of the boundary. If the boundary shape is regarded as arbitrary within some requirements of smoothness, then the full generality of shapes cannot be defined with a finite number of parameters, and one must use the concept of the Frechet derivative of the cost with respect to a function. Clearly such a derivative cannot be determined directly by separate variation of each design parameter, because there are now an infinite number of these.

Using techniques of control theory, however, the gradient can be determined indirectly by solving an adjoint equation which has coefficients determined by the solution of the flow equations. This directly corresponds to the gradient technique for trajectory optimization pioneered by Bryson [4]. The cost of solving the adjoint equation is comparable to the cost of solving the flow equations, with the consequence that the gradient with respect to an arbitrarily large number of parameters can be calculated with roughly the same computational cost as two flow solutions. Once the gradient has been calculated, a descent method can be used to determine a shape change which will make an improvement in the design. The gradient can then be recalculated, and the whole process can be repeated until the design converges to an optimum solution, usually within 50 to 100 cycles. The fast calculation of the gradients makes optimization computationally feasible even for designs in three-dimensional viscous flow. There is a possibility that the descent method could converge to a local minimum rather than the global optimum solution. In practice this has not proved a difficulty, provided care is taken in the choice of a cost function which properly reflects the design requirements. Conceptually, with this approach the problem is viewed as infinitely dimensional, with the control being the shape of the bounding surface. Eventually the equations must be discretized for a numerical implementation of the method. For this purpose the flow and adjoint equations may either be separately discretized from their representations as differential equations, or, alternatively, the flow equations may be discretized first, and the discrete adjoint equations then derived directly from the discrete flow equations.

The effectiveness of optimization as a tool for aerodynamic design also depends crucially on the proper choice of cost functions and constraints. One popular approach is to define a target pressure distribution, and then solve the inverse problem of finding the shape that will produce that pressure distribution. Since such a shape does not necessarily exist, direct inverse methods may be ill-posed. The problem of designing a two-dimensional profile to attain a desired pressure distribution was studied by Lighthill, who solved it for the case of incompressible flow with a conformal mapping of the profile to a unit circle [5]. The speed over the profile is

$$q = \frac{1}{h} |\nabla \phi|,$$

where ϕ is the potential which is known for incompressible flow and h is the modulus of the mapping function. The surface value of h can be obtained by setting $q = q_d$, where q_d is the desired speed, and since the mapping function is analytic, it is uniquely determined by the value of h on the boundary. A solution exists for a given speed q_∞ at infinity only if

$$\frac{1}{2\pi} \oint q \, d\theta = q_\infty,$$

and there are additional constraints on q if the profile is required to be closed.

The difficulty that the target pressure may be unattainable may be circumvented by treating the inverse problem as a special case of the optimization problem, with a cost function which measures the error in the solution of the inverse problem. For example, if p_d is the desired surface pressure, one may take the cost function to be an integral over the the body surface of the square of the pressure error,

$$I = \frac{1}{2} \int_B (p - p_d)^2 dB,$$

or possibly a more general Sobolev norm of the pressure error. This has the advantage of converting a possibly ill posed problem into a well posed one. It has the disadvantage that it incurs the computational costs associated with optimization procedures.

The inverse problem still leaves the definition of an appropriate pressure architecture to the designer. One may prefer to directly improve suitable performance parameters, for example, to minimize the drag at a given lift and Mach number. In this case it is important to introduce appropriate constraints. For example, if the span is not fixed the vortex drag can be made arbitrarily small by sufficiently increasing the span. In practice, a useful approach is to fix the planform, and optimize the wing sections subject to constraints on minimum thickness.

2 Formulation of the Design Problem as a Control Problem

The simplest approach to optimization is to define the geometry through a set of design parameters, which may, for example, be the weights α_i applied to a set of shape functions $b_i(x)$ so that the shape is represented as

$$f(x) = \sum \alpha_i b_i(x).$$

Then a cost function I is selected which might, for example, be the drag coefficient or the lift to drag ratio, and I is regarded as a function of the parameters α_i. The sensitivities $\frac{\partial I}{\partial \alpha_i}$ may now be estimated by making a small variation $\delta \alpha_i$ in each design parameter in turn and recalculating the flow to obtain the change in I. Then

$$\frac{\partial I}{\partial \alpha_i} \approx \frac{I(\alpha_i + \delta \alpha_i) - I(\alpha_i)}{\delta \alpha_i}.$$

The gradient vector $\frac{\partial I}{\partial \alpha}$ may now be used to determine a direction of improvement. The simplest procedure is to make a step in the negative gradient direction by setting

$$\alpha^{n+1} = \alpha^n - \lambda \delta \alpha,$$

so that to first order

$$I + \delta I = I - \frac{\partial I^T}{\partial \alpha} \delta \alpha = I - \lambda \frac{\partial I^T}{\partial \alpha} \frac{\partial I}{\partial \alpha}.$$

More sophisticated search procedures may be used such as quasi-Newton methods, which attempt to estimate the second derivative $\frac{\partial^2 I}{\partial \alpha_i \partial \alpha_j}$ of the cost function from changes in the gradient $\frac{\partial I}{\partial \alpha}$ in successive optimization steps. These methods also generally introduce line searches to find the minimum in the search direction which is defined at each step. The main disadvantage of this approach is the need for a number of flow calculations proportional to the number of design variables to estimate the gradient. The computational costs can thus become prohibitive as the number of design variables is increased.

Using techniques of control theory, however, the gradient can be determined indirectly by solving an adjoint equation which has coefficients defined by the solution of the flow equations. The cost of solving the adjoint equation is comparable to that of solving the flow equations. Thus the gradient can be determined with roughly the computational costs of two flow solutions, independently of the number of design variables, which may be infinite if the boundary is regarded as a free surface. The underlying concepts are clarified by the following abstract description of the adjoint method.

For flow about an airfoil or wing, the aerodynamic properties which define the cost function are functions of the flow-field variables (w) and the physical location of the boundary, which may be represented by the function $\mathcal{F}$, say. Then

$$I = I\left(w, \mathcal{F}\right),$$

and a change in $\mathcal{F}$ results in a change

$$\delta I = \left[\frac{\partial I^T}{\partial w}\right]_I \delta w + \left[\frac{\partial I^T}{\partial \mathcal{F}}\right]_{II} \delta \mathcal{F} \tag{1}$$

in the cost function. Here, the subscripts I and II are used to distinguish the contributions due to the variation δw in the flow solution from the change associated directly with the modification $\delta \mathcal{F}$ in the shape. This notation assists in grouping the numerous terms that arise during the derivation of the full Navier–Stokes adjoint operator, outlined later, so that the basic structure of the approach as it is sketched in the present section can easily be recognized.

Suppose that the governing equation R which expresses the dependence of w and $\mathcal{F}$ within the flowfield domain D can be written as

$$R\left(w, \mathcal{F}\right) = 0. \tag{2}$$

Then δw is determined from the equation

$$\delta R = \left[\frac{\partial R}{\partial w}\right]_I \delta w + \left[\frac{\partial R}{\partial \mathcal{F}}\right]_{II} \delta \mathcal{F} = 0. \tag{3}$$

Since the variation δR is zero, it can be multiplied by a Lagrange Multiplier ψ and subtracted from the variation δI without changing the result. Thus equation (1) can be replaced by

$$\begin{aligned}
\delta I &= \frac{\partial I^T}{\partial w}\delta w + \frac{\partial I^T}{\partial \mathcal{F}}\delta \mathcal{F} - \psi^T \left(\left[\frac{\partial R}{\partial w}\right]\delta w + \left[\frac{\partial R}{\partial \mathcal{F}}\right]\delta \mathcal{F}\right) \\
&= \left\{\frac{\partial I^T}{\partial w} - \psi^T \left[\frac{\partial R}{\partial w}\right]\right\}_I \delta w + \left\{\frac{\partial I^T}{\partial \mathcal{F}} - \psi^T \left[\frac{\partial R}{\partial \mathcal{F}}\right]\right\}_{II} \delta \mathcal{F}. \tag{4}
\end{aligned}$$

Choosing ψ to satisfy the adjoint equation

$$\left[\frac{\partial R}{\partial w}\right]^T \psi = \frac{\partial I}{\partial w} \tag{5}$$

the first term is eliminated, and we find that

$$\delta I = \mathcal{G}\delta\mathcal{F}, \tag{6}$$

where

$$\mathcal{G} = \frac{\partial I^T}{\partial \mathcal{F}} - \psi^T \left[\frac{\partial R}{\partial \mathcal{F}}\right].$$

The advantage is that (6) is independent of δw, with the result that the gradient of I with respect to an arbitrary number of design variables can be determined without the need for additional flow-field evaluations. In the case that (2) is a partial differential equation, the adjoint equation (5) is also a partial differential equation and determination of the appropriate boundary conditions requires careful mathematical treatment.

In reference [6] Jameson derived the adjoint equations for transonic flows modeled by both the potential flow equation and the Euler equations. The theory was developed in terms of partial differential equations, leading to an adjoint partial differential equation. In order to obtain numerical solutions both the flow and the adjoint equations must be discretized. The control theory might be applied directly to the discrete flow equations which result from the numerical approximation of the flow equations by finite element, finite volume or finite difference procedures. This leads directly to a set of discrete adjoint equations with a matrix which is the transpose of the Jacobian matrix of the full set of discrete nonlinear flow equations. On a three-dimensional mesh with indices i, j, k the individual adjoint equations may be derived by collecting together all the terms multiplied by the variation $\delta w_{i,j,k}$ of the discrete flow variable $w_{i,j,k}$. The resulting discrete adjoint equations represent a possible discretization of the adjoint partial differential equation. If these equations are solved exactly they can provide an exact gradient of the inexact cost function which results from the discretization of the flow equations. The discrete adjoint equations derived directly from the discrete flow equations become very complicated when the flow equations are discretized with higher order upwind biased schemes using flux limiters. On the other hand any consistent discretization of the adjoint partial differential equation will yield the exact gradient in the limit as the mesh is refined. The trade-off between the complexity of the adjoint discretization, the accuracy of the resulting estimate of the gradient, and its impact on the computational cost to approach an optimum solution is a subject of ongoing research.

The true optimum shape belongs to an infinitely dimensional space of design parameters. One motivation for developing the theory for the partial differential equations of the flow is to provide an indication in principle of how such a solution could be approached if sufficient computational resources were available. Another motivation is that it highlights the possibility of generating ill posed formulations of the problem. For example, if one attempts to calculate the sensitivity of the pressure at a particular location to changes in the boundary shape, there is the possibility that a shape modification could cause a shock wave to pass over that location. Then the sensitivity could become unbounded. The movement of the shock, however, is continuous as the shape changes. Therefore a quantity such as the drag coefficient, which is determined by integrating the pressure over the surface, also depends continuously on the shape. The adjoint equation allows the sensitivity of the drag coefficient to be determined without the explicit evaluation of pressure sensitivities which would be ill posed.

The discrete adjoint equations, whether they are derived directly or by discretization of the adjoint partial differential equation, are linear. Therefore they could be solved by direct numerical inversion. In three-dimensional problems on a mesh with, say, n intervals in each coordinate direction, the number of unknowns is proportional to n^3 and the bandwidth to n^2. The complexity of direct inversion is proportional to the number of unknowns multiplied by the square of the bandwidth, resulting in a complexity proportional to n^7. The cost of direct inversion can thus become prohibitive as the mesh is refined, and it becomes more efficient to use iterative solution methods. Moreover, because of the similarity of the adjoint equations to the flow equations, the same iterative methods which have been proved to be efficient for the solution of the flow equations are efficient for the solution of the adjoint equations.

Studies of the use of control theory for optimum shape design of systems governed by elliptic equations were initiated by Pironneau [7]. The control theory approach to optimal aerodynamic design was first applied to transonic flow by Jameson [8, 6, 9, 10, 11, 12]. He formulated the method for inviscid compressible flows with shock waves governed by both the potential flow and the Euler equations [6]. Numerical results showing the method to be extremely effective for the design of airfoils in transonic potential flow were presented in [13], and for three-dimensional wing design using the Euler equations in [14]. More recently the method has been employed for the shape design of complex aircraft configurations [15, 16], using a grid perturbation approach to accommodate the geometry modifications. The method has been used to support the aerodynamic design studies of several industrial projects, including the Beech Premier and the McDonnell Douglas MDXX and Blended Wing-Body projects. The application to the MDXX is described in [10]. The experience gained in these industrial applications made it clear that the viscous effects cannot be ignored in transonic wing design, and the method has therefore been extended to treat the Reynolds Averaged Navier-Stokes equations [12]. Adjoint methods have also been the subject of studies by a number of other authors, including Baysal and Eleshaky [17], Huan and Modi [18], Desai and Ito [19], Anderson and Venkatakrishnan [20], and Peraire and Elliot [21]. Ta'asan, Kuruvila and Salas [22], who have implemented a one shot approach in which the constraint represented by the flow equations is only required to be satisfied by the final converged solution. In their work, computational costs are also reduced by applying multigrid techniques to the geometry modifications as well as the solution of the flow and adjoint equations.

The next section presents the formulation for the case of airfoils in transonic flow. The governing equation is taken to be the transonic potential flow equation, and the profile is generated by conformal mapping from a unit circle. Thus the control is taken to be the modulus of the mapping function on the boundary. This leads to a generalization of Lighthill's method both to compressible flow, and to design for more general criteria. The subsequent sections discuss the application of the method to automatic wing design with the flow modeled by the three-dimensional Euler and Navier-Stokes equations. The computational costs are low enough that it has proved possible to determine optimum wing designs using the Euler equations in a few hours on workstations such as the IBM590 or the Silicon Graphics Octane.

3 Airfoil Design for Potential Flow using Conformal Mapping

Consider the case of two-dimensional compressible inviscid flow. In the absence of shock waves, an initially irrotational flow will remain irrotational, and we can assume that the velocity vector $\mathbf{q}$ is the gradient of a potential ϕ. In the presence of weak shock waves this remains a fairly good approximation.

1a: z-Plane.   1b: σ-Plane.

Figure 1: Conformal Mapping.

Let p, ρ, c, and M be the pressure, density, speed-of-sound, and Mach number q/c. Then the potential flow equation is

$$\nabla \cdot (\rho \nabla \phi) = 0, \tag{7}$$

where the density is given by

$$\rho = \left\{ 1 + \frac{\gamma - 1}{2} M_\infty^2 \left(1 - q^2 \right) \right\}^{\frac{1}{(\gamma-1)}}, \tag{8}$$

while

$$p = \frac{\rho^\gamma}{\gamma M_\infty^2}, \quad c^2 = \frac{\gamma p}{\rho}. \tag{9}$$

Here M_∞ is the Mach number in the free stream, and the units have been chosen so that p and q have a value of unity in the far field.

Suppose that the domain D exterior to the profile C in the z-plane is conformally mapped on to the domain exterior to a unit circle in the σ-plane as sketched in Figure 1. Let R and θ be polar coordinates in the σ-plane, and let r be the inverted radial coordinate $\frac{1}{R}$. Also let h be the modulus of the derivative of the mapping function

$$h = \left| \frac{dz}{d\sigma} \right|. \tag{10}$$

Now the potential flow equation becomes

$$\frac{\partial}{\partial \theta} (\rho \phi_\theta) + r \frac{\partial}{\partial r} (r \rho \phi_r) = 0 \text{ in } D, \tag{11}$$

where the density is given by equation (8), and the circumferential and radial velocity components are

$$u = \frac{r \phi_\theta}{h}, \quad v = \frac{r^2 \phi_r}{h}, \tag{12}$$

while

$$q^2 = u^2 + v^2. \tag{13}$$

The condition of flow tangency leads to the Neumann boundary condition

$$v = \frac{1}{h} \frac{\partial \phi}{\partial r} = 0 \text{ on } C. \tag{14}$$

In the far field, the potential is given by an asymptotic estimate, leading to a Dirichlet boundary condition at $r = 0$ [23].

Suppose that it is desired to achieve a specified velocity distribution q_d on C. Introduce the cost function

$$I = \frac{1}{2} \int_C (q - q_d)^2 \, d\theta,$$

The design problem is now treated as a control problem where the control function is the mapping modulus h, which is to be chosen to minimize I subject to the constraints defined by the flow equations (7–14).

A modification δh to the mapping modulus will result in variations $\delta\phi$, δu, δv, and $\delta\rho$ to the potential, velocity components, and density. The resulting variation in the cost will be

$$\delta I = \int_C (q - q_d)\, \delta q\, d\theta, \tag{15}$$

where, on C, $q = u$. Also,

$$\delta u = r\frac{\delta\phi_\theta}{h} - u\frac{\delta h}{h}, \quad \delta v = r^2\frac{\delta\phi_r}{h} - v\frac{\delta h}{h},$$

while according to equation (8)

$$\frac{\partial\rho}{\partial u} = -\frac{\rho u}{c^2}, \quad \frac{\partial\rho}{\partial v} = -\frac{\rho v}{c^2}.$$

It follows that $\delta\phi$ satisfies

$$L\delta\phi = -\frac{\partial}{\partial\theta}\left(\rho M^2\phi_\theta\frac{\delta h}{h}\right) - r\frac{\partial}{\partial r}\left(\rho M^2 r\phi_r\frac{\delta h}{h}\right)$$

where

$$L \equiv \frac{\partial}{\partial\theta}\left\{\rho\left(1 - \frac{u^2}{c^2}\right)\frac{\partial}{\partial\theta} - \frac{\rho uv}{c^2}r\frac{\partial}{\partial r}\right\} + r\frac{\partial}{\partial r}\left\{\rho\left(1 - \frac{v^2}{c^2}\right)r\frac{\partial}{\partial r} - \frac{\rho uv}{c^2}\frac{\partial}{\partial\theta}\right\}. \tag{16}$$

Then, if ψ is any periodic differentiable function which vanishes in the far field,

$$\int_D \frac{\psi}{r^2}L\delta\phi\, dS = \int_D \rho M^2\nabla\phi\cdot\nabla\psi\frac{\delta h}{h}\, dS, \tag{17}$$

where dS is the area element $r\, dr\, d\theta$, and the right hand side has been integrated by parts.

Now we can augment equation (15) by subtracting the constraint (17). The auxiliary function ψ then plays the role of a Lagrange multiplier. Thus,

$$\begin{aligned}
\delta I = & \int_C (q - q_d)q\frac{\delta h}{h}\, d\theta - \int_C \delta\phi\frac{\partial}{\partial\theta}\left(\frac{q - q_d}{h}\right)d\theta \\
& - \int_D \frac{\psi}{r^2}L\delta\phi\, dS + \int_D \rho M^2\nabla\phi\cdot\nabla\psi\frac{\delta h}{h}\, dS.
\end{aligned}$$

Now suppose that ψ satisfies the adjoint equation

$$L\psi = 0 \quad \text{in } D \tag{18}$$

with the boundary condition

$$\frac{\partial \psi}{\partial r} = \frac{1}{\rho} \frac{\partial}{\partial \theta} \left(\frac{q - q_d}{h} \right) \quad \text{on } C. \tag{19}$$

Then, integrating by parts,

$$\int_D \frac{\psi}{r^2} L \delta\phi \, dS = -\int_C \rho \psi_r \delta\phi \, d\theta,$$

and

$$\delta I = -\int_C (q - q_d) \, q \frac{\delta h}{h} \, d\theta + \int_D \rho M^2 \nabla\phi \cdot \nabla\psi \frac{\delta h}{h} \, dS. \tag{20}$$

Here the first term represents the direct effect of the change in the metric, while the area integral represents a correction for the effect of compressibility. When the second term is deleted the method reduces to a variation of Lighthill's method [5].

Equation (20) can be further simplified to represent δI purely as a boundary integral because the mapping function is fully determined by the value of its modulus on the boundary. Set

$$\log \frac{dz}{d\sigma} = \mathcal{F} + i\beta,$$

where

$$\mathcal{F} = \log \left| \frac{dz}{d\sigma} \right| = \log h,$$

and

$$\delta\mathcal{F} = \frac{\delta h}{h}.$$

Then $\mathcal{F}$ satisfies Laplace's equation

$$\Delta\mathcal{F} = 0 \text{ in } D,$$

and if there is no stretching in the far field, $\mathcal{F} \to 0$. Also $\delta\mathcal{F}$ satisfies the same conditions. Introduce another auxiliary function P which satisfies

$$\Delta P = \rho M^2 \nabla\psi \cdot \nabla\psi \text{ in } D, \tag{21}$$

and

$$P = 0 \text{ on } C.$$

Then, the area integral in equation (20) is

$$\int_D \Delta P \, \delta\mathcal{F} \, dS = \int_C \delta\mathcal{F} \frac{\partial P}{\partial r} \, d\theta - \int_D P \Delta \delta\mathcal{F} \, dS,$$

and finally

$$\delta I = \int_C \mathcal{G} \, \delta\mathcal{F} \, d\theta,$$

where $\mathcal{F}_c$ is the boundary value of $\mathcal{F}$, and

$$\mathcal{G} = \frac{\partial P}{\partial r} - (q - q_d)\, q. \tag{22}$$

This suggests setting

$$\delta \mathcal{F}_c = -\lambda \mathcal{G}$$

so that if λ is a sufficiently small positive quantity

$$\delta I = -\int_C \lambda \mathcal{G}^2 \, d\theta < 0.$$

Arbitrary variations in $\mathcal{F}$ cannot, however, be admitted. The condition that $\mathcal{F} \to 0$ in the far field, and also the requirement that the profile should be closed, imply constraints which must be satisfied by $\mathcal{F}$ on the boundary C. Suppose that $\log\left(\frac{dz}{d\sigma}\right)$ is expanded as a power series

$$\log\left(\frac{dz}{d\sigma}\right) = \sum_{n=o}^{\infty} \frac{c_n}{\sigma^n}, \tag{23}$$

where only negative powers are retained, because otherwise $\left(\frac{dz}{d\sigma}\right)$ would become unbounded for large σ. The condition that $\mathcal{F} \to 0$ as $\sigma \to \infty$ implies

$$c_o = 0.$$

Also, the change in z on integration around a circuit is

$$\Delta z = \int \frac{dz}{d\sigma}\, d\sigma = 2\pi i\, c_1,$$

so the profile will be closed only if

$$c_1 = 0.$$

In order to satisfy these constraints, we can project $\mathcal{G}$ onto the admissible subspace for $\mathcal{F}_c$ by setting

$$c_o = c_1 = 0. \tag{24}$$

Then the projected gradient $\tilde{\mathcal{G}}$ is orthogonal to $\mathcal{G} - \tilde{\mathcal{G}}$, and if we take

$$\delta \mathcal{F}_c = -\lambda \tilde{\mathcal{G}},$$

it follows that to first order

$$\begin{aligned}
\delta I &= -\int_C \lambda \mathcal{G} \tilde{\mathcal{G}} \, d\theta = -\int_C \lambda \left(\tilde{\mathcal{G}} + \mathcal{G} - \tilde{\mathcal{G}}\right) \mathcal{G} \, d\theta \\
&= -\int_C \lambda \tilde{\mathcal{G}}^2 \, d\theta < 0.
\end{aligned}$$

If the flow is subsonic, this procedure should converge toward the desired speed distribution since the solution will remain smooth, and no unbounded derivatives will appear. If, however, the flow is transonic, one must allow for the appearance of shock waves in the trial solutions, even if q_d is smooth. Then $q - q_d$ is not differentiable. This difficulty can be circumvented by a more sophisticated choice of the cost function. Consider the choice

$$I = \frac{1}{2} \int_C \left(\lambda_1 \mathcal{Z}^2 + \lambda_2 \left(\frac{d\mathcal{Z}}{d\theta} \right)^2 \right) d\theta, \tag{25}$$

where λ_1 and λ_2 are parameters, and the periodic function $\mathcal{Z}(\theta)$ satisfies the equation

$$\lambda_1 \mathcal{Z} - \frac{d}{d\theta} \lambda_2 \frac{d\mathcal{Z}}{d\theta} = q - q_d. \tag{26}$$

Then,

$$
\delta I = \int_C \left(\lambda_1 \mathcal{Z} \delta \mathcal{Z} + \lambda_2 \frac{d\mathcal{Z}}{d\theta} \frac{d}{d\theta} \delta \mathcal{Z} \right) d\theta = \int_C \mathcal{Z} \left(\lambda_1 \delta \mathcal{Z} - \frac{d}{d\theta} \lambda_2 \frac{d}{d\theta} \delta \mathcal{Z} \right) d\theta
$$

$$
= \int_C \mathcal{Z} \, \delta q \, d\theta.
$$

Thus, $\mathcal{Z}$ replaces $q - q_d$ in the previous formulas, and if one modifies the boundary condition (19) to

$$\frac{\partial \psi}{\partial r} = \frac{1}{\rho} \frac{\partial}{\partial \theta} \left(\frac{\mathcal{Z}}{h} \right) \quad \text{on } C, \tag{27}$$

the formula for the gradient becomes

$$\mathcal{G} = \frac{\partial P}{\partial r} - \mathcal{Z} q \tag{28}$$

instead of equation (22). Smoothing can also be introduced directly in the descent procedure by choosing $\delta \mathcal{F}_c$ to satisfy

$$\delta \mathcal{F}_c - \frac{\partial}{\partial \theta} \beta \frac{\partial}{\partial \theta} \delta \mathcal{F}_c = -\lambda \mathcal{G}, \tag{29}$$

where β is a smoothing parameter. Then to first order

$$
\int \mathcal{G} \, \delta \mathcal{F} = -\frac{1}{\lambda} \int \left(\delta \mathcal{F}_c^2 - \delta \mathcal{F}_c \frac{\partial}{\partial \theta} \beta \frac{\partial}{\partial \theta} \delta \mathcal{F}_c \right) d\theta
$$

$$
= -\frac{1}{\lambda} \int \left(\delta \mathcal{F}_c^2 + \beta \left(\frac{\partial}{\partial \theta} \delta \mathcal{F}_c \right)^2 \right) d\theta < 0.
$$

The smoothed correction should now be projected onto the admissible subspace.

The final design procedure is thus as follows. Choose an initial profile and corresponding mapping function $\mathcal{F}$. Then:

1. Solve the flow equations (7–14) for ϕ, u, v, q, ρ.

2. Solve the ordinary differential equation (26) for $\mathcal{Z}$.

3. Solve the adjoint equation (16 and 18) or ψ subject to the boundary condition (27).

4. Solve the auxiliary Poisson equation (21) for P.

5. Evaluate $\mathcal{G}$ by equation (28)

6. Correct the boundary mapping function $\mathcal{F}_c$ by $\delta\mathcal{F}_c$ calculated from equation (29), projected onto the admissible subspace defined by (24).

7. Return to step 1.

Numerical Tests of Optimal Airfoil Design

The practical realization of the design procedure depends on the availability of sufficiently fast and accurate numerical procedures for the implementation of the essential steps, in particular the solution of both the flow and the adjoint equations. If the numerical procedures are not accurate enough, the resulting errors in the gradient may impair or prevent the convergence of the descent procedure. If the procedures are too slow, the cumulative computing time may become excessive. In this case, it was possible to build the design procedure around the author's computer program FLO36, which solves the transonic potential flow equation in conservation form in a domain mapped to the unit disk. The solution is obtained by a very rapid multigrid alternating direction method. The original scheme is described in Reference [24]. The program has been much improved since it was originally developed, and well converged solutions of transonic flows on a mesh with 128 cells in the circumferential direction and 32 cells in the radial direction are typically obtained in 5-20 multigrid cycles. The scheme uses artificial dissipative terms to introduce upwind biasing which simulates the rotated difference scheme [23], while preserving the conservation form. The alternating direction method is a generalization of conventional alternating direction methods, in which the scalar parameters are replaced by upwind difference operators to produce a scheme which remains stable when the type changes from elliptic to hyperbolic as the flow becomes locally supersonic [24]. The conformal mapping is generated by a power series of the form of equation (23) with an additional term

$$\left(1 - \frac{\epsilon}{\phi}\right) \log\left(1 - \frac{1}{\sigma}\right)$$

to allow for a wedge angle ϵ at the trailing edge. The coefficients are determined by an iterative process with the aid of fast Fourier transforms [23].

The adjoint equation has a form very similar to the flow equation. While it is linear in its dependent variable, it also changes type from elliptic in subsonic zones of the flow to hyperbolic in supersonic zones of the flow. Thus, it was possible to adapt exactly the same algorithm to solve both the adjoint and the flow equations, but with reverse biasing of the difference operators in the downwind direction in the adjoint equation, corresponding to the reversed direction of the zone of dependence. The Poisson equation (21) is solved by the Buneman algorithm.

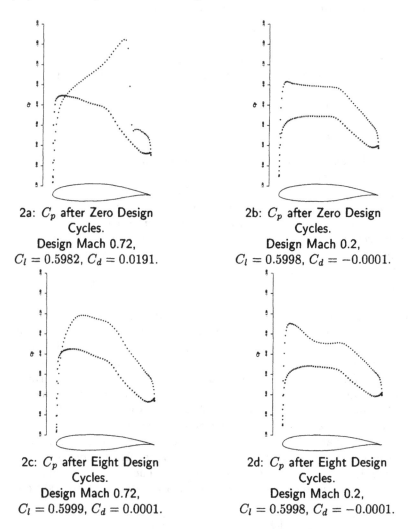

2a: C_p after Zero Design
Cycles.
Design Mach 0.72,
$C_l = 0.5982$, $C_d = 0.0191$.

2b: C_p after Zero Design
Cycles.
Design Mach 0.2,
$C_l = 0.5998$, $C_d = -0.0001$.

2c: C_p after Eight Design
Cycles.
Design Mach 0.72,
$C_l = 0.5999$, $C_d = 0.0001$.

2d: C_p after Eight Design
Cycles.
Design Mach 0.2,
$C_l = 0.5998$, $C_d = -0.0001$.

Figure 2: Optimization of an Airfoil at Two Design Points.

The efficiency of the present approach, which uses separate discretizations of the flow and adjoint equations, depends on the fact that in the limit of zero mesh width the discrete adjoint solution converges to the true adjoint solution. This allows the use of a rather simple discretization of the adjoint equation modeled after the discretization of the flow equation. Numerical experiments confirm that in practice separate discretizations of the flow and adjoint equations yields good convergence to an optimum solution.

As an example of the application of the method, Figure 2 presents a calculation in which an airfoil was redesigned to improve its transonic performance by reducing the pressure drag induced by the appearance of a shock wave. The drag coefficient was therefore included in the cost function so that equation (25) is replaced by

$$I = \frac{1}{2} \int_C \left(\lambda_1 \mathcal{Z}^2 + \lambda_2 \left(\frac{d\mathcal{Z}}{d\theta} \right)^2 \right) d\theta + \lambda_3 \, C_d,$$

where λ_3 is a parameter which may be varied to alter the trade-off between drag reduction and deviation from the desired pressure distribution. Representing the drag as

$$D = \int_C (p - p_\infty) \frac{dy}{d\theta} \, d\theta,$$

the procedure of Section 3 may be used to determine the gradient by solving the adjoint equation with a modified boundary condition. A penalty on the desired pressure distribution is still needed to avoid a situation in which the optimum shape is a flat plate with no lift and no drag.

It was also desired to preserve the subsonic characteristics of the airfoil. Therefore two design points were specified, Mach 0.20 and Mach 0.720, and in each case the lift coefficient was forced to be 0.6. The composite cost function was taken to be the sum of the values of the cost function at the two design points. The transonic drag coefficient was reduced from 0.0191 to 0.0001 in 8 design cycles. In order to achieve this reduction the airfoil had to be modified so that its subsonic pressure distribution became more peaky at the leading edge. This is consistent with the results of experimental research on transonic airfoils, in which it has generally been found necessary to have a peaky subsonic presure distribution in order to delay the onset of the transonic drag rise. It is also important to control the adverse pressure gradient on the rear upper surface, which can lead to premature separation of the viscous boundary layer. It can be seen that there is no steepening of this gradient due to the redesign.

4   The Navier-Stokes Equations

For the derivations that follow, it is convenient to use Cartesian coordinates (x_1, x_2, x_3) and to adopt the convention of indicial notation where a repeated index "i" implies summation over $i = 1$ to 3. The three-dimensional Navier-Stokes equations then take the form

$$\frac{\partial w}{\partial t} + \frac{\partial f_i}{\partial x_i} = \frac{\partial f_{vi}}{\partial x_i} \quad \text{in } \mathcal{D}, \tag{30}$$

where the state vector w, inviscid flux vector f and viscous flux vector f_v are described respectively by

$$w = \left\{ \begin{array}{c} \rho \\ \rho u_1 \\ \rho u_2 \\ \rho u_3 \\ \rho E \end{array} \right\}, \quad f_i = \left\{ \begin{array}{c} \rho u_i \\ \rho u_i u_1 + p\delta_{i1} \\ \rho u_i u_2 + p\delta_{i2} \\ \rho u_i u_3 + p\delta_{i3} \\ \rho u_i H \end{array} \right\}, \quad f_{vi} = \left\{ \begin{array}{c} 0 \\ \sigma_{ij}\delta_{j1} \\ \sigma_{ij}\delta_{j2} \\ \sigma_{ij}\delta_{j3} \\ u_j\sigma_{ij} + k\frac{\partial T}{\partial x_i} \end{array} \right\}.$$

(31)

In these definitions, ρ is the density, u_1, u_2, u_3 are the Cartesian velocity components, E is the total energy and δ_{ij} is the Kronecker delta function. The pressure is determined by the equation of state

$$p = (\gamma - 1)\rho \left\{ E - \frac{1}{2}(u_i u_i) \right\},$$

and the stagnation enthalpy is given by

$$H = E + \frac{p}{\rho},$$

where γ is the ratio of the specific heats. The viscous stresses may be written as

$$\sigma_{ij} = \mu \left(\frac{\partial u_i}{\partial x_j} + \frac{\partial u_j}{\partial x_i} \right) + \lambda \delta_{ij} \frac{\partial u_k}{\partial x_k}, \tag{32}$$

where μ and λ are the first and second coefficients of viscosity. The coefficient of thermal conductivity and the temperature are computed as

$$k = \frac{c_p \mu}{Pr}, \quad T = \frac{p}{R\rho}, \tag{33}$$

where Pr is the Prandtl number, c_p is the specific heat at constant pressure, and R is the gas constant.

For discussion of real applications using a discretization on a body conforming structured mesh, it is also useful to consider a transformation to the computational coordinates (ξ_1, ξ_2, ξ_3) defined by the metrics

$$K_{ij} = \left[\frac{\partial x_i}{\partial \xi_j} \right], \quad J = \det(K), \quad K_{ij}^{-1} = \left[\frac{\partial \xi_i}{\partial x_j} \right].$$

The Navier-Stokes equations can then be written in computational space as

$$\frac{\partial(Jw)}{\partial t} + \frac{\partial(F_i - F_{vi})}{\partial \xi_i} = 0 \quad \text{in } \mathcal{D}, \tag{34}$$

where the inviscid and viscous flux contributions are now defined with respect to the computational cell faces by $F_i = S_{ij} f_j$ and $F_{vi} = S_{ij} f_{vj}$, and the quantity $S_{ij} = JK_{ij}^{-1}$ represents the projection of the ξ_i cell face along the x_j axis. In obtaining equation (34) we have made use of the property that

$$\frac{\partial S_{ij}}{\partial \xi_i} = 0 \tag{35}$$

which represents the fact that the sum of the face areas over a closed volume is zero, as can be readily verified by a direct examination of the metric terms.

5 Formulation of the Optimal Design Problem for the Navier-Stokes Equations

Aerodynamic optimization is based on the determination of the effect of shape modifications on some performance measure which depends on the flow. For convenience, the coordinates ξ_i describing the fixed computational domain are chosen so that each boundary conforms to a constant value of one of these coordinates. Variations in the shape then result in corresponding variations in the mapping derivatives defined by K_{ij}.

Suppose that the performance is measured by a cost function

$$I = \int_B \mathcal{M}(w, S)\, dB_\xi + \int_\mathcal{D} \mathcal{P}(w, S)\, dD_\xi,$$

containing both boundary and field contributions where dB_ξ and dD_ξ are the surface and volume elements in the computational domain. In general, $\mathcal{M}$ and $\mathcal{P}$ will depend on both the flow variables w and the metrics S defining the computational space. The design problem is now treated as a control problem where the boundary shape represents the control function, which is chosen to minimize I subject to the constraints defined by the flow equations (34). A shape change produces a variation in the flow solution δw and the metrics δS which in turn produce a variation in the cost function

$$\delta I = \int_B \delta \mathcal{M}(w, S)\, dB_\xi + \int_\mathcal{D} \delta \mathcal{P}(w, S)\, dD_\xi. \tag{36}$$

This can be split as

$$\delta I = \delta I_I + \delta I_{II}, \tag{37}$$

with

$$\begin{aligned}
\delta \mathcal{M} &= [\mathcal{M}_w]_I\, \delta w + \delta \mathcal{M}_{II}, \\
\delta \mathcal{P} &= [\mathcal{P}_w]_I\, \delta w + \delta \mathcal{P}_{II},
\end{aligned} \tag{38}$$

where we continue to use the subscripts I and II to distinguish between the contributions associated with the variation of the flow solution δw and those associated with the metric variations δS. Thus $[\mathcal{M}_w]_I$ and $[\mathcal{P}_w]_I$ represent

$\frac{\partial \mathcal{M}}{\partial w}$ and $\frac{\partial \mathcal{P}}{\partial w}$ with the metrics fixed, while $\delta \mathcal{M}_{II}$ and $\delta \mathcal{P}_{II}$ represent the contribution of the metric variations δS to $\delta \mathcal{M}$ and $\delta \mathcal{P}$.

In the steady state, the constraint equation (53) specifies the variation of the state vector δw by

$$\delta R = \frac{\partial}{\partial \xi_i} \delta (F_i - F_{vi}) = 0. \tag{39}$$

Here, also, δR, δF_i and δF_{vi} can be split into contributions associated with δw and δS using the notation

$$\begin{aligned}
\delta R &= \delta R_I + \delta R_{II} \\
\delta F_i &= [F_{iw}]_I \, \delta w + \delta F_{i\,II} \\
\delta F_{vi} &= [F_{viw}]_I \, \delta w + \delta F_{vi\,II}.
\end{aligned} \tag{40}$$

The inviscid contributions are easily evaluated as

$$[F_{iw}]_I = S_{ij} \frac{\partial f_i}{\partial w}, \quad \delta F_{vi\,II} = \delta S_{ij} f_j.$$

The details of the viscous contributions are complicated by the additional level of derivatives in the stress and heat flux terms.

Multiplying by a co-state vector ψ, which will play an analogous role to the Lagrange multiplier introduced in equation (4), and integrating over the domain produces

$$\int_{\mathcal{D}} \psi^T \frac{\partial}{\partial \xi_i} \delta (F_i - F_{vi}) \, d\mathcal{D}_\xi = 0. \tag{41}$$

Assuming that ψ is differentiable the terms with subscript I may be integrated by parts to give

$$\int_B n_i \psi^T \delta (F_i - F_{vi})_I \, d\mathcal{B}_\xi - \int_{\mathcal{D}} \frac{\partial \psi^T}{\partial \xi_i} \delta (F_i - F_{vi})_I \, d\mathcal{D}_\xi + \int_{\mathcal{D}} \psi^T \delta R_{II} \, d\mathcal{D}_\xi = 0. \tag{42}$$

This equation results directly from taking the variation of the weak form of the flow equations, where ψ is taken to be an arbitrary differentiable test function. Since the left hand expression equals zero, it may be subtracted from the variation in the cost function (36) to give

$$\begin{aligned}
\delta I &= \delta I_{II} - \int_{\mathcal{D}} \psi^T \delta R_{II} \, d\mathcal{D}_\xi - \int_B \left[\delta \mathcal{M}_I - n_i \psi^T \delta (F_i - F_{vi})_I \right] d\mathcal{B}_\xi \\
&\quad + \int_{\mathcal{D}} \left[\delta \mathcal{P}_I + \frac{\partial \psi^T}{\partial \xi_i} \delta (F_i - F_{vi})_I \right] d\mathcal{D}_\xi.
\end{aligned} \tag{43}$$

Now, since ψ is an arbitrary differentiable function, it may be chosen in such a way that δI no longer depends explicitly on the variation of the state vector δw. The gradient of the cost function can then be evaluated directly from the metric variations without having to recompute the variation δw resulting from the perturbation of each design variable.

Comparing equations (38) and (40), the variation δw may be eliminated from (43) by equating all field terms with subscript "I" to produce a differential adjoint system governing ψ

$$\frac{\partial \psi^T}{\partial \xi_i} [F_{iw} - F_{viw}]_I + [\mathcal{P}_w]_I = 0 \quad \text{in } \mathcal{D}. \tag{44}$$

The corresponding adjoint boundary condition is produced by equating the subscript "I" boundary terms in equation (43) to produce

$$n_i \psi^T [F_{iw} - F_{viw}]_I = [\mathcal{M}_w]_I \quad \text{on } \mathcal{B}. \tag{45}$$

The remaining terms from equation (43) then yield a simplified expression for the variation of the cost function which defines the gradient

$$\delta I = \delta I_{II} + \int_{\mathcal{D}} \psi^T \delta R_{II} \, d\mathcal{D}_\xi, \tag{46}$$

which consists purely of the terms containing variations in the metrics with the flow solution fixed. Hence an explicit formula for the gradient can be derived once the relationship between mesh perturbations and shape variations is defined.

Comparing equations (38) and (40), the variation δw may be eliminated from (43) by equating all field terms with subscript "I" to produce a differential adjoint system governing ψ

$$\frac{\partial \psi^T}{\partial \xi_i} [F_{iw} - F_{viw}]_I + \mathcal{P}_w = 0 \quad \text{in } \mathcal{D}. \tag{47}$$

The corresponding adjoint boundary condition is produced by equating the subscript "I" boundary terms in equation (43) to produce

$$n_i \psi^T [F_{iw} - F_{viw}]_I = \mathcal{M}_w \quad \text{on } \mathcal{B}. \tag{48}$$

The remaining terms from equation (43) then yield a simplified expression for the variation of the cost function which defines the gradient

$$\delta I = \int_{\mathcal{B}} \left\{ \delta \mathcal{M}_{II} - n_i \psi^T [\delta F_i - \delta F_{vi}]_{II} \right\} d\mathcal{B}_\xi$$

$$+ \int_{\mathcal{D}} \left\{ \delta \mathcal{P}_{II} + \frac{\partial \psi^T}{\partial \xi_i} [\delta F_i - \delta F_{vi}]_{II} \right\} d\mathcal{D}_\xi. \tag{49}$$

The details of the formula for the gradient depend on the way in which the boundary shape is parameterized as a function of the design variables, and the way in which the mesh is deformed as the boundary is modified. Using the relationship between the mesh deformation and the surface modification, the field integral is reduced to a surface integral by integrating along the coordinate lines emanating from the surface. Thus the expression for δI is finally reduced to the form of equation (6)

$$\delta I = \int_B \mathcal{G} \delta \mathcal{F} \, d\mathcal{B}_\xi$$

where $\mathcal{F}$ represents the design variables, and $\mathcal{G}$ is the gradient, which is a function defined over the boundary surface.

The boundary conditions satisfied by the flow equations restrict the form of the left hand side of the adjoint boundary condition (48). Consequently, the boundary contribution to the cost function $\mathcal{M}$ cannot be specified arbitrarily. Instead, it must be chosen from the class of functions which allow cancellation of all terms containing δw in the boundary integral of equation (43). On the other hand, there is no such restriction on the specification of the field contribution to the cost function $\mathcal{P}$, since these terms may always be absorbed into the adjoint field equation (47) as source terms.

It is convenient to develop the inviscid and viscous contributions to the adjoint equations separately. Also, for simplicity, it will be assumed that the portion of the boundary that undergoes shape modifications is restricted to the coordinate surface $\xi_2 = 0$. Then equations (43) and (45) may be simplified by incorporating the conditions

$$n_1 = n_3 = 0, \quad n_2 = 1, \quad d\mathcal{B}_\xi = d\xi_1 d\xi_3,$$

so that only the variations δF_2 and δF_{v2} need to be considered at the wall boundary.

6   Derivation of the Inviscid Adjoint Terms

The inviscid contributions have been previously derived in [13, 25] but are included here for completeness. Taking the transpose of equation (44), the inviscid adjoint equation may be written as

$$C_i^T \frac{\partial \psi}{\partial \xi_i} = 0 \quad \text{in } \mathcal{D}, \tag{50}$$

where the inviscid Jacobian matrices in the transformed space are given by

$$C_i = S_{ij} \frac{\partial f_j}{\partial w}.$$

The transformed velocity components have the form

$$U_i = S_{ij} u_j,$$

and the condition that there is no flow through the wall boundary at $\xi_2 = 0$ is equivalent to

$$U_2 = 0,$$

so that

$$\delta U_2 = 0$$

when the boundary shape is modified. Consequently the variation of the inviscid flux at the boundary reduces to

$$\delta F_2 = \delta p \left\{ \begin{array}{c} 0 \\ S_{21} \\ S_{22} \\ S_{23} \\ 0 \end{array} \right\} + p \left\{ \begin{array}{c} 0 \\ \delta S_{21} \\ \delta S_{22} \\ \delta S_{23} \\ 0 \end{array} \right\}. \tag{51}$$

Since δF_2 depends only on the pressure, it is now clear that the performance measure on the boundary $\mathcal{M}(w, S)$ may only be a function of the pressure and metric terms. Otherwise, complete cancellation of the terms containing δw in the boundary integral would be impossible. One may, for example, include arbitrary measures of the forces and moments in the cost function, since these are functions of the surface pressure.

In order to design a shape which will lead to a desired pressure distribution, a natural choice is to set

$$I = \frac{1}{2} \int_B (p - p_d)^2 \, dS$$

where p_d is the desired surface pressure, and the integral is evaluated over the actual surface area. In the computational domain this is transformed to

$$I = \frac{1}{2} \int \int_{B_w} (p - p_d)^2 \, |S_2| \, d\xi_1 \, d\xi_3,$$

where the quantity

$$|S_2| = \sqrt{S_{2j} S_{2j}}$$

denotes the face area corresponding to a unit element of face area in the computational domain. Now, to cancel the dependence of the boundary integral on δp, the adjoint boundary condition reduces to

$$\psi_j n_j = p - p_d \tag{52}$$

where n_j are the components of the surface normal

$$n_j = \frac{S_{2j}}{|S_2|}.$$

This amounts to a transpiration boundary condition on the co-state variables corresponding to the momentum components. Note that it imposes no restriction on the tangential component of ψ at the boundary.

In the presence of shock waves, neither p nor p_d are necessarily continuous at the surface. The boundary condition is then in conflict with the assumption that ψ is differentiable. This difficulty can be circumvented by the use of a smoothed boundary condition [25].

7 Derivation of the Viscous Adjoint Terms

In computational coordinates, the viscous terms in the Navier–Stokes equations have the form

$$\frac{\partial F_{vi}}{\partial \xi_i} = \frac{\partial}{\partial \xi_i} \left(S_{ij} f_{vj} \right).$$

Computing the variation δw resulting from a shape modification of the boundary, introducing a co-state vector ψ and integrating by parts following the steps outlined by equations (39) to (42) produces

$$\int_B \psi^T \left(\delta S_{2j} f_{vj} + S_{2j} \delta f_{vj} \right) dB_\xi - \int_D \frac{\partial \psi^T}{\partial \xi_i} \left(\delta S_{ij} f_{vj} + S_{ij} \delta f_{vj} \right) dD_\xi,$$

where the shape modification is restricted to the coordinate surface $\xi_2 = 0$ so that $n_1 = n_3 = 0$, and $n_2 = 1$. Furthermore, it is assumed that the boundary contributions at the far field may either be neglected or else eliminated by a proper choice of boundary conditions as previously shown for the inviscid case [13, 25].

The viscous terms will be derived under the assumption that the viscosity and heat conduction coefficients μ and k are essentially independent of the flow, and that their variations may be neglected. This simplification has been successfully used for may aerodynamic problems of interest. In the case of some turbulent flows, there is the possibility that the flow variations could result in significant changes in the turbulent viscosity, and it may then be necessary to account for its variation in the calculation.

Transformation to Primitive Variables

The derivation of the viscous adjoint terms is simplified by transforming to the primitive variables

$$\tilde{w}^T = (\rho, u_1, u_2, u_3, p)^T,$$

because the viscous stresses depend on the velocity derivatives $\frac{\partial u_i}{\partial x_j}$, while the heat flux can be expressed as

$$\kappa \frac{\partial}{\partial x_i} \left(\frac{p}{\rho} \right).$$

where $\kappa = \frac{k}{R} = \frac{\gamma\mu}{Pr(\gamma-1)}$. The relationship between the conservative and primitive variations is defined by the expressions

$$\delta w = M\delta\tilde{w}, \quad \delta\tilde{w} = M^{-1}\delta w$$

which make use of the transformation matrices $M = \frac{\partial w}{\partial \tilde{w}}$ and $M^{-1} = \frac{\partial \tilde{w}}{\partial w}$. These matrices are provided in transposed form for future convenience

$$M^T = \begin{bmatrix} 1 & u_1 & u_2 & u_3 & \frac{u_i u_i}{2} \\ 0 & \rho & 0 & 0 & \rho u_1 \\ 0 & 0 & \rho & 0 & \rho u_2 \\ 0 & 0 & 0 & \rho & \rho u_3 \\ 0 & 0 & 0 & 0 & \frac{1}{\gamma-1} \end{bmatrix}$$

$$M^{-1^T} = \begin{bmatrix} 1 & -\frac{u_1}{\rho} & -\frac{u_2}{\rho} & -\frac{u_3}{\rho} & \frac{(\gamma-1)u_i u_i}{2} \\ 0 & \frac{1}{\rho} & 0 & 0 & -(\gamma-1)u_1 \\ 0 & 0 & \frac{1}{\rho} & 0 & -(\gamma-1)u_2 \\ 0 & 0 & 0 & \frac{1}{\rho} & -(\gamma-1)u_3 \\ 0 & 0 & 0 & 0 & \gamma-1 \end{bmatrix}.$$

The conservative and primitive adjoint operators L and $\tilde{L}$ corresponding to the variations δw and $\delta\tilde{w}$ are then related by

$$\int_{\mathcal{D}} \delta w^T L\psi \, d\mathcal{D}_\xi = \int_{\mathcal{D}} \delta\tilde{w}^T \tilde{L}\psi \, d\mathcal{D}_\xi,$$

with

$$\tilde{L} = M^T L,$$

so that after determining the primitive adjoint operator by direct evaluation of the viscous portion of (44), the conservative operator may be obtained by the transformation $L = M^{-1^T}\tilde{L}$. Since the continuity equation contains no viscous terms, it makes no contribution to the viscous adjoint system. Therefore, the derivation proceeds by first examining the adjoint operators arising from the momentum equations.

Contributions from the Momentum Equations

In order to make use of the summation convention, it is convenient to set $\psi_{j+1} = \phi_j$ for $j = 1, 2, 3$. Then the contribution from the momentum equations is

$$\int_B \phi_k \left(\delta S_{2j}\sigma_{kj} + S_{2j}\delta\sigma_{kj} \right) d\mathcal{B}_\xi \int_{\mathcal{D}} \frac{\partial \phi_k}{\partial \xi_i} \left(\delta S_{ij}\sigma_{kj} + S_{ij}\delta\sigma_{kj} \right) d\mathcal{D}_\xi. \tag{53}$$

The velocity derivatives in the viscous stresses can be expressed as

$$\frac{\partial u_i}{\partial x_j} = \frac{\partial u_i}{\partial \xi_l}\frac{\partial \xi_l}{\partial x_j} = \frac{S_{lj}}{J}\frac{\partial u_i}{\partial \xi_l}$$

with corresponding variations

$$\delta \frac{\partial u_i}{\partial x_j} = \left[\frac{S_{lj}}{J} \right]_I \frac{\partial}{\partial \xi_l} \delta u_i + \left[\frac{\partial u_i}{\partial \xi_l} \right]_{II} \delta \left(\frac{S_{lj}}{J} \right).$$

The variations in the stresses are then

$$\begin{aligned}
\delta \sigma_{kj} &= \left\{ \mu \left[\frac{S_{lj}}{J} \frac{\partial}{\partial \xi_l} \delta u_k + \frac{S_{lk}}{J} \frac{\partial}{\partial \xi_l} \delta u_j \right] + \lambda \left[\delta_{jk} \frac{S_{lm}}{J} \frac{\partial}{\partial \xi_l} \delta u_m \right] \right\}_I \\
&+ \left\{ \mu \left[\delta \left(\frac{S_{lj}}{J} \right) \frac{\partial u_k}{\partial \xi_l} + \delta \left(\frac{S_{lk}}{J} \right) \frac{\partial u_j}{\partial \xi_l} \right] + \lambda \left[\delta_{jk} \delta \left(\frac{S_{lm}}{J} \right) \frac{\partial u_m}{\partial \xi_l} \right] \right\}_{II}.
\end{aligned}$$

As before, only those terms with subscript I, which contain variations of the flow variables, need be considered further in deriving the adjoint operator. The field contributions that contain δu_i in equation (53) appear as

$$- \int_D \frac{\partial \phi_k}{\partial \xi_i} S_{ij} \left\{ \mu \left(\frac{S_{lj}}{J} \frac{\partial}{\partial \xi_l} \delta u_k + \frac{S_{lk}}{J} \frac{\partial}{\partial \xi_l} \delta u_j \right) + \lambda \delta_{jk} \frac{S_{lm}}{J} \frac{\partial}{\partial \xi_l} \delta u_m \right\} d\mathcal{D}_\xi.$$

This may be integrated by parts to yield

$$\begin{aligned}
&\int_D \delta u_k \frac{\partial}{\partial \xi_l} \left(S_{lj} S_{ij} \frac{\mu}{J} \frac{\partial \phi_k}{\partial \xi_i} \right) d\mathcal{D}_\xi \\
+ &\int_D \delta u_j \frac{\partial}{\partial \xi_l} \left(S_{lk} S_{ij} \frac{\mu}{J} \frac{\partial \phi_k}{\partial \xi_i} \right) d\mathcal{D}_\xi \\
+ &\int_D \delta u_m \frac{\partial}{\partial \xi_l} \left(S_{lm} S_{ij} \frac{\lambda \delta_{jk}}{J} \frac{\partial \phi_k}{\partial \xi_i} \right) d\mathcal{D}_\xi,
\end{aligned}$$

where the boundary integral has been eliminated by noting that $\delta u_i = 0$ on the solid boundary. By exchanging indices, the field integrals may be combined to produce

$$\int_D \delta u_k \frac{\partial}{\partial \xi_l} S_{lj} \left\{ \mu \left(\frac{S_{ij}}{J} \frac{\partial \phi_k}{\partial \xi_i} + \frac{S_{ik}}{J} \frac{\partial \phi_j}{\partial \xi_i} \right) + \lambda \delta_{jk} \frac{S_{im}}{J} \frac{\partial \phi_m}{\partial \xi_i} \right\} d\mathcal{D}_\xi,$$

which is further simplified by transforming the inner derivatives back to Cartesian coordinates

$$\int_D \delta u_k \frac{\partial}{\partial \xi_l} S_{lj} \left\{ \mu \left(\frac{\partial \phi_k}{\partial x_j} + \frac{\partial \phi_j}{\partial x_k} \right) + \lambda \delta_{jk} \frac{\partial \phi_m}{\partial x_m} \right\} d\mathcal{D}_\xi. \qquad (54)$$

The boundary contributions that contain δu_i in equation (53) may be simplified using the fact that

$$\frac{\partial}{\partial \xi_l} \delta u_i = 0 \quad \text{if} \quad l = 1, 3$$

on the boundary $\mathcal{B}$ so that they become

$$\int_B \phi_k S_{2j} \left\{ \mu \left(\frac{S_{2j}}{J} \frac{\partial}{\partial \xi_2} \delta u_k + \frac{S_{2k}}{J} \frac{\partial}{\partial \xi_2} \delta u_j \right) + \lambda \delta_{jk} \frac{S_{2m}}{J} \frac{\partial}{\partial \xi_2} \delta u_m \right\} d\mathcal{B}_\xi. \qquad (55)$$

Together, (54) and (55) comprise the field and boundary contributions of the momentum equations to the viscous adjoint operator in primitive variables.

Contributions from the Energy Equation

In order to derive the contribution of the energy equation to the viscous adjoint terms it is convenient to set

$$\psi_5 = 0, \quad Q_j = u_i \sigma_{ij} + \kappa \frac{\partial}{\partial x_j} \left(\frac{p}{\rho} \right),$$

where the temperature has been written in terms of pressure and density using (33). The contribution from the energy equation can then be written as

$$\int_B \theta \left(\delta S_{2j} Q_j + S_{2j} \delta Q_j \right) dB_\xi - \int_D \frac{\partial \theta}{\partial \xi_i} \left(\delta S_{ij} Q_j + S_{ij} \delta Q_j \right) dD_\xi. \tag{56}$$

The field contributions that contain $\delta u_i, \delta p$, and $\delta \rho$ in equation (56) appear as

$$-\int_D \frac{\partial \theta}{\partial \xi_i} S_{ij} \delta Q_j dD_\xi = -\int_D \frac{\partial \theta}{\partial \xi_i} S_{ij} \left\{ \delta u_k \sigma_{kj} + u_k \delta \sigma_{kj} \right.$$

$$\left. + \kappa \frac{S_{lj}}{J} \frac{\partial}{\partial \xi_l} \left(\frac{\delta p}{\rho} - \frac{p}{\rho} \frac{\delta \rho}{\rho} \right) \right\} dD_\xi. \tag{57}$$

The term involving $\delta \sigma_{kj}$ may be integrated by parts to produce

$$\int_D \delta u_k \frac{\partial}{\partial \xi_l} S_{lj} \left\{ \mu \left(u_k \frac{\partial \theta}{\partial x_j} + u_j \frac{\partial \theta}{\partial x_k} \right) + \lambda \delta_{jk} u_m \frac{\partial \theta}{\partial x_m} \right\} dD_\xi, \tag{58}$$

where the conditions $u_i = \delta u_i = 0$ are used to eliminate the boundary integral on B. Notice that the other term in (57) that involves δu_k need not be integrated by parts and is merely carried on as

$$-\int_D \delta u_k \sigma_{kj} S_{ij} \frac{\partial \theta}{\partial \xi_i} dD_\xi \tag{59}$$

The terms in expression (57) that involve δp and $\delta \rho$ may also be integrated by parts to produce both a field and a boundary integral. The field integral becomes

$$\int_D \left(\frac{\delta p}{\rho} - \frac{p}{\rho} \frac{\delta \rho}{\rho} \right) \frac{\partial}{\partial \xi_l} \left(S_{lj} S_{ij} \frac{\kappa}{J} \frac{\partial \theta}{\partial \xi_i} \right) dD_\xi$$

which may be simplified by transforming the inner derivative to Cartesian coordinates

$$\int_D \left(\frac{\delta p}{\rho} - \frac{p}{\rho} \frac{\delta \rho}{\rho} \right) \frac{\partial}{\partial \xi_l} \left(S_{lj} \kappa \frac{\partial \theta}{\partial x_j} \right) dD_\xi. \tag{60}$$

The boundary integral becomes

$$\int_B \kappa \left(\frac{\delta p}{\rho} - \frac{p}{\rho} \frac{\delta \rho}{\rho} \right) \frac{S_{2j} S_{ij}}{J} \frac{\partial \theta}{\partial \xi_i} dB_\xi. \tag{61}$$

This can be simplified by transforming the inner derivative to Cartesian coordinates

$$\int_B \kappa \left(\frac{\delta p}{\rho} - \frac{p}{\rho} \frac{\delta \rho}{\rho} \right) \frac{S_{2j}}{J} \frac{\partial \theta}{\partial x_j} dB_\xi, \tag{62}$$

and identifying the normal derivative at the wall

$$\frac{\partial}{\partial n} = S_{2j} \frac{\partial}{\partial x_j}, \tag{63}$$

and the variation in temperature

$$\delta T = \frac{1}{R} \left(\frac{\delta p}{\rho} - \frac{p}{\rho} \frac{\delta \rho}{\rho} \right),$$

to produce the boundary contribution

$$\int_B k \delta T \frac{\partial \theta}{\partial n} dB_\xi. \tag{64}$$

This term vanishes if T is constant on the wall but persists if the wall is adiabatic.

There is also a boundary contribution left over from the first integration by parts (56) which has the form

$$\int_B \theta \delta \left(S_{2j} Q_j \right) dB_\xi, \tag{65}$$

where

$$Q_j = k \frac{\partial T}{\partial x_j},$$

since $u_i = 0$. Notice that for future convenience in discussing the adjoint boundary conditions resulting from the energy equation, both the δw and δS terms corresponding to subscript classes I and II are considered simultaneously. If the wall is adiabatic

$$\frac{\partial T}{\partial n} = 0,$$

so that using (63),

$$\delta \left(S_{2j} Q_j \right) = 0,$$

and both the δw and δS boundary contributions vanish.

On the other hand, if T is constant $\frac{\partial T}{\partial \xi_l} = 0$ for $l = 1, 3$, so that

$$Q_j = k \frac{\partial T}{\partial x_j} = k \left(\frac{S_{lj}}{J} \frac{\partial T}{\partial \xi_l} \right) = k \left(\frac{S_{2j}}{J} \frac{\partial T}{\partial \xi_2} \right).$$

Thus, the boundary integral (65) becomes

$$\int_B k\theta \left\{ \frac{S_{2j}{}^2}{J} \frac{\partial}{\partial \xi_2} \delta T + \delta \left(\frac{S_{2j}{}^2}{J} \right) \frac{\partial T}{\partial \xi_2} \right\} dB_\xi .\qquad (66)$$

Therefore, for constant T, the first term corresponding to variations in the flow field contributes to the adjoint boundary operator and the second set of terms corresponding to metric variations contribute to the cost function gradient.

All together, the contributions from the energy equation to the viscous adjoint operator are the three field terms (58), (59) and (60), and either of two boundary contributions (64) or (66), depending on whether the wall is adiabatic or has constant temperature.

The Viscous Adjoint Field Operator

Collecting together the contributions from the momentum and energy equations, the viscous adjoint operator in primitive variables can be expressed as

$$
\begin{aligned}
(\tilde{L}\psi)_1 &= -\frac{p}{\rho^2} \frac{\partial}{\partial \xi_l} \left(S_{lj} \kappa \frac{\partial \theta}{\partial x_j} \right) \\
(\tilde{L}\psi)_{i+1} &= \frac{\partial}{\partial \xi_l} \left\{ S_{lj} \left[\mu \left(\frac{\partial \phi_i}{\partial x_j} + \frac{\partial \phi_j}{\partial x_i} \right) + \lambda \delta_{ij} \frac{\partial \phi_k}{\partial x_k} \right] \right\} \\
&+ \frac{\partial}{\partial \xi_l} \left\{ S_{lj} \left[\mu \left(u_i \frac{\partial \theta}{\partial x_j} + u_j \frac{\partial \theta}{\partial x_i} \right) + \lambda \delta_{ij} u_k \frac{\partial \theta}{\partial x_k} \right] \right\} \quad \text{for} \quad i = 1, 2, 3 \\
&- \sigma_{ij} S_{lj} \frac{\partial \theta}{\partial \xi_l} \\
(\tilde{L}\psi)_5 &= \frac{1}{\rho} \frac{\partial}{\partial \xi_l} \left(S_{lj} \kappa \frac{\partial \theta}{\partial x_j} \right) .
\end{aligned}
$$

The conservative viscous adjoint operator may now be obtained by the transformation

$$L = M^{-1}{}^T \tilde{L}.$$

8   Viscous Adjoint Boundary Conditions

It was recognized in Section 5 that the boundary conditions satisfied by the flow equations restrict the form of the performance measure that may be chosen for the cost function. There must be a direct correspondence between the flow variables for which variations appear in the variation of the cost function, and those variables for which variations appear in the boundary terms arising during the derivation of the adjoint field equations. Otherwise it would be impossible to eliminate the dependence of δI on δw through proper specification of the adjoint boundary condition. As in the derivation of the field equations, it proves convenient to consider the contributions from the momentum equations and the energy equation separately.

Boundary Conditions Arising from the Momentum Equations

The boundary term that arises from the momentum equations including both the δw and δS components (53) takes the form

$$\int_B \phi_k \delta \left(S_{2j} \sigma_{kj} \right) dB_\xi.$$

Replacing the metric term with the corresponding local face area S_2 and unit normal n_j defined by

$$|S_2| = \sqrt{S_{2j} S_{2j}}, \quad n_j = \frac{S_{2j}}{|S_2|}$$

then leads to

$$\int_B \phi_k \delta \left(|S_2| \, n_j \sigma_{kj} \right) dB_\xi.$$

Defining the components of the surface stress as

$$\tau_k = n_j \sigma_{kj}$$

and the physical surface element

$$dS = |S_2| \, dB_\xi,$$

the integral may then be split into two components

$$\int_B \phi_k \tau_k \, |\delta S_2| \, dB_\xi + \int_B \phi_k \delta \tau_k dS, \tag{67}$$

where only the second term contains variations in the flow variables and must consequently cancel the δw terms arising in the cost function. The first term will appear in the expression for the gradient.

A general expression for the cost function that allows cancellation with terms containing $\delta \tau_k$ has the form

$$I = \int_B \mathcal{N}(\tau) dS, \tag{68}$$

corresponding to a variation

$$\delta I = \int_B \frac{\partial \mathcal{N}}{\partial \tau_k} \delta \tau_k dS,$$

for which cancellation is achieved by the adjoint boundary condition

$$\phi_k = \frac{\partial \mathcal{N}}{\partial \tau_k}.$$

Natural choices for $\mathcal{N}$ arise from force optimization and as measures of the deviation of the surface stresses from desired target values.

For viscous force optimization, the cost function should measure friction drag. The friction force in the x_i direction is

$$CD_{fi} = \int_B \sigma_{ij} dS_j = \int_B S_{2j} \sigma_{ij} dB_\xi$$

so that the force in a direction with cosines q_i has the form

$$C_{qf} = \int_B q_i S_{2j} \sigma_{ij} dB_\xi .$$

Expressed in terms of the surface stress τ_i, this corresponds to

$$C_{qf} = \int_B q_i \tau_i dS,$$

so that basing the cost function (68) on this quantity gives

$$\mathcal{N} = q_i \tau_i .$$

Cancellation with the flow variation terms in equation (67) therefore mandates the adjoint boundary condition

$$\phi_k = n_k .$$

Note that this choice of boundary condition also eliminates the first term in equation (67) so that it need not be included in the gradient calculation.

In the inverse design case, where the cost function is intended to measure the deviation of the surface stresses from some desired target values, a suitable definition is

$$\mathcal{N}(\tau) = \frac{1}{2} a_{lk} (\tau_l - \tau_{dl})(\tau_k - \tau_{dk}),$$

where τ_d is the desired surface stress, including the contribution of the pressure, and the coefficients a_{lk} define a weighting matrix. For cancellation

$$\phi_k \delta \tau_k = a_{lk} (\tau_l - \tau_{dl}) \delta \tau_k .$$

This is satisfied by the boundary condition

$$\phi_k = a_{lk} (\tau_l - \tau_{dl}) . \tag{69}$$

Assuming arbitrary variations in $\delta \tau_k$, this condition is also necessary.

In order to control the surface pressure and normal stress one can measure the difference

$$n_j \{ \sigma_{kj} + \delta_{kj} (p - p_d) \},$$

where p_d is the desired pressure. The normal component is then

$$\tau_n = n_k n_j \sigma_{kj} + p - p_d,$$

so that the measure becomes

$$\mathcal{N}(\tau) \;=\; \frac{1}{2}\tau_n^2$$

$$=\; \frac{1}{2}n_l n_m n_k n_j \left\{\sigma_{lm} + \delta_{lm}\left(p - p_d\right)\right\}\left\{\sigma_{kj} + \delta_{kj}\left(p - p_d\right)\right\}.$$

This corresponds to setting

$$a_{lk} = n_l n_k$$

in equation (69). Defining the viscous normal stress as

$$\tau_{vn} = n_k n_j \sigma_{kj},$$

the measure can be expanded as

$$\mathcal{N}(\tau) \;=\; \frac{1}{2}n_l n_m n_k n_j \sigma_{lm}\sigma_{kj} + \frac{1}{2}\left(n_k n_j \sigma_{kj} + n_l n_m \sigma_{lm}\right)\left(p - p_d\right)$$

$$+\frac{1}{2}\left(p - p_d\right)^2$$

$$=\; \frac{1}{2}\tau_{vn}^2 + \tau_{vn}\left(p - p_d\right) + \frac{1}{2}\left(p - p_d\right)^2 .$$

For cancellation of the boundary terms

$$\phi_k\left(n_j\delta\sigma_{kj} + n_k\delta p\right) = \left\{n_l n_m \sigma_{lm} + n_l^2\left(p - p_d\right)\right\}n_k\left(n_j\delta\sigma_{kj} + n_k\delta p\right)$$

leading to the boundary condition

$$\phi_k = n_k\left(\tau_{vn} + p - p_d\right).$$

In the case of high Reynolds number, this is well approximated by the equations

$$\phi_k = n_k\left(p - p_d\right), \tag{70}$$

which should be compared with the single scalar equation derived for the inviscid boundary condition (52). In the case of an inviscid flow, choosing

$$\mathcal{N}(\tau) = \frac{1}{2}\left(p - p_d\right)^2$$

requires

$$\phi_k n_k\delta p = \left(p - p_d\right)n_k^2\delta p = \left(p - p_d\right)\delta p$$

which is satisfied by equation (70), but which represents an overspecification of the boundary condition since only the single condition (52) need be specified to ensure cancellation.

Boundary Conditions Arising from the Energy Equation

The form of the boundary terms arising from the energy equation depends on the choice of temperature boundary condition at the wall. For the adiabatic case, the boundary contribution is (64)

$$\int_B k\delta T \frac{\partial \theta}{\partial n} d\mathcal{B}_\xi,$$

while for the constant temperature case the boundary term is (66). One possibility is to introduce a contribution into the cost function which depends on T or $\frac{\partial T}{\partial n}$ so that the appropriate cancellation would occur. Since there is little physical intuition to guide the choice of such a cost function for aerodynamic design, a more natural solution is to set

$$\theta = 0$$

in the constant temperature case or

$$\frac{\partial \theta}{\partial n} = 0$$

in the adiabatic case. Note that in the constant temperature case, this choice of θ on the boundary would also eliminate the boundary metric variation terms in (65).

9 Implementation of Navier-Stokes Design

The design procedures can be summarized as follows:

1. Solve the flow equations for ρ, u_1, u_2, u_3, p.

2. Solve the adjoint equations for ψ subject to appropriate boundary conditions.

3. Evaluate $\mathcal{G}$.

4. Project $\mathcal{G}$ into an allowable subspace that satisfies any geometric constraints.

5. Update the shape based on the direction of steepest descent.

6. Return to 1 until convergence is reached.

Practical implementation of the viscous design method relies heavily upon fast and accurate solvers for both the state (w) and co-state (ψ) systems. This work uses well-validated software for the solution of the Euler and Navier-Stokes equations developed over the course of many years [26, 27, 28].

For inverse design the lift is fixed by the target pressure. In drag minimization it is also appropriate to fix the lift coefficient, because the induced drag is a major fraction of the total drag, and this could be reduced simply by reducing the lift. Therefore the angle of attack is adjusted during the flow solution to force a specified lift coefficient to be attained, and the influence of variations of the angle of attack is included in the calculation of the gradient. The vortex drag also depends on the span loading, which may be constrained by other considerations such as structural loading or buffet onset. Consequently, the option is provided to force the span loading by adjusting the twist distribution as well as the angle of attack during the flow solution.

Discretization

Both the flow and the adjoint equations are discretized using a semi-discrete cell-centered finite volume scheme. The convective fluxes across cell interfaces are represented by simple arithmetic averages of the fluxes computed using values from the cells on either side of the face, augmented by artificial diffusive terms to prevent numerical oscillations in the vicinity of shock waves. Continuing to use the summation convention for repeated indices, the numerical convective flux across the interface between cells A and B in a three dimensional mesh has the form

$$h_{AB} = \frac{1}{2} S_{AB_j} \left(f_{A_j} + f_{B_j} \right) - d_{AB},$$

where S_{AB_j} is the component of the face area in the jth Cartesian coordinate direction, $\left(f_{A_j} \right)$ and $\left(f_{B_j} \right)$ denote the flux f_j as defined by equation (12) and d_{AB} is the diffusive term. Variations of the computer program provide options for alternate constructions of the diffusive flux.

The simplest option implements the Jameson-Schmidt-Turkel scheme [26, 29], using scalar diffusive terms of the form

$$d_{AB} = \epsilon^{(2)} \Delta w - \epsilon^{(4)} \left(\Delta w^+ - 2\Delta w + \Delta w^- \right),$$

where

$$\Delta w = w_B - w_A$$

and Δw^+ and Δw^- are the same differences across the adjacent cell interfaces behind cell A and beyond cell B in the AB direction. By making the coefficient $\epsilon^{(2)}$ depend on a switch proportional to the undivided second difference of a flow quantity such as the pressure or entropy, the diffusive flux becomes a third order quantity, proportional to the cube of the mesh width in regions where the solution is smooth. Oscillations are suppressed near a shock wave because $\epsilon^{(2)}$ becomes of order unity, while $\epsilon^{(4)}$ is reduced to zero by the same switch. For a scalar conservation law, it is shown in reference [29] that $\epsilon^{(2)}$ and $\epsilon^{(4)}$ can be constructed to make the scheme satisfy the local extremum diminishing (LED) principle that local maxima cannot increase while local minima cannot decrease.

The second option applies the same construction to local characteristic variables. There are derived from the eigenvectors of the Jacobian matrix A_{AB} which exactly satisfies the relation

$$A_{AB}(w_B - w_A) = S_{AB_j}\left(f_{B_j} - f_{A_j}\right).$$

This corresponds to the definition of Roe [30]. The resulting scheme is LED in the characteristic variables. The third option implements the H-CUSP scheme proposed by Jameson [31] which combines differences $f_B - f_A$ and $w_B - w_A$ in a manner such that stationary shock waves can be captured with a single interior point in the discrete solution. This scheme minimizes the numerical diffusion as the velocity approaches zero in the boundary layer, and has therefore been preferred for viscous calculations in this work.

Similar artificial diffusive terms are introduced in the discretization of the adjoint equation, but with the opposite sign because the wave directions are reversed in the adjoint equation. Satisfactory results have been obtained using scalar diffusion in the adjoint equation while characteristic or H-CUSP constructions are used in the flow solution.

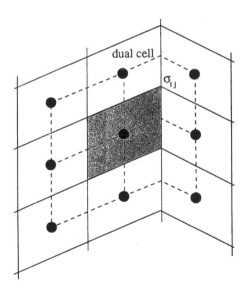

Figure 3: Cell-centered scheme. σ_{ij} evaluated at vertices of the primary mesh

The discretization of the viscous terms of the Navier Stokes equations requires the evaluation of the velocity derivatives $\frac{\partial u_i}{\partial x_j}$ in order to calculate the viscous stress tensor σ_{ij} defined in equation (32). These are most conveniently evaluated at the cell vertices of the primary mesh by introducing a dual mesh which connects the cell centers of the primary mesh, as depicted in Figure (3). According to the Gauss formula for a control volume V with boundary S

$$\int_V \frac{\partial v_i}{\partial x_j} dv = \int_S u_i n_j dS$$

where n_j is the outward normal. Applied to the dual cells this yields the estimate

$$\frac{\partial v_i}{\partial x_j} = \frac{1}{\text{vol}} \sum_{\text{faces}} \bar{u}_i n_j S$$

where S is the area of a face, and $\bar{u}_i$ is an estimate of the average of u_i over that face. In order to determine the viscous flux balance of each primary cell, the viscous flux across each of its faces is then calculated from the average of the viscous stress tensor at the four vertices connected by that face. This leads to a compact scheme with a stencil connecting each cell to its 26 nearest neighbors.

The semi-discrete schemes for both the flow and the adjoint equations are both advanced to steady state by a multi-stage time stepping scheme. This is a generalized Runge-Kutta scheme in which the convective and diffusive terms are treated differently to enlarge the stability region [29, 32]. Convergence to a steady state is accelerated by residual averaging and a multi-grid procedure [33]. These algorithms have been implemented both for single and multiblock meshes and for operation on parallel computers with message passing using the MPI (Message Passing Interface) protocol [9, 34, 35].

In this work, the adjoint and flow equations are discretized separately. The alternative approach of deriving the discrete adjoint equations directly from the discrete flow equations yields another possible discretization of the adjoint partial differential equation which is more complex. If the resulting equations were solved exactly, they could provide the exact gradient of the inexact cost function which results from the discretization of the flow equations. On the other hand, any consistent discretization of the adjoint partial differential equation will yield the exact gradient as the mesh is refined, and separate discretization has proved to work perfectly well in practice. It should also be noted that the discrete gradient includes both mesh effects and numerical errors such as spurious entropy production which may not reflect the true cost function of the continuous problem.

Mesh Generation and Geometry Control

Meshes for both viscous optimization and for the treatment of complex configurations are externally generated in order to allow for their inspection and careful quality control. Single block meshes with a C-H topology have been used for viscous optimization of wing-body combinations, while multiblock meshes have been generated for complex configurations using GRIDGEN [36].

In either case geometry modifications are accommodated by a grid perturbation scheme. For viscous wing-body design using single block meshes, the wing surface mesh points themselves are taken as the design variables. A simple mesh perturbation scheme is then used, in which the mesh points lying on a mesh line projecting out from the wing surface are all shifted in the same sense as the surface mesh point, with a decay factor proportional to the arc length along the mesh line. The resulting perturbation in the face areas of the neighboring cells are then included in the gradient calculation. For complex configurations the geometry is controlled by superposition of analytic "bump" functions defined over the surfaces which are to be modified. The grid is then perturbed to conform to modifications of the surface shape by the WARP3D and WARP-MB algorithms described in [34].

Optimization

Two main search procedures have been used in our applications to date. The first is a simple descent method in which small steps are taken in the negative gradient direction. Let $\mathcal{F}$ represent the design variable, and $\mathcal{G}$ the gradient. Then the iteration

$$\delta \mathcal{F} = -\lambda \mathcal{G}$$

can be regarded as simulating the time dependent process

$$\frac{d\mathcal{F}}{dt} = -\mathcal{G}$$

where λ is the time step Δt. Let A be the Hessian matrix with elements

$$A_{ij} = \frac{\partial \mathcal{G}_i}{\partial \mathcal{F}_j} = \frac{\partial^2 I}{\partial \mathcal{F}_i \partial \mathcal{F}_j}.$$

Suppose that a locally minimum value of the cost function $I^* = I(\mathcal{F}^*)$ is attained when $\mathcal{F} = \mathcal{F}^*$. Then the gradient $\mathcal{G}^* = \mathcal{G}(\mathcal{F}^*)$ must be zero, while the Hessian matrix $A^* = A(\mathcal{F}^*)$ must be positive definite. Since $\mathcal{G}^*$ is zero, the cost function can be expanded as a Taylor series in the neighborhood of $\mathcal{F}^*$ with the form

$$I(\mathcal{F}) = I^* + \frac{1}{2}(\mathcal{F} - \mathcal{F}^*) A (\mathcal{F} - \mathcal{F}^*) + \dots$$

Correspondingly,

$$\mathcal{G}(\mathcal{F}) = A(\mathcal{F} - \mathcal{F}^*) + \dots$$

As $\mathcal{F}$ approaches $\mathcal{F}^*$, the leading terms become dominant. Then, setting $\hat{\mathcal{F}} = (\mathcal{F} - \mathcal{F}^*)$, the search process approximates

$$\frac{d\hat{\mathcal{F}}}{dt} = -A^* \hat{\mathcal{F}}.$$

Also, since A^* is positive definite it can be expanded as

$$A^* = RMR^T,$$

where M is a diagonal matrix containing the eigenvalues of A^*, and

$$RR^T = R^T R = I.$$

Setting

$$v = R^T \hat{\mathcal{F}},$$

the search process can be represented as

$$\frac{dv}{dt} = -Mv.$$

The stability region for the simple forward Euler stepping scheme is a unit circle centered at -1 on the negative real axis. Thus for stability we must choose

$$\mu_{\max}\Delta t = \mu_{\max}\lambda < 2,$$

while the asymptotic decay rate, given by the smallest eigenvalue, is proportional to

$$e^{-\mu_{\min}t}.$$

In order to make sure that each new shape in the optimization sequence remains smooth, it proves essential to smooth the gradient and to replace $\mathcal{G}$ by its smoothed value $\bar{\mathcal{G}}$ in the descent process. This also acts as a preconditioner which allows the use of much larger steps. To apply smoothing in the ξ_1 direction, for example, the smoothed gradient $\bar{\mathcal{G}}$ may be calculated from a discrete approximation to

$$\bar{\mathcal{G}} - \frac{\partial}{\partial \xi_1}\epsilon\frac{\partial}{\partial \xi_1}\bar{\mathcal{G}} = \mathcal{G} \tag{71}$$

where ϵ is the smoothing parameter. If one sets $\delta\mathcal{F} = -\lambda\bar{\mathcal{G}}$, then, assuming the modification is applied on the surface $\xi_2 = $ constant, the first order change in the cost function is

$$\begin{aligned}
\delta I &= -\iint \mathcal{G}\delta\mathcal{F}\, d\xi_1 d\xi_3 \\
&= -\lambda\iint \left(\bar{\mathcal{G}} - \frac{\partial}{\partial \xi_1}\epsilon\frac{\partial\bar{\mathcal{G}}}{\partial \xi_1}\right)\bar{\mathcal{G}}\, d\xi_1 d\xi_3 \\
&= -\lambda\iint \left(\bar{\mathcal{G}}^2 + \epsilon\left(\frac{\partial\bar{\mathcal{G}}}{\partial \xi_1}\right)^2\right) d\xi_1 d\xi_3 \\
&< 0,
\end{aligned}$$

assuring an improvement if λ is sufficiently small and positive, unless the process has already reached a stationary point at which $\mathcal{G} = 0$.

It turns out that this approach is tolerant to the use of approximate values of the gradient, so that neither the flow solution nor the adjoint solution need be fully converged before making a shape change. This results in very large savings in the computational cost. For inviscid optimization it is necessary to use only 15 multigrid cycles for the flow solution and the adjoint solution in each design iteration. For viscous optimization, about 20-30 multigrid cycles are needed.

Our second main search procedure incorporates a quasi-Newton method for general constrained optimization. In this class of methods the step is defined as

$$\delta \mathcal{F} = -\lambda P \mathcal{G},$$

where P is a preconditioner for the search. An ideal choice is $P = A^{*-1}$, so that the corresponding time dependent process reduces to

$$\frac{d\hat{\mathcal{F}}}{dt} = -\hat{\mathcal{F}},$$

for which all the eigenvalues are equal to unity, and $\hat{\mathcal{F}}$ is reduced to zero in one time step by the choice $\Delta t = 1$ if the Hessian, A, is constant. The full Newton method takes $P = A^{-1}$, requiring the evaluation of the Hessian matrix, A, at each step. It corresponds to the use of the Newton-Raphson method to solve the non-linear equation $\mathcal{G} = 0$. Quasi-Newton methods estimate A^* from the change in the gradient during the search process. This requires accurate estimates of the gradient at each time step. In order to obtain these, both the flow solution and the adjoint equation must be fully converged. Most quasi-Newton methods also require a line search in each search direction, for which the flow equations and cost function must be accurately evaluated several times. They have proven quite robust in practice for aerodynamic optimization [37].

In the applications to complex configurations presented below the optimization was carried out using the existing, well validated software NPSOL. This software, which implements a quasi-Newton method for optimization with both linear and non-linear constraints, has proved very reliable but is generally more expensive than the simple search method with smoothing.

10 Industrial Experience and Results

The methods described in this paper have been quite thoroughly tested in industrial applications in which they were used as a tool for aerodynamic design. They have proved useful both in inverse mode to find shapes that would produce desired pressure distributions, and for direct minimization of the drag. They have been applied both to well understood configurations that have gradually evolved through incremental improvements guided by wind tunnel tests and computational simulation, and to new concepts for which there is a limited knowledge base. In either case they have enabled engineers to produce improved designs.

Substantial improvements are usually obtained with 20−200 design cycles, depending on the difficulty of the case. One concern is the possibility of getting trapped in a local minimum. In practice this has not proved to be a source of difficulty. In inverse mode, it often proves possible to come very close to realizing the target pressure distribution, thus effectively demonstrating convergence. In drag minimization, the result of the optimization is usually a shock-free wing. If one considers drag minimization of airfoils in two-dimensional inviscid transonic flow, it can be seen that every shock-free airfoil produces zero drag, and thus optimization based solely on drag has a highly non-unique solution. Different shock-free airfoils can be obtained by starting from different initial profiles. One may also influence the character of the final design by blending a target pressure distribution with the drag in the definition of the cost function.

Similar considerations apply to three-dimensional wing design. Since the vortex drag can be reduced simply by reducing the lift, the lift coefficient must be fixed for a meaningful drag minimization. In order to do this the angle of attack α is adjusted during the flow solution. It has proved most effective to make a small change $\delta\alpha$ proportional to the difference between the actual and the desired lift coefficient at every iteration in the flow calculation. A typical wing of a transport aircraft is designed for a lift coefficient in the range of 0.4 to 0.6. The total wing drag may be broken down into vortex drag, drag due to viscous effects, and shock drag. The vortex drag coefficient is typically in the range of 0.0100 (100 counts), while the friction drag coefficient is in the range of 45 counts, and the shock drag at a Mach number just before the onset of severe drag rise is of the order of 15 counts. With a fixed span, typically dictated by structural limits or a constraint imposed by airport gates, the vortex drag is entirely a function of span loading, and is minimized by an elliptic loading unless winglets are added. Transport aircraft usually have highly tapered wings with very large root chords to accommodate retraction of the undercarriage. An elliptic loading may lead to excessively large section lift coefficients on the outboard wing, leading to premature shock stall or buffet when the load is increased. The structure weight is also reduced by a more inboard loading which reduces the root bending moment. Thus the choice of span loading is influenced by other considerations. The skin friction of transport aircraft is typically very close to flat plate skin friction in turbulent flow, and is very insensitive to section variations. An exception to this is the case of smaller executive jet aircraft, for which the Reynolds number may be small enough to allow a significant run of laminar flow if the suction peak of the pressure distribution is moved back on the section. This leaves the shock drag as the primary target for wing section optimization. This is reduced to zero if the wing is shock-free, leaving no room for further improvement. Thus the attainment of a shock-free flow is a demonstration of a successful drag minimization. In practice range is maximized by maximizing $M\frac{L}{D}$, and this is likely to be increased by increasing the lift coefficient to the point where a weak shock appears. One may also use optimization to find the maximum Mach number at which the shock drag can be eliminated or significantly reduced for a wing with a given sweepback angle and thickness. Alternatively one may try to find the largest wing thickness or the minimum sweepback angle for which the shock drag can be eliminated at a given Mach number. This can yield both savings in structure weight and increased fuel volume . If there is no fixed limit for the wing span, such as a gate constraint, increased thickness can be used to allow an increase in aspect ratio for a wing of equal weight, in turn leading to a reduction in vortex drag. Since the vortex drag is usually the largest component of the total wing drag, this is probably the most effective design strategy, and it may pay to increase the wing thickness to the point where the optimized section produces a weak shock wave rather than a shock-free flow [25].

The first major industrial application of an adjoint based aerodynamic optimization method was the wing design of the Beech Premier [38] in 1995. The method was successfully used in inverse mode as a tool to obtain pressure distributions favorable to the maintenance of natural laminar flow over a range of cruise Mach numbers. Wing contours were obtained which yielded the desired pressure distribution in the presence of closely coupled engine nacelles on the fuselage above the wing trailing edge.

During 1996 some preliminary studies indicated that the wings of both the McDonnell Douglas MD-11 and the Boeing 747-200 could be made shock-free in a representative cruise condition by using very small shape modifications, with consequent drag savings which could amount to several percent of the total drag. This led to a decision to evaluate adjoint-based design methods in the design of the McDonnell Douglas MDXX during the summer and fall of 1996. In initial studies wing redesigns were carried out for inviscid transonic flow modeled by the Euler equations. A redesign to minimize the drag at a specified lift and Mach number required about 40 design cycles, which could be completed overnight on a workstation.

Three main lessons were drawn from these initial studies: (i) the fuselage effect is to large to be ignored and must be included in the optimization, (ii) single-point designs could be too sensitive to small variations in the flight condition, typically producing a shock-free flow at the design point with a tendency to break up into a severe double shock pattern below the design point, and (iii) the shape changes necessary to optimize a wing in transonic flow are smaller than the boundary layer displacement thickness, with the consequence that viscous effects must be included in the final design.

In order to meet the first two of these considerations, the second phase of the study was concentrated on the optimization of wing-body combinations with multiple design points. These were still performed with inviscid flow to reduce computational cost and allow for fast turnaround. It was found that comparatively insensitive designs could be obtained by minimizing the drag at a fixed Mach number for three fairly closely spaced lift coefficients such as 0.5, 0.525, and 0.55, or alternatively three nearby Mach numbers with a fixed lift coefficient.

The third phase of the project was focused on the design with viscous effects using as a starting point wings which resulted from multipoint inviscid optimization. While the full viscous adjoint method was still under development, it was found that useful improvements could be realized, particularly in inverse mode, using the inviscid result to provide the target pressure, by coupling an inviscid adjoint solver to a viscous flow solver. Computer costs are many times larger, both because finer meshes are needed to resolve the boundary layer, and because more iterations are needed in the flow and adjoint solutions. In order to force the specified lift coefficient the number of iterations in each flow solution had to be increased from 15 to 100. To achieve overnight turnaround a fully parallel implementation of the software had to be developed. Finally it was found that in order to produce sufficiently accurate results, the number of mesh points had to be increased to about 1.8 million. In the final phase of this project it was planned to carry out a propulsion integration study using the multiblock versions of the software. This study was not completed due to the cancellation of the entire MDXX project.

In the next subsections we present examples of the use of the adjoint method for viscous inverse and drag minimization in two dimensional flow. We then show a three-dimensional wing design using the Euler equations and a wing design using the full viscous adjoint method in its current form, implemented in the computer program SYN107. These calculations were all performed using the simple descent method with smoothing of the gradient. This has proved to be very efficient: in all cases the final optimum design was achieved with a total computational cost equivalent to the cost of from 2 to 10 converged flow solutions. The remaining subsections present results of optimizations for complete configurations in inviscid transonic and supersonic flow using the multiblock parallel design program, SYN107-MB.

Inverse design of an airfoil in transonic viscous flow

Our first example shows an inverse design in two dimensional viscous transonic flow obtained using the two-dimensional design code SYN103. The target pressure is that of the section of the ONERA M6 wing at Mach .75 and a lift coefficient of .50. It was calculated using SYN103 in analysis mode, thus it should be exactly realizable. A C-type mesh was used which contained 256 intervals in the chordwise and 96 cells in the normal direction for a total of 24, 576 cells. The design calculation was started with the NACA 0012 airfoil as the initial profile, and the ONERA M6 pressure distribution was almost exactly recovered in 25 design cycles. In the first cycle 120 iterations were used in both the flow and the adjoint solutions. In the subsequent cycles only 30 iterations were used in both the flow and adjoint solutions. Figure 4 shows the initial profile and pressure distribution with the pressure coefficient plotted vertically in the negative direction. It then shows the results after one, five and twenty five design cycles, with the target represented by circles. It also superposes on each redesigned profile the smoothed gradient plotted in the direction of the shape modification. A fixed scale is used so that it is possible to observe the decrease in the magnitude of the gradient as the calculation converges enough to ensure that they were fairly close to convergence. The root mean square error between the target and actual pressure was reduced from .0530 to .0016 in the course of the entire calculation which took 3569 seconds using a single Silicon Graphics R10000 processor. A fully converged flow solution using 500 iterations on the same mesh took 936 seconds, so the cost of the entire design calculation was about that of three flow solutions.

Drag reduction of an airfoil in viscous flow

The next example shows a redesign of the RAE2822 airfoil to reduce the drag at a fixed lift coefficient of .65 in transonic flow at Mach .75. In this case a shock free flow was obtained after 10 design cycles, in each of which both the flow and the adjoint solutions were calculated with 25 multigrid cycles. A grid with 512 × 64 cells was used. The pressure drag was reduced from .0091 to .0041, while the viscous drag remained essentially constant. The constraint was imposed that the thickness of the profile could not be reduced by only permitting outward movement from the initial profile. Figure 5 displays the sequence of pressure distributions, showing the elimination of the shock wave. It also shows the initial profile, and the smoothed gradient superposed on the subsequent profiles. It can be seen that the gradient continues to have an inward component, indicating that the drag might be further reduced if a thickness reduction were permitted. It should be noted that the unsmoothed gradient is in the sense of crossing over the trailing edge, because the resulting non physical shape would correspond to a sink in the free stream which would have a negative drag. The solution of the smoothing equation (71) with a two point boundary condition allow the trailing edge to be frozen.

Three point inviscid redesign of the Boeing 747 wing

The third example shows a redesign of the wing of the Boeing 747 to reduce its drag in a typical cruising condition. It has been our experience that drag minimization at a single point tends to produce a wing which is shock free at its design point, but tends to display undesirable characteristics off its design point. Typically, a double shock pattern forms below the design lift coefficient and Mach number, and a single fairly strong shock above the design point. To alleviate this tendency the calculation was performed with three design points. In carrying out multipoint designs of this kind a composite gradient is calculated as a weighted average of the gradients calculated for each design point separately. In this case the design points were selected as lift coefficients of .38, .42 and .46 for the exposed wing at Mach .85. Because the fuselage has a significant effect on the flow over the wing, the calculations were performed for the wing body combination, but the shape modifications were restricted to the wing alone. The fuselage also contributes to the lift, so that the total lift coefficient at the mid design point was estimated to be .50.

The results are displayed in Figures 6- 8 and in Table 1 which shows the drag at three design points of the initial wing, and the final wing after 30 design cycles. It can been seen that a drag reduction was obtained over the entire range of lift coefficients, and at the mid design point the redesigned wing is almost shock free. Figure 9 shows the modification in the wing section about half way out the span. It can be seen that a useful drag reduction can be obtained by a very small change in the wing shape. This is because of the extreme sensitivity of the transonic flow. Also, it is clear that without a tool of this kind it would be almost impossible to find an optimum shape.

| Design Conditions | | Initial | Three Point Design | |
Mach	C_L	C_D Original	C_D Redesign	C_D Reduction (%)
0.85	0.38	0.0071	0.0064	9.8
0.85	0.42	0.0086	0.0077	10.4
0.85	0.46	0.0106	0.0095	10.3

Table 1: Drag Reduction for Multipoint Design.

Transonic Viscous Wing-Body Design

A typical result for drag minimization of a wing body combination in transonic viscous flow is presented next. The viscous adjoint optimization method was used with a Baldwin-Lomax turbulence model. The initial wing is similar to one produced during the MDXX design studies. Figures 10-12 show the result of the wing-body redesign on a C-H mesh with $288 \times 96 \times 64$ cells. The wing has sweep back of about 38 degrees at the 1/4 chord. A total of 44 iterations of the viscous optimization procedure resulted in a shock-free wing at a cruise design point of Mach 0.86, with a lift coefficient of 0.61 for the wing-body combination at a Reynolds number of 101 million based on the root chord. Using 48 processors of an SGI Origin2000 parallel computer, each design iteration takes about 22 minutes so that overnight turnaround for such a calculation is possible. Figure 10 compares the pressure distribution of the final design with that of the initial wing. The final wing is quite thick, with a thickness to chord ratio of about 14 percent at the root and 9 percent at the tip. The optimization was performed with a constraint that the section modifications were not allowed to decrease the thickness anywhere. The design offers excellent performance at the nominal cruise point. A drag reduction of 2.2 counts was achieved from the initial wing which had itself been derived by inviscid optimization. Figures 11 and 12 show the results of a Mach number sweep to determine the drag rise. The drag coefficients shown in the figures represent the total wing drag including shock, vortex, and skin friction contributions. It can be seen that a double shock pattern forms below the design point, while there is actually a slight increase in the drag coefficient at Mach 0.85. This wing has a low drag coefficient, however, over a wide range of conditions. Above the design point a single shock forms and strengthens as the Mach number increases.

Transonic Multipoint Constrained Aircraft Design

As a first example of the automatic design capability for complex configurations, we show drag reduction for a typical business jet configuration. The objective of the design is to alter the geometry of the wing in order to minimize the configuration inviscid drag at three different flight conditions simultaneously. Realistic geometric spar thickness constraints are enforced. The geometry chosen for this analysis is a full configuration business jet composed of wing, fuselage, pylon, nacelle, and empennage. The inviscid multiblock mesh around this configuration follows a general C-O topology with special blocking to capture the geometric details of the nacelles, pylons and empennage. A total of 240 point-to-point matched blocks with 4,157,440 cells (including halos) are used to grid the complete configuration. This mesh allows the use of 4 multigrid levels obtained through recursive coarsening of the initial fine mesh. The upstream, downstream, upper and lower far field boundaries are located at an approximate distance of 15 wing semispans, while the far field boundary beyond the wing tip is located at a distance approximately equal to 5 semispans. An engineering-accuracy solution (with a decrease of 4 orders of magnitude in the average density residual) can be obtained in 100 multigrid cycles. This kind of solution can be routinely accomplished in under 20 minutes of wall clock time using 32 processors of an SGI Origin2000 computer.

The initial configuration was designed for Mach $= 0.8$ and $C_L = 0.3$. The three operating points chosen for this design are Mach $= 0.81$ with $C_L = 0.35$, Mach $= 0.82$ with $C_L = 0.30$, and Mach $= 0.83$ with $C_L = 0.25$. For each of the design points, both Mach number and lift coefficient are held fixed. In order to demonstrate the advantage of a multipoint design approach, the final solution at the middle design point will be compared with a single point design at the same conditions. As the geometry of the wing is modified, the design algorithm computes new wing-fuselage intersections. The wing component is made up of six airfoil defining sections. Eighteen Hicks-Henne design variables are applied to five of these sections for a total of 90 design variables. The sixth section at the symmetry plane is not modified. Spar thickness constraints were also enforced on each defining station at the $x/c = 0.2$ and $x/c = 0.8$ locations. Maximum thickness was forced to be preserved at $x/c = 0.4$ for all six defining sections. To ensure an adequate included angle at the trailing edge, each section was also constrained to preserve thickness at $x/c = 0.95$. Finally, to preserve leading edge bluntness, the upper surface of each section was forced to maintain its height above the camber line at $x/c = 0.02$. Combined, a total of 30 linear geometric constraints were imposed on the configuration.

Figures 13 – 15 show the initial and final airfoil geometries and C_p distributions after 5 NPSOL design iterations. It is evident that the new design has significantly reduced the shock strengths on both upper and lower wing surfaces at all design points. The transitions between design points are also quite smooth. For comparison purposes, a single point drag minimization study (Mach $= 0.81$ and $C_L = 0.25$) is carried out starting from the same initial configuration and using the same design variables and geometric constraints.

Figures 16 – 18 show comparisons of the solutions from the three-point design with those of the single point design. Interestingly, the upper surface shapes for both final designs are very similar. However, in the case of the single point design, a strong lower surface shock appears at the Mach $= 0.83$, $C_L = 0.25$ design point. The three-point design is able to suppress the formation of this lower surface shock and achieves a 9 count drag benefit over the single point design at this condition. However, it has a 1 count penalty at the single point design condition. The three-point design features a weak single shock for one of the three design points and a very weak double shock at another design point. Table 2 summarizes the drag results for the two designs. The C_D values have been normalized by the drag of the initial configuration at the second design point. Figure 19 shows the surface of

Design Mach	Conditions C_L	Initial Relative C_D	Single Point Design Relative C_D	Three Point Design Relative C_D
0.81	0.35	1.00257	0.85003	0.85413
0.82	0.30	1.00000	0.77350	0.77915
0.83	0.25	1.08731	0.81407	0.76836

Table 2: Drag Reduction for Single and Multipoint Designs.

the configuration colored by the local coefficient of pressure, C_p, before and after redesign for the middle design point. One can clearly observe that the strength of the shock wave on the upper surface of the configuration has been considerably reduced.

Finally, Figure 20 shows the parallel scalability of the multiblock design method for the mesh in question using up to 32 processors of an SGI Origin2000 parallel computer. Despite the fact that the multigrid technique is used in both the flow and adjoint solvers, the demonstrated parallel speedups are outstanding.

Supersonic Constrained Aircraft Design

For supersonic design, provided that turbulent flow is assumed over the entire configuration, the inviscid Euler equations suffice for aerodynamic design since the pressure drag is not greatly affected by the inclusion of viscous effects. Moreover, flat plate skin friction estimates of viscous drag are often very good approximations. In this study we show drag reduction of a generic supersonic transport configuration used in reference [39] using the inviscid Euler equations to model the flow.

The baseline supersonic transport configuration was sized to accommodate 300 passengers with a gross take-off weight of 750,000 lbs. The supersonic cruise point is Mach 2.2 with a C_L of 0.105. Figure 21 shows that the planform is a cranked-delta configuration with a break in the leading edge sweep. The inboard leading edge sweep is 68.5 degrees while the outboard is 49.5 degrees. Since the Mach angle at $M = 2.2$ is 63 degrees it is clear that some leading edge bluntness may be used inboard without a significant wave drag penalty. Blunt leading edge airfoils were created with thickness ranging from 4% at the root to 2.5% at the leading edge break point. These symmetric airfoils were chosen to accommodate thick spars at roughly the 5% and 80% chord locations over the span up to the leading edge break. Outboard of the leading edge break where the wing sweep is ahead of the Mach cone, a sharp leading edge was used to avoid unnecessary wave drag. The airfoils were chosen to be symmetric, biconvex shapes modified to have a region of constant thickness over the mid-chord. The four-engine configuration features axisymmetric nacelles tucked close to the wing lower surface. This layout favors reduced wave drag by minimizing the exposed boundary layer diverter area. However, in practice it may be problematic because of the channel flows occurring in the juncture region of the diverter, wing, and nacelle at the wing trailing edge.

The computational mesh on which the design is run has 180 blocks and 1,500,000 mesh cells (including halos), while the underlying geometry entities define the wing with 16 sectional cuts and the body with 200 sectional cuts. In this case, where we hope to optimize the shape of the wing, care must be taken to ensure that the nacelles remain properly attached with diverter heights being maintained.

The objective of the design is to reduce the total drag of the configuration at a single design point (Mach = 2.2, $C_L = 0.105$) by modifying the wing shape. Just as in the transonic case, 18 design variables of the Hicks-Henne type are chosen for each wing defining section. Similarly, instead of applying them to all 16 sections, they are applied to 8 of the sections and then lofted linearly to the neighboring sections. Spar thickness constraints are imposed for all wing defining sections at $x/c = 0.05$ and $x/c = 0.8$. An additional maximum thickness constraint is specified along the span at $x/c = 0.5$. A final thickness constraint is enforced at $x/c = 0.95$ to ensure a reasonable trailing edge included angle. An iso-C_p representation of the initial and final designs is depicted in Figure 21 for both the upper and lower surfaces.

It is noted that the strong oblique shock evident near the leading edge of the upper surface on the initial configuration is largely eliminated in the final design after 5 NPSOL design iterations. Also, it is seen that the upper surface pressure distribution in the vicinity of the nacelles has formed an unexpected pattern. However, recalling that thickness constraints abound in this design, these upper surface pressure patterns are assumed to be the result of sculpting of the lower surface near the nacelles which affects the upper surface shape via the thickness constraints. For the lower surface, the leading edge has developed a suction region while the shocks and expansions around the nacelles have been somewhat reduced. Figure 22 shows the pressure coefficients and (scaled) airfoil sections for four sectional cuts along the wing. These cuts further demonstrate the removal of the oblique shock on the upper surface and the addition of a suction region on the leading edge of the lower surface. The airfoil sections have been scaled by a factor of 2 so that shape changes may be seen more easily. Most notably, the section at 38.7% span has had the lower surface drastically modified such that a large region of the aft airfoil has a forward-facing portion near where the pressure spike from the nacelle shock impinges on the surface. The final overall pressure drag was reduced by 8%, from $C_D = 0.0088$ to $C_D = 0.0081$.

11   Conclusions

The adjoint design method presented in these notes is now well established and has proved effective in a variety of industrial applications. The method combines the versatility of numerical optimization methods with the efficiency of inverse design methods. The geometry is modified by a grid perturbation technique which is applicable to arbitrary configurations. Both the wing-body and multiblock version of the design algorithms have been implemented in parallel using the MPI (Message Passing Interface) Standard, and they both yield excellent parallel speedups. The combination of computational efficiency with geometric flexibility provides a powerful tool, with the final goal being to create practical aerodynamic shape design methods for complete aircraft configurations.

Acknowledgment This work has benefited from the generous support of AFOSR, ONR, the NASA-IBM Cooperative Research Agreement, and the DoD under the Grand Challenge Projects of the High Performance Computing Modernization Program.

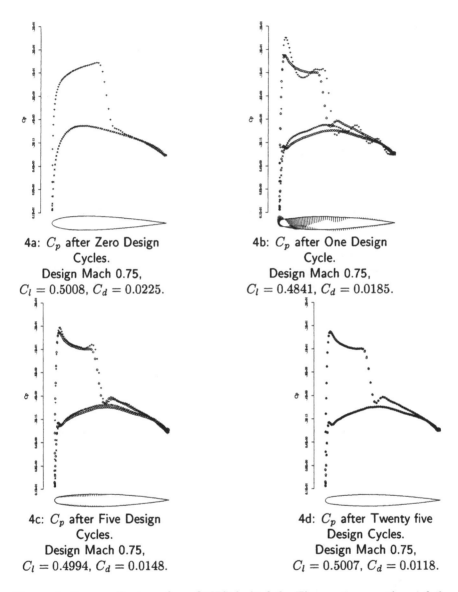

4a: C_p after Zero Design
Cycles.
Design Mach 0.75,
$C_l = 0.5008$, $C_d = 0.0225$.

4b: C_p after One Design
Cycle.
Design Mach 0.75,
$C_l = 0.4841$, $C_d = 0.0185$.

4c: C_p after Five Design
Cycles.
Design Mach 0.75,
$C_l = 0.4994$, $C_d = 0.0148$.

4d: C_p after Twenty five
Design Cycles.
Design Mach 0.75,
$C_l = 0.5007$, $C_d = 0.0118$.

Figure 4: Inverse Design of an ONERA Airfoil. The vectors on the airfoil surface represent the direction and magnitude of the gradient.

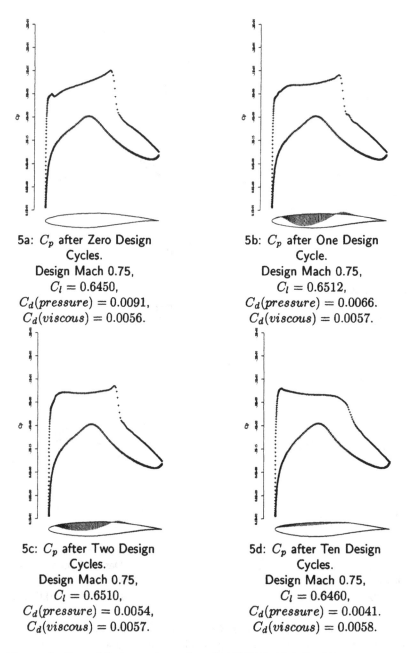

5a: C_p after Zero Design
Cycles.
Design Mach 0.75,
$C_l = 0.6450$,
$C_d(pressure) = 0.0091$,
$C_d(viscous) = 0.0056$.

5b: C_p after One Design
Cycle.
Design Mach 0.75,
$C_l = 0.6512$,
$C_d(pressure) = 0.0066$.
$C_d(viscous) = 0.0057$.

5c: C_p after Two Design
Cycles.
Design Mach 0.75,
$C_l = 0.6510$,
$C_d(pressure) = 0.0054$,
$C_d(viscous) = 0.0057$.

5d: C_p after Ten Design
Cycles.
Design Mach 0.75,
$C_l = 0.6460$,
$C_d(pressure) = 0.0041$.
$C_d(viscous) = 0.0058$.

Figure 5: Drag Minimization of an RAE2822 Airfoil. The vectors on the
airfoil surface represent the direction and magnitude of the gradient.

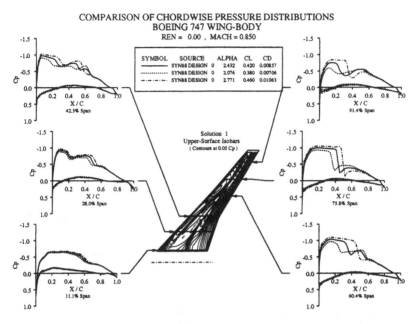

Figure 6: Pressure distribution of the Boeing 747 Wing-Body before optimization.

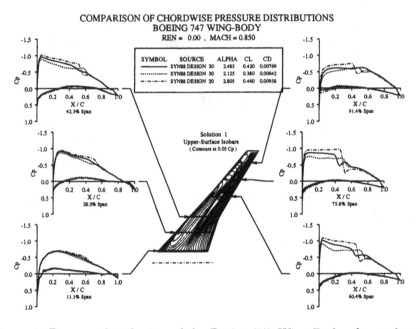

Figure 7: Pressure distribution of the Boeing 747 Wing-Body after a three point optimization.

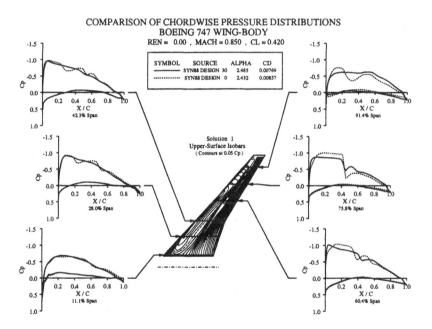

Figure 8: Comparison of Original and Optimized Boeing 747 Wing-Body at the mid design point

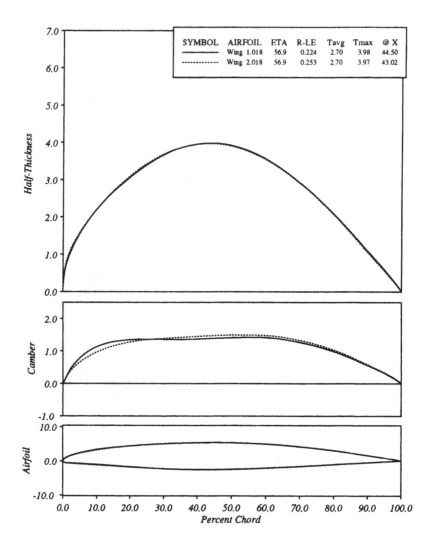

Figure 9: Original and Re-designed Wing section for the Boeing 747 Wing-Body at mid-span.

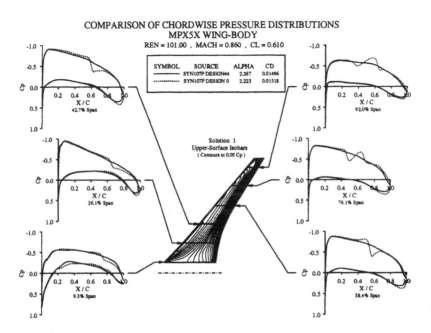

Figure 10: Pressure distribution of the MPX5X before and after optimization.

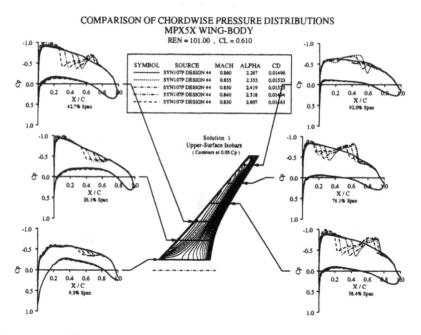

Figure 11: Off design performance of the MPX5X below the design point.

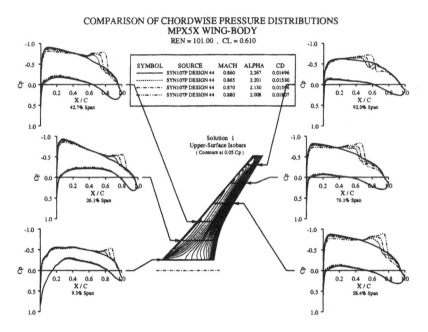

Figure 12: Off design performance of the MPX5X above the design point.

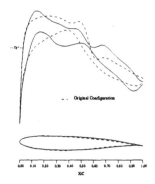

13a: span station $z = 0.190$

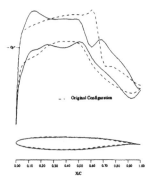

13b: span station $z = 0.475$

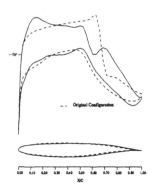

13c: span station $z = 0.665$

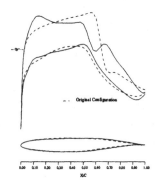

13d: span station $z = 0.856$

Figure 13: Business Jet Configuration. Multipoint Drag Minimization at Fixed Lift.
Design Point 1, $M = 0.81$, $C_L = 0.35$
90 Hicks-Henne variables. Spar Constraints Active.
$- - -$, Initial Pressures
————, Pressures After 5 Design Cycles.

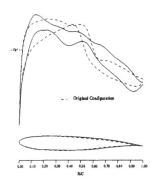

14a: span station $z = 0.190$

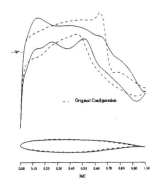

14b: span station $z = 0.475$

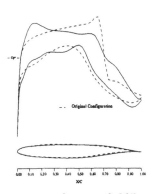

14c: span station $z = 0.665$

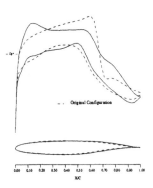

14d: span station $z = 0.856$

Figure 14: Business Jet Configuration. Multipoint Drag Minimization at Fixed Lift.
Design Point 2, $M = 0.82$, $C_L = 0.30$
90 Hicks-Henne variables. Spar Constraints Active.
- - -, Initial Pressures
———, Pressures After 5 Design Cycles.

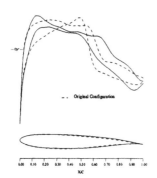

15a: span station $z = 0.190$

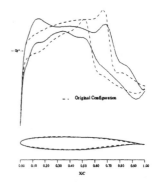

15b: span station $z = 0.475$

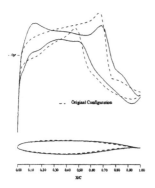

15c: span station $z = 0.665$

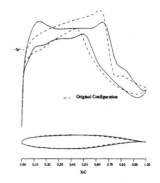

15d: span station $z = 0.856$

Figure 15: Business Jet Configuration. Multipoint Drag Minimization at Fixed Lift.
Design Point 3, $M = 0.83$, $C_L = 0.25$
90 Hicks-Henne variables. Spar Constraints Active.
$- - -$, Initial Pressures
————, Pressures After 5 Design Cycles.

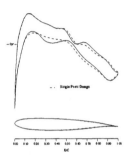

16a: span station $z = 0.190$

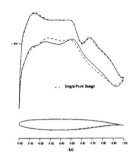

16b: span station $z = 0.475$

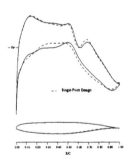

16c: span station $z = 0.665$

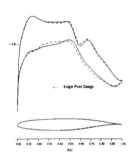

16d: span station $z = 0.856$

Figure 16: Business Jet Configuration. Single Point vs. Multipoint Drag Minimization at Fixed Lift.
Design Point 1, $M = 0.81$, $C_L = 0.35$
90 Hicks-Henne variables. Spar Constraints Active.
– – –, Single Point Design Pressures.
———, Multipoint Design Pressures.

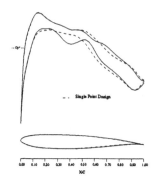

17a: span station $z = 0.190$

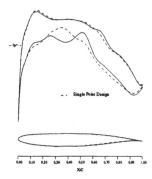

17b: span station $z = 0.475$

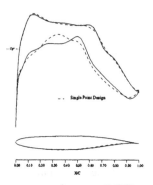

17c: span station $z = 0.665$

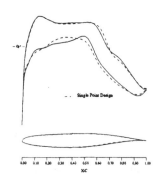

17d: span station $z = 0.856$

Figure 17: Business Jet Configuration. Single Point vs. Multipoint Drag Minimization at Fixed Lift.
Design Point 2, $M = 0.82$, $C_L = 0.30$
90 Hicks-Henne variables. Spar Constraints Active.
$-\ -\ -$, Single Point Design Pressure.
————, Multipoint Design Pressures.

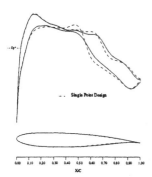

18a: span station $z = 0.190$

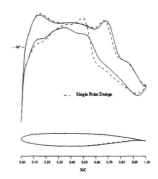

18b: span station $z = 0.475$

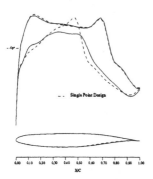

18c: span station $z = 0.665$

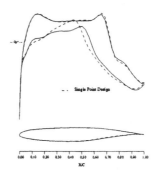

18d: span station $z = 0.856$

Figure 18: Business Jet Configuration. Single Point vs. Multipoint Drag Minimization at Fixed Lift.
Design Point 3, $M = 0.83$, $C_L = 0.25$
90 Hicks-Henne variables. Spar Constraints Active.
– – –, Single Point Design Pressures.
———, Multipoint Design Pressure.

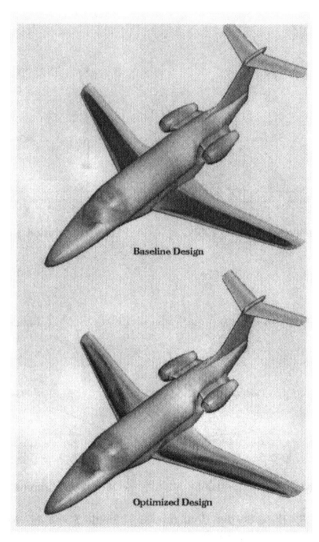

Figure 19: Geometry Surface Colored by local C_p Before and After Redesign.

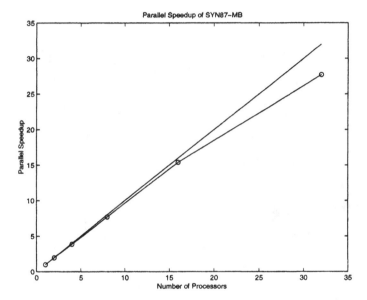

Figure 20: Scalability Study for Multiblock Design Method.

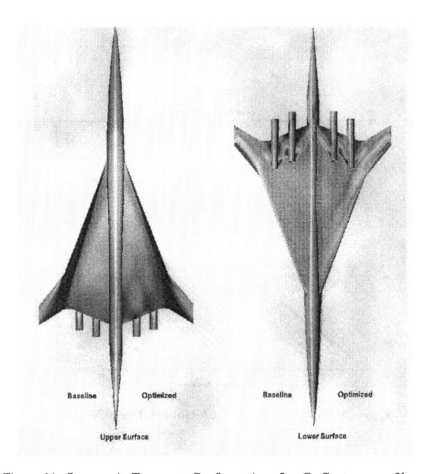

Figure 21: Supersonic Transport Configuration. Iso-C_p Contours on Upper and Lower Surfaces. Baseline and Optimized Designs. $M = 2.2$, $C_L = 0.105$.

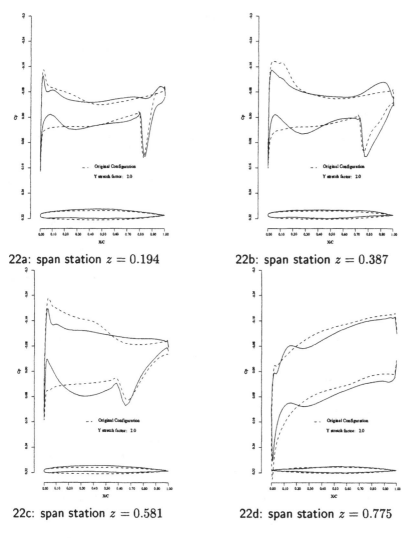

22a: span station $z = 0.194$

22b: span station $z = 0.387$

22c: span station $z = 0.581$

22d: span station $z = 0.775$

Figure 22: Supersonic Transport Configuration. Drag Minimization at Fixed Lift.
$M = 2.20$, $C_L = 0.105$
144 Hicks-Henne variables. Spar Constraints Active.
$- - -$, Initial Pressures
————, Pressures After 5 Design Cycles.

References

[1] R. M. Hicks, E. M. Murman, and G. N. Vanderplaats. An assessment of airfoil design by numerical optimization. *NASA TM X-3092*, Ames Research Center, Moffett Field, California, July 1974.

[2] R. M. Hicks and P. A. Henne. Wing design by numerical optimization. *Journal of Aircraft*, 15:407–412, 1978.

[3] J. L. Lions. *Optimal Control of Systems Governed by Partial Differential Equations*. Springer-Verlag, New York, 1971. Translated by S.K. Mitter.

[4] A. E. Bryson and Y. C. Ho. *Applied Optimal Control*. Hemisphere, Washington, DC, 1975.

[5] M. J. Lighthill. A new method of two-dimensional aerodynamic design. *R & M 1111*, Aeronautical Research Council, 1945.

[6] A. Jameson. Aerodynamic design via control theory. *Journal of Scientific Computing*, 3:233–260, 1988.

[7] O. Pironneau. *Optimal Shape Design for Elliptic Systems*. Springer-Verlag, New York, 1984.

[8] A. Jameson. Optimum aerodynamic design using CFD and control theory. *AIAA paper 95-1729*, AIAA 12th Computational Fluid Dynamics Conference, San Diego, CA, June 1995.

[9] A. Jameson and J.J. Alonso. Automatic aerodynamic optimization on distributed memory architectures. *AIAA paper 96-0409*, 34th Aerospace Sciences Meeting and Exhibit, Reno, Nevada, January 1996.

[10] A. Jameson. Re-engineering the design process through computation. *AIAA paper 97-0641*, 35th Aerospace Sciences Meeting and Exhibit, Reno, Nevada, January 1997.

[11] A. Jameson, N. Pierce, and L. Martinelli. Optimum aerodynamic design using the Navier-Stokes equations. *AIAA paper 97-0101*, 35th Aerospace Sciences Meeting and Exhibit, Reno, Nevada, January 1997.

[12] A. Jameson, L. Martinelli, and N. A. Pierce. Optimum aerodynamic design using the Navier-Stokes equations. *Theoret. Comput. Fluid Dynamics*, 10:213–237, 1998.

[13] A. Jameson. Automatic design of transonic airfoils to reduce the shock induced pressure drag. In *Proceedings of the 31st Israel Annual Conference on Aviation and Aeronautics, Tel Aviv*, pages 5–17, February 1990.

[14] A. Jameson. Optimum aerodynamic design via boundary control. In *AGARD-VKI Lecture Series, Optimum Design Methods in Aerodynamics*. von Karman Institute for Fluid Dynamics, 1994.

[15] J. Reuther, A. Jameson, J. J. Alonso, M. J. Rimlinger, and D. Saunders. Constrained multipoint aerodynamic shape optimization using an adjoint formulation and parallel computers. *AIAA paper 97-0103*, 35th Aerospace Sciences Meeting and Exhibit, Reno, Nevada, January 1997.

[16] J. Reuther, J. J. Alonso, J. C. Vassberg, A. Jameson, and L. Martinelli. An efficient multiblock method for aerodynamic analysis and design on distributed memory systems. *AIAA paper 97-1893*, June 1997.

[17] O. Baysal and M. E. Eleshaky. Aerodynamic design optimization using sensitivity analysis and computational fluid dynamics. *AIAA Journal*, 30(3):718–725, 1992.

[18] J.C. Huan and V. Modi. Optimum design for drag minimizing bodies in incompressible flow. *Inverse Problems in Engineering*, 1:1–25, 1994.

[19] M. Desai and K. Ito. Optimal controls of Navier-Stokes equations. *SIAM J. Control and Optimization*, 32(5):1428–1446, 1994.

[20] W. K. Anderson and V. Venkatakrishnan. Aerodynamic design optimization on unstructured grids with a continuous adjoint formulation. *AIAA paper 97-0643*, 35th Aerospace Sciences Meeting and Exhibit, Reno, Nevada, January 1997.

[21] J. Elliott and J. Peraire. 3-D aerodynamic optimization on unstructured meshes with viscous effects. *AIAA paper 97-1849*, June 1997.

[22] S. Ta'asan, G. Kuruvila, and M. D. Salas. Aerodynamic design and optimization in one shot. *AIAA paper 92-0025*, 30th Aerospace Sciences Meeting and Exhibit, Reno, Nevada, January 1992.

[23] A. Jameson. Iterative solution of transonic flows over airfoils and wings, including flows at Mach 1. *Communications on Pure and Applied Mathematics*, 27:283–309, 1974.

[24] A. Jameson. Acceleration of transonic potential flow calculations on arbitrary meshes by the multiple grid method. *AIAA paper 79-1458*, Fourth AIAA Computational Fluid Dynamics Conference, Williamsburg, Virginia, July 1979.

[25] A. Jameson. Optimum aerodynamic design using control theory. *Computational Fluid Dynamics Review*, pages 495–528, 1995.

[26] A. Jameson, W. Schmidt, and E. Turkel. Numerical solutions of the Euler equations by finite volume methods with Runge-Kutta time stepping schemes. *AIAA paper 81-1259*, January 1981.

[27] L. Martinelli and A. Jameson. Validation of a multigrid method for the Reynolds averaged equations. *AIAA paper 88-0414*, 1988.

[28] S. Tatsumi, L. Martinelli, and A. Jameson. A new high resolution scheme for compressible viscous flows with shocks. *AIAA paper* To Appear, AIAA 33nd Aerospace Sciences Meeting, Reno, Nevada, January 1995.

[29] A. Jameson. Analysis and design of numerical schemes for gas dynamics 1, artificial diffusion, upwind biasing, limiters and their effect on multigrid convergence. *Int. J. of Comp. Fluid Dyn.*, 4:171–218, 1995.

[30] P.L. Roe. Approximate Riemann solvers, parameter vectors, and difference schemes. *Journal of Computational Physics*, 43:357–372, 1981.

[31] A. Jameson. Analysis and design of numerical schemes for gas dynamics 2, artificial diffusion and discrete shock structure. *Int. J. of Comp. Fluid Dyn.*, 5:1–38, 1995.

[32] L. Martinelli. Calculations of viscous flows with a multigrid method. *Princeton University Thesis*, May 1987.

[33] A. Jameson. Multigrid algorithms for compressible flow calculations. In W. Hackbusch and U. Trottenberg, editors, *Lecture Notes in Mathematics, Vol. 1228*, pages 166–201. Proceedings of the 2nd European Conference on Multigrid Methods, Cologne, 1985, Springer-Verlag, 1986.

[34] J. J. Reuther, A. Jameson, J. J. Alonso, M. Rimlinger, and D. Saunders. Constrained multipoint aerodynamic shape optimization using an adjoint formulation and parallel computers: Part i. *Journal of Aircraft*, 1998. Accepted for publication.

[35] J. J. Reuther, A. Jameson, J. J. Alonso, M. Rimlinger, and D. Saunders. Constrained multipoint aerodynamic shape optimization using an adjoint formulation and parallel computers: Part ii. *Journal of Aircraft*, 1998. Accepted for publication.

[36] J.P. Steinbrenner, J.R. Chawner, and C.L. Fouts. The GRIDGEN 3D multiple block grid generation system. Technical report, Flight Dynamics Laboratory, Wright Research and Development Center, Wright-Patterson Air Force Base, Ohio, July 1990.

[37] J. Reuther and A. Jameson. Aerodynamic shape optimization of wing and wing-body configurations using control theory. *AIAA paper 95-0123*, AIAA 33rd Aerospace Sciences Meeting, Reno, Nevada, January 1995.

[38] J. Gallman, J. Reuther, N. Pfeiffer, W. Forrest, and D. Bernstorf. Business jet wing design using aerodynamic shape optimization. *AIAA paper 96-0554*, 34th Aerospace Sciences Meeting and Exhibit, Reno, Nevada, January 1996.

[39] J. Reuther, J.J. Alonso, M.J. Rimlinger, and A. Jameson. Aerodynamic shape optimization of supersonic aircraft configurations via an adjoint formulation on parallel computers. *AIAA paper 96-4045*, 6th AIAA/NASA/ISSMO Symposium on Multidisciplinary Analysis and Optimization, Bellevue, WA, September 1996.

Complexity in Industrial Problems
Some remarks

J.L. LIONS

College de France
3. Rue d'Ulm
75231 Paris Cedex 05

Abstract

When working on the numerical approximation of the solution of the **state equations** modeling industrial (complex) systems, one meets a number of problems which are rather universal.

We present here an **introduction to some** of these problems and to **some** of the methods which can be used, some of these methods being already quite old (we give then a short introduction to them). Others are on the contrary in development stage, and we then present them with more details.

We begin with **stiff problems**, i.e. models where one has several orders of magnitude of difference in the **size** of the coefficients. We introduce in such situations **stiff asymptotic expansions**. Questions of this type are met very often, for instance in electromagnetism.

In Section 3 we consider **multi scales problems**, where the "multi scales" refer to **time** scales and to **space** scales as well. In these cases, associated with multi physics with intrinsic mixing properties of the system under study, one can use expansions based on **slow and fast variables, in time and, or in space**. The main ideas are presented for rapidly rotating fluids and for flows in porous media, where the expansion leads to Darcy's law.

Section 4 deals with the "universal" difficulty of **the complexity of the geometry** of the system under study. A classical method used to tackle this type of question is the DDM (**domain decomposition method**).

An introduction is presented in Section 4, an introduction to the **decomposition of the energy space** being presented in Section 5. The Section 4 and 5 are based on the **virtual control method**, introduced by O. PIRONNEAU and the A. in several conferences in 1998 and in the notes, referred to in Section 4. It contains as a particular case artificial or fictitions domains methods. Many applications and extensions are now under progress, in joint papers by O. PIRONNEAU and the A. The Decomposition of the Energy space has been introduced in a note of R. GLOWINSKI, O. PIRONNEAU and the A. (CRAS, 1999), developments being under development with T.W. PAN Others are in progress with J.P. PERIAUX

The method of **virtual control** applies in a very efficient manner to problems where there is an **effective control** (a **real** control !). Extensive developments are then needed. They will be presented elsewhere, a short introduction being given in the second note of O. PIRONNEAU and the A. referred to above.

Contents

1 Introduction

One of the characteristics of the mathematical problems arising in Industry is their complexity.

We present in this set of lectures **some** of the reasons why such is the case together with **some** of the methods that can be used to tackle (or to try to...) these problems. □

First examples arise in the **size** of the coefficients which may vary by **several orders of magnitude** in various connected regions. Examples of such situations abound in electronics (junctions), in electromagnetism, in composite materials, etc... From a mathematical and computational point of view, these questions enter in the family of **stiff problems**.

For these problems, a natural method is to try **asymptotic expansions**. This is what we present in Section 2, following the A. [1] (cf. the Bibliography of Section 2). This section can be considered as some kind of "warming up" for the following sections.

Of course the "real" stiff industrial problems are **much** more complicated than the ones presented here, in particular due to the systematic presence of non linearities. But the asymptotic expansions presented in Section 2 can serve as a bottom line for more complicated (and partly formal) computations. □

Another often met difficulty in Industrial problems comes from the various **scales**, both with respect to **space variables** and with respect to **time**, which enter in the questions to be solved. This is due, among other things, to the **multi-physics aspects of most of the industrial problems**.

The scales (both in time and in space) can differ again by several orders of magnitude depending on the "physic" which is considered.

A simple (more precisely : a simple looking) example of such a situation is presented in Section 3, for **rapidly rotating fluids**. Many problems of this type arise in all sorts of engines and they do arise in Climatology as well. We introduce **asymptotic expansions** where more and more complicated "ansatz" are needed. We also consider rotating fluids in porous media, which leads, when adequate scaling is made and a proper ansatz is introduced, to the Darcy's law.

Of course the source of complexity which is probably the most frequent is the **complexity of the geometry**. Examples abound : planes, ships, engines, etc... A classical way to address these questions is to "cut the domain in pieces". These are the famous DDM, Domain Decomposition Methods, an Industry by itself !

With O. PIRONNEAU, we have introduced in 1998, in several lectures and in the Notes refered to in the Bibliography of Section 4, a family of methods based on the idea of **virtual controls**. Several systematic publications are in preparation on these topics. We present in Section 4 an **introduction** to these publications.

As we show in the text, the method of artificial domain or of fictitions domain, is related to a particular case of the general methodology of virtual controls. We refer to the work of R. GLOWINSKI, T.W. PAN and J. PERIAUX (cf. the Bibliography of Section 4).   □

Of course the geometrical domain is not the only component of the problem than can be "cut in pieces". In most stationnary problems there is a natural "energy space", a space where some estimates can be obtained. One can then think of **decomposing the energy space.**

It turns out that here again the idea of **virtual controls** (used in a different fashion) leads to a **completely general family** of decompositions of the energy space. This is presented in Section 5 which can be thought of as an introduction to the paper of R. GLOWINSKI, O. PIRONNEAU and the A., refered to in Section 5.   □

The technique of **virtual control** seems to be particularly useful for problems where is **an effective control**. We confine ourselves here to refer to the notes of O. PIRONNEAU and the A. given in Section 4. Several papers are now in preparation on these questions.

2   Size of coefficients and stiff problems

2.1   A second order elliptic problem.

Let Ω be a bounded open set of $\mathbf{R}^d$, d=2 or 3 in the applications), defined (cf. Figure 1) by

$$\Omega = \overline{\Omega}_1 \cup \Omega_0$$

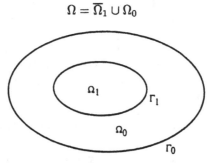

Figure 1

The boundary $\partial\Omega_1$ of Ω_1 is given by

$$\partial\Omega_1 = \Gamma_1$$

and

$$\partial\Omega_0 = \Gamma_0 \cup \Gamma_1.$$

We are given the operators

$$A_0 = -\sum_{i,j=1}^{d} \frac{\partial}{\partial x_i}\left(a_{ij}^0 \frac{\partial}{\partial x_j}\right) \quad \text{in } \Omega_0,$$

$$A_1 = -\sum_{i,j=1}^{d} \frac{\partial}{\partial x_i}\left(a_{ij}^1 \frac{\partial}{\partial x_j}\right) \quad \text{in } \Omega_1, \tag{2.1}$$

where

$$a_{ij}^k \in L^\infty(\Omega_k), \qquad k = 0,1,$$

$$\sum_{i,j=1}^{d} a_{ij}^k(x)\xi_i\xi_j \ge \alpha^k \sum_{i=1}^{d} \xi_i^2 \quad, \quad \alpha^k > 0, \text{ a.e. in } \Omega_k. \tag{2.2}$$

For $u, v \in H^1(\Omega_k)$, $k = 0,1$, we set

$$a_k(u,v) = \sum_{i,j=1}^{d} \int_{\Omega_k} a_{ij}^k \frac{\partial u}{\partial x_j} \frac{\partial v}{\partial x_i} dx. \tag{2.3}$$

We then introduce

$$a_\varepsilon(u,v) = a_0(u,v) + \varepsilon a_1(u,v) \tag{2.4}$$

where $u, v \in H^1(\Omega)$ and where in, say, $a_0(u,v)$, one takes for u and v the restriction of u and v to Ω_0 (idem for Ω_1).

In (2.4) ε is > 0 and "small".

We denote by u_ε **the unique solution** of

$$a_\varepsilon(u_\varepsilon, v) = (f, v) \qquad \forall\, v \in H_0^1(\Omega),$$
$$u_\varepsilon \in H_0^1(\Omega) \tag{2.5}$$

where f is given in $L^2(\Omega)$ and $(f,v) = \int_\Omega fv\, dx$.

Remark 2.1

Because of (1.3), the bilinear form $a_\varepsilon(u,v)$ is coercive on $H_0^1(\Omega)$ (subspace of $H^1(\Omega)$ of those functions v in $H^1(\Omega)$ which are zero on Γ), so that (1.6) admits a unique solution. □

Remark 2.2

Let f_1 = restriction of f to Ω_i, $u_{\varepsilon i}$ the restriction of u_ε.

It follows from (2.5) that

$$A_0 u_{\varepsilon 0} = f_0 \qquad \text{in } \Omega_0,$$
$$\varepsilon A_1\, u_{\varepsilon 1} = f_1 \qquad \text{in } \Omega_1. \tag{2.6}$$

Since $u_\varepsilon \in H_0^1(\Omega)$, one has

$$u_{\varepsilon 0} = 0 \qquad \text{on } \Gamma_0,$$
$$u_{\varepsilon 0} = u_{\varepsilon 1} \quad \text{on } \Gamma_1, \tag{2.7}$$

There is another condition on the interface Γ_1, namely

$$\frac{\partial u_{\varepsilon 0}}{\partial \eta_{A_0}} = \varepsilon \frac{\partial u_{\varepsilon 1}}{\partial \eta_{A_1}} \tag{2.8}$$

where $\dfrac{\partial}{\partial \eta_{A_1}}$ = conormal derivative attached to A_i, the normal to Γ_1 being directed towards the exterior of Ω_0 to fix ideas. Of course (2.8) is defined in a **weak** form by the variational formulation (2.5). It becomes true in a **strong** form if the coefficients a_{ij}^k are sufficiently smooth.

The verification of (2.8) is standard. One multiplies both equations in (2.6) by v (we use the same notation for $v \in H_0^1(\Omega)$ and for its restrictions to Ω_0, Ω_1). One obtains

$$-\int_{\Gamma_1} \frac{\partial u_{\varepsilon 0}}{\partial \eta_{A_0}} v \, d\Gamma_1 + a_0(u_{\varepsilon 0}, v) = (f_0, v) \quad \left(= \int_{\Omega_0} f_0 \, v \, dx \right),$$

$$\varepsilon \int_{\Gamma_1} \frac{\partial u_{\varepsilon 1}}{\partial \eta_{A_1}} v \, d\Gamma_1 + \varepsilon a_1(u_{\varepsilon 1}, v) = (f_1, v).$$

Adding up and using (2.5), it follows that

$$\int_{\Gamma_1} \left(-\frac{\partial u_{\varepsilon 0}}{\partial \eta_{A_0}} + \varepsilon \frac{\partial u_{\varepsilon 1}}{\partial \eta_{A_1}} \right) v \, d\Gamma_1 = 0 \qquad \forall \, v \in H_0^1(\Omega)$$

hence (2.8) follows. $\qquad\qquad\qquad\qquad\qquad\qquad\qquad\qquad\qquad\qquad$ □

Remark 2.3

The operator A_ε attached to the bilinear form $a_\varepsilon(u, v)$ is

$$A_\varepsilon = A_0 + \varepsilon A_1 \text{ in } \Omega$$

(this writing is **formal**, since A_i is defined only in Ω_i), i.e. an operator whose coefficients **are of different orders of magnitude on Ω_0 and on Ω_1**□

We are looking for an asymptotic expansion of u_ε. We try the "ansatz"

$$u_\varepsilon = \frac{u^{-1}}{\varepsilon} + u^0 + \varepsilon \, a^1 + \dots. \tag{2.9}$$

Using (2.9) in (2.5) and identifying the powers of ε, we obtain :

$$a_0(u^{-1}, v) = 0,$$
$$a_0(u^0, v) + a_1(u^{-1}, v) = (f, v),$$
$$a_0(u^{-1}, v) + a_1(u^0, v) = 0,$$
$$\dots. \tag{2.10}$$

We now solve these equations.

It follows from $(2.10)_1$ and (2.2) (with $k = 0$) that $u_0^{-1} = $ constant in Ω_0 and since $u_\varepsilon = 0$ on Γ_0, we take $u_0^{-1} = 0$ on Γ_0 so that

$$u_0^{-1} = 0. \tag{2.11}$$

We consider now $(2.10)_2$ where we take $v \in H_0^1(\Omega_1), v = 0$ outside Ω_1. It follows that

$$a_1(u^{-1}, v) = (f_1, v) \qquad \forall v \in H_0^1(\Omega_1)$$

and $u_1^{-1} = u_0^{-1}$ on Γ_1 i.e.

$$A_1 u_1^{-1} = f_1 \quad \text{in } \Omega_1,$$
$$u_1^{-1} = 0 \qquad \text{on } \Gamma_1. \tag{2.12}$$

Equations (2.11), (2.12) uniquely define u^{-1}.
It follows from (2.12) that

$$\int_{\gamma_1} \frac{\partial u_1^{-1}}{\partial \eta_{A_1}} v \, d\Gamma_1 + a_1(u^{-1}, v) = (f_1, v) \qquad \forall v \in H^1(\Omega_1). \tag{2.13}$$

This is **not** formal if the coefficients of A_1 are in $W^{1,\infty}(\Omega_1)$.
We use (2.13) in $(2.10)_2$. We obtain

$$a_0(u^0, v) = (f_0, v) + \int_{\Gamma_1} \frac{\partial u_1^{-1}}{\partial \eta_{A_1}} v \, d\Gamma_1 . \tag{2.14}$$

This is equivalent to

$$A_0 \, u_0^0 = f_0 \qquad \text{in } \Omega_0,$$
$$u_0^0 = 0 \qquad \text{on } \Gamma_0,$$
$$\frac{\partial u_0^0}{\partial \eta_{A_0}} = \frac{\partial u_1^{-1}}{\partial \eta_{A_1}} \quad \text{on } \Gamma_1 . \tag{2.15}$$

We proceed with $(2.10)_3$ where we take $v \in H_0^1(\Omega_1)$.
Then $a_1(u^0, v) = 0 \quad \forall v \in H_0^1(\Omega_1)$, hence

$$A_1 u_1^0 = 0 \qquad \text{in } \Omega_1,$$
$$u_1^0 = u_0^0 \qquad \text{on } \Gamma_1. \tag{2.16}$$

The equations (2.15), (2.16) uniquely define u^0.
We proceed in this way.
The expansion (2.9) is uniquely defined in this way. $\square$

Remark 2.4.

Problem (2.5) is a **stiff problem**. Expansion (2.9) is a **stiff expansion** (cf. J.L. LIONS [1], Chapter 1). **One has the following estimate**

$$\|u_\varepsilon - (\frac{u^{-1}}{\varepsilon} + u^0 + \cdots + \varepsilon^j u^j)\|_{H_0^1(\Omega)} \leq c\, \varepsilon^{j+1}. \qquad (2.17)$$

□

Remark 2.5.

The application of different finite element approximations in Ω_0 and in Ω_1 **and** of the asymptotic expansion is straightforward.   □

Remark 2.6.

Everything extends to the cases where we have **several** orders of magnitude in the coefficients : $a_\varepsilon(u,v) = a_0(u,v) + \varepsilon a_1(u,v) + \cdots + \varepsilon^m a_m(u,v)$, $a_k(u,v)$ defined in

$$\Omega_k, \quad \Omega_k \cap \Omega_l = \phi \quad \text{for } k \neq l, \quad \Omega = \Omega_0 \cup (\bigcup_{k=1}^{m} \bar{\Omega}_k).$$

□

We now show how the previous remarks can be applied to Stokes equations.

2.2   Stiff Stokes equations.

The geometry is the same than in Section 2.1.

We introduce :

$$V = \{v | v \in H_0^1(\Omega)^d, \ \operatorname{div} v = 0\},$$
$$V_0 = \{v | v \in H^1(\Omega_0)^d, \ \operatorname{div} v = 0, \quad v = 0 \text{ on } \Gamma_0\},$$
$$V_1 = \{v | v \in H^1(\Omega_1)^d, \ \operatorname{div} v = 0\},$$

and

$$a_k(u,v) = \sum \int_{\Omega_k} \frac{\partial u_i}{\partial x_j} \frac{\partial v_i}{\partial x_j} dx$$

where $u = \{u_1, \ldots, u_d\}$. One has to be careful here with the notations. If $u_1 \in V_1$ then $u_1 = \{u_{11}, u_{12}, \ldots, u_{1d}\}$.

Let f be given in $(L^2(\Omega))^d$. We consider the problem

$$a_\varepsilon(u_\varepsilon, v) = (f, v) \quad \forall\, v \in V,$$
$$u_\varepsilon \in V \qquad (2.18)$$

where

$$a_\varepsilon(u,v) = a_0(u,v) + \varepsilon a_1(u,v) \qquad (2.19)$$

and where $(f, v) = \int_\Omega \sum_{j=1}^{d} f_j v_j \, dx$.

The interpretation in classical terms of (2.18) is

$$- \Delta u_{\varepsilon 0} = f_0 - \nabla p_{\varepsilon 0},$$
$$\text{div } u_{\varepsilon 0} = 0 \quad \text{in } \Omega_0 ,$$
$$u_{\varepsilon 0} = 0 \text{ on } \Gamma_0 , \quad \int_{\Omega_0} p_0 \, dx = 0 \qquad (2.20)$$

We choose here and below this normalization.

$$- \varepsilon u_{\varepsilon 1} = f_1 - \varepsilon \nabla p_{\varepsilon 1},$$
$$\text{div } u_{\varepsilon 1} = 0 \quad \text{in } \Omega_1 ,$$
$$u_{\varepsilon 1} = u_{\varepsilon 0} \text{ on } \Gamma_1 , \quad \int_{\Omega_1} p_1 \, dx = 0 \qquad (2.21)$$

and

$$\varepsilon \frac{\partial u_{\varepsilon 1}}{\partial n} - \frac{\partial u_{\varepsilon 0}}{\partial n} = (\varepsilon p_{\varepsilon 1} - p_{\varepsilon 0})n \quad \text{on } \Gamma_1 . \qquad (2.22)$$

It is a **stiff Stokes problem.**   □

Remark 2.7

In (2.21) we have **normalized** $p_{\varepsilon 1}$ by introducing the coefficient ε. Of course this does not change the expansion, provided we use the proper powers of ε.   □

We use the same ansatz than in Section 2.1 and we have of course the same equations (2.10). We still have

$$u_0^{-1} = 0 \qquad \text{in } \Omega_0. \qquad (2.23)$$

Then

$$a_1(u^{-1}, v) = (f_1, v) \qquad \forall\, v \in V_1 \quad \text{and } v = 0 \quad \text{on } \Gamma_1$$

gives

$$- \Delta u_1^{-1} = f_1 - \nabla p_1^{-1} \quad \text{in } \Omega_1 ,$$
$$\text{div } u_1^{-1} = 0$$
$$u_1^{-1} = 0 \qquad \text{on } \Gamma_1 . \qquad (2.24)$$

Then

$$\int_{\Gamma_1} \frac{\partial u_1^{-1}}{\partial n} v + a_1(u^{-1}, v) = (f_1, v) + \int_{\Gamma_1} p_1^{-1} vn \, d\Gamma_1$$

so that $(2.10)_2$ gives

$$a_0(u^0, v) = (f_0, v) + \int_{\Gamma_1} (\frac{\partial u_1^{-1}}{\partial n} - n\, p_1^{-1}) v \; d\Gamma_1$$

so that

$- \Delta u_0^0 = f_0 - \nabla p_0^0$

div $u_0^0 = 0$

$u_0^0 = 0$ on Γ_0 ,

$\dfrac{\partial u_0^0}{\partial n} - n\, p_0^0 = \dfrac{\partial u_1^{-1}}{\partial n} - n\, p_1^{-1}$ on Γ_1 up to an additive vector   (2.25)

and we proceed in this way. (Variational formulation is better than (2.25)).

2.3   Remarks and problems.

All what has been presented extends (at least formally) to non linear problems. cf. J.L. LIONS, loc. cit.

The situation becomes more complicated for **evolution problems**. Let us consider, with the notations of Section 2.1, the parabolic equation

$$(\frac{\partial u_\varepsilon}{\partial t}, v) + a_\varepsilon(u_\varepsilon, v) = (f, v) \qquad \forall v \in H_0^1(\Omega),$$

$$u_\varepsilon \in L^2(0, T; H_0^1(\Omega)), \frac{\partial u_\varepsilon}{\partial t} \in L^2(0, T; H^{-1}(\Omega)),$$

$$\text{where } H^{-1}(\Omega) = \text{ (dual of } H_0^1(\Omega))$$

$$u_\varepsilon|_{t=0} = 0 \quad , \quad f \in L^2(0, T; L^2(\Omega)). \tag{2.26}$$

One can as well **rescale the time** so that (3.1) becomes

$$\varepsilon(\frac{\partial u_\varepsilon}{\partial \tau}, v) + a_\varepsilon(u_\varepsilon, v) = (f_\varepsilon, v) \qquad \forall v \in H_0^1(\Omega),$$

$$u_\varepsilon \in L^2(0, T^*; H_0^1(\Omega)), \frac{\partial u_\varepsilon}{\partial \tau} \in L^2(0, T^*; H^{-1}(\Omega)),$$

$$u_\varepsilon|_{\tau=0} = 0 \quad , \quad \text{where } f_\varepsilon \text{ is } f \text{ after time rescaling, and } T^* = T/\varepsilon. \tag{2.27}$$

An expansion entirely analogous to the one presented above is valid for **another** evolution equation, namely

$$(\frac{\partial u_\varepsilon}{\partial t}, v)_{\Omega_0} + \varepsilon(\frac{\partial u_\varepsilon}{\partial t}, v)_{\Omega_1} + a_\varepsilon(u_\varepsilon, v) = (f, v)$$

$$u_\varepsilon|_{t=0} = 0 \quad , \quad \text{where } (f, g)_{\Omega_i} = \int_{\Omega_1} fg \; dx. \tag{2.28}$$

Therefore for (2.26) (or the equivalent equation (2.27)) we need to introduce **boundary layers at the interface**. The reason is that the time scales are **not** the same on both sides of Γ_1. □

Remark 2.8
From a numerical point of view, the **time steps** should be of the same size on both sides, but the adjustment on Γ_1 does not seem to have been adressed in the litterature.

2.4 Bibliography

J.L. LIONS [1] *Perturbations singulières dans les problèmes aux limites et en contrôle optimal.* Lecture Notes in Math. Springer, 323 (1973).

3 Time scales and space scales

3.1 Rotating fluids.

Let Ω be a bounded open set of $\mathbf{R}^3$. We consider the Navier Stokes equations with a Coriolis like term

$$\frac{\partial u}{\partial t} + u\nabla u - \mu\Delta u + \frac{1}{\varepsilon}\, k \times u = f - \nabla p \qquad \text{in } \Omega,$$
$$\text{div } u = 0\,,$$
$$u = 0 \qquad\qquad \text{on } \Gamma = \partial\Omega\,,$$
$$u(x,0) = u^0(x) \qquad \text{in } \Omega. \tag{3.1}$$

In $(3.1)_1$, $k = \{0,0,1\}$, and ε is "small".
The coefficient μ is given > 0.
Let u_ε be a̲ solution of (3.1) (t̲h̲e̲ solution if $\Omega \subset \mathbf{R}^2$). Cf. J. LERAY [1][2], J.L. LIONS [1], J.L. LIONS and G. PRODI [1] for the uniqueness in 2D. We address here the following question :

what can be said on the behaviour of u_ε as $\varepsilon \to 0$? (3.2)

Remark 3.1
This is a long standing problem of fundamental importance in many applications. The classical reference is H.P. GREENSPAN [1]. □

We are going to proceed in a formal fashion to begin with. We "simply" look for

$$u_\varepsilon = u + \varepsilon u^1 + \dots \tag{3.3}$$

and we try to define $u = u(x,t)$ as a reasonable limit. This is what we attempt in Section 3.2.

3.2   The first term in the expansion

We **rescale** p (as in Section 2.2. This is always possible !) so that we rewrite the equation $(3.1)_1$ in the **equivalent form**

$$\frac{\partial u}{\partial t} + u\nabla u - \mu\Delta u + \frac{1}{\varepsilon} k \times u = f - \frac{1}{\varepsilon} \nabla p,$$

the other terms being inchanged.   (3.4)

Then if $u_\varepsilon, v_\varepsilon$ is a couple of solutions, we look for (3.3) and $p_\varepsilon = p + \varepsilon p^1 + \ldots$

By identification of the terms of power ε^{-1}, we obtain

$$k \times u = -\nabla p \tag{3.5}$$

and of course

$$\text{div } u = 0 . \tag{3.6}$$

Equation (3.5) is equivalent to

$$u_2 = \partial_1 p \qquad (\partial_i = \frac{\partial}{\partial x_i})$$
$$u_1 = \partial_2 p$$
$$0 = -\partial_3 p . \tag{3.7}$$

It follows from (3.7) that $\partial_1 u_1 + \partial_2 u_2 = 0$ so that (3.6) reduces to

$$\partial_3 u_3 = 0. \tag{3.8}$$

It follows from (3.7), (3.8) that

$$u_1, u_2, u_3 \text{ and } p \text{ are independant of } x_3. \tag{3.9}$$

Remark 3.2
It is clear from (3.9) that in general **neither the initial condition nor the boundary conditions can be satisfied by** u.   □

Remark 3.3
Let us assume that (cf. Figure 1)

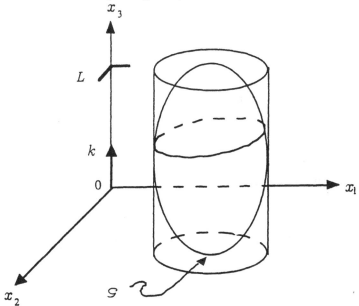

Figure 1

$$\Omega \subset \mathcal{G} \times]0, L[\tag{3.10}$$

where $\mathcal{G}$ is an open set of $\mathbf{R}^2$

We introduce the space $H_0^2(\mathcal{G}) = \{p | p \in H^2(\mathcal{G}), p = 0, \dfrac{\partial p}{\partial n} = 0 \quad \text{on} \quad \partial \mathcal{G}\}$
and we look for a function

$$u = \{u_1, u_2, u_3\} = \nabla^\perp p = \{-\partial_2 p, \partial_1 p, 0\}.$$

Of course with these choices it is **impossible** to satisfy for u all the
boundary conditions of u_ε. $\qquad\qquad\square$

In order to proceed, we introduce the variational formulation of (3.1).
We define

$$(u, v) = \int_\Omega (u_1 v_1 + u_2 v_2 + u_3 v_3) \, dx,$$

$$a(u, v) = \mu \int_\Omega \sum_{i,j=1}^{3} \frac{\partial u_i}{\partial x_j} \frac{\partial v_i}{\partial x_j} dx,$$

$$b(u, v, w) = \int_\Omega \sum_{i,j} u_j \frac{\partial v_i}{\partial x_j} w_i \, dx.$$

Then (3.1) is equivalent (by definition of the weak solutions !) to

$$(\frac{\partial u}{\partial t}, v) + a(u, v) + b(u, u, v) + \frac{1}{\varepsilon}(k \times u, v) = (f, v) \qquad \forall v \in V,$$

$$u \in L^2(0, T; V),$$

$$u|_{t=0} = u^0, \tag{3.11}$$

where

$$V = \{v | v \in (H_0^1(\Omega))^3, \ \text{div } v = 0\}.$$

We are looking for a "solution" (of some kind) of (3.11), of the form

$$u = \{-\partial_2 p, \partial_1 p, 0\}, \ p \in H_0^2(\mathcal{G}). \tag{3.12}$$

We take (this is formal !) in (3.11)

$$v = \{-\partial_2 q, \partial_1 q, 0\}, \ q \in H_0^2(\mathcal{G})$$

(this is formal because v does not satisfy the boundary conditions for being in V). For this choice $(k \times u, v) = 0$.

We introduce :

$$l(x') = \text{length of } \Omega \text{ above } x' = \{x_1, x_2\} \tag{3.13}$$

and we define

$$\rho(p, q) = \int_{\mathcal{G}} l(x')(\partial_1 p \partial_1 q + \partial_2 p \partial_2 q) dx',$$

$$\pi(p, q) = \int_{\mathcal{G}} l(x')(\partial_1^2 p \partial_1^2 q + \partial_2^2 p \partial_2^2 q + 2\partial_1 \partial_2 p \partial_1 \partial_2 q) dx',$$

$$\beta(p, p, q) = \int_{\mathcal{G}} l(x')((\partial_1 p \partial_1 \partial_2 p - \partial_2 p \partial_1^2 p)\partial_1 q - (\partial_2 p \partial_1 \partial_2 p - \partial_1 p \partial_2^2 p)\partial_2 q) dx'.$$
$$\tag{3.14}$$

Moreover we define

$$\sigma(f, q) = -\int_{\mathcal{G}} \bar{f}_1(x')\partial_2 q dx' + \int_{\mathcal{G}} \bar{f}_2(x')\partial_1 q dx' \tag{3.15}$$

where

$$\bar{f}_i(x') = \int f_i(x', x_3) dx_3$$

where the integral is taken on $\Omega \cap$ the "vertical" line in x_3 over x'. With these notations, a reasonable (?) approximation of u_ε is given by

$$u = \nabla^\perp p \tag{3.16}$$

where p is solution of

$$\rho(\frac{\partial p}{\partial t}, q) + \pi(p, q) + \beta(p, p, q) = \sigma(f, q) \qquad \forall q \in H_0^2(\mathcal{G}),$$

$$p \in L^2(0, T; H_0^2(\mathcal{G}))$$

$$p(0) = p^0 \text{ given, } \textbf{to be defined in } H_0^1(\mathcal{G}). \tag{3.17}$$

Before we define p^0, a few remarks are in order.

Remark 3.4.

If Ω is such that

$$l(x') \geq l_0 > 0 \tag{3.18}$$

one can show the existence and uniqueness of a solution of (3.17).

This is still the case when $l(x')$ can be zero on the boundary of $\mathcal{G}$, by introducing Sobolev spaces **with weights**. (The formulation in (3.17) has to be modified accordingly). □

Remark 3.5.

If $\Omega = \mathcal{G} \times (0, L)$ then $l(x') = L$. □

Remark 3.6.

In the case $\Omega = \mathcal{G} \times (0, L)$, the situation is slightly simpler if the boundary conditions on u_ε are of periodicity in x_3. □

We have now to define p^0 in the best possible way. It does not make sense to take $p^0 = u^0(x)$ since $u^0(x)$ depends (in general) of x_3.

The natural choice is to **define** p^0 as the solution of

$$\rho(p^0, q) = -\int_{\mathcal{G}} \bar{u}_1^0 \partial_2 q dx' + \int_{\mathcal{G}} \bar{u}_2^0 \partial_1 q dx' \qquad \forall q. \tag{3.19}$$

□

Remark 3.7.

In the case $\Omega = \mathcal{G} \times]0, L[$ one can prove the weak convergence of a set of solutions u_ε towards u given by (3.16) where p is given by the solution of (3.17), (3.19).

The general case seems to be an open question. □

Remark 3.8.

There are **necessarily boundary layers** at time $t = 0$ and on the boundary of Ω. We begin with the "boundary layer in time". □

3.3 Slow and fast time variables.

Let us introduce **a fast time**

$$\tau = t/\varepsilon . \tag{3.20}$$

Let us look for a solution (ansatz) in the form

$$u = u_\varepsilon = \tilde{u}(x, \tau) \ . \tag{3.21}$$

Since $\dfrac{\partial}{\partial t} = \dfrac{1}{\varepsilon} \dfrac{\partial}{\partial \tau}$, equation (3.4) gives, by identifying the terms in ε^{-1} :

$$\frac{\partial \tilde{u}}{\partial \tau} + k \times \tilde{u} = -\nabla \tilde{p}$$
$$\operatorname{div} \tilde{u} = 0 \tag{3.22}$$

$$\tilde{u}(0) = u^0 - \nabla^\perp p^0 \tag{3.23}$$

One can show (this is a very interesting set of equations) that (3.22), (3.23) **admits a unique solution, subject to the boundary conditions**

$$\tilde{u}n = 0 \qquad \text{on} \quad \partial \Omega \ . \tag{3.24}$$

□

We have now a "better" approximation of u_ε by taking, for the fast approximation

$$u + \tilde{u}. \tag{3.25}$$

But the boundary conditions are **not** satisfied

Therefore a boundary layer has to be introduced. For the case where $\Omega = \Pi^2 \times]0, L[$ cf. N. MASMOUDI [1] and the Bibliography therein.

3.4   Further remarks

Remark 3.9
Scaling **in space variables** appears in the above considerations in the **boundary layers**.   □

Remark 3.10
The content of the previous sections (without the boundary layers introduced by N. MASMOUDI, loc. cit.) has been presented in J.L. LIONS [3], where a more general situation (but geometrically simpler) was studied, in connection with climatology, following J.L. LIONS, R. TEMAM and S. WANG [1], [2].

Remark 3.11
The problem of passing to the limit from the **incompressible case** to the **compressible** situation for viscous fluids has been solved by P.L. LIONS and N. MASMOUDI [1].   □

Remark 3.12. - Darcy's law for rotating fluids.

Let Ω be a bounded open set of $\mathbf{R}^3$. We introduce a domain Ω_ε obtained from Ω after taking out "many" tiny holes arranged in a periodic manner.

More precisely we introduce

$$Y =]0,1[^3 \quad \mathcal{O} \subset Y, \partial\mathcal{O} = S \, , \mathcal{Y} = Y \backslash \bar{\mathcal{O}},$$

$\chi =$ characteristic function of $\mathcal{Y}$, **extended in $\mathbf{R}^3$ in a periodic manner.**

We then define

$$\Omega_\varepsilon = \{x | x \in \Omega, \quad \chi(x/\varepsilon) = 1\} \tag{3.26}$$

(we can arrange things so that Ω_ε is open).

It amounts to the same thing to take out of Ω the sets $\varepsilon\mathcal{O}$ and all their translations.

We may think of Ω_ε as a **porous material**.

We now consider the Stokes equation in Ω_ε, with rotation around $k = \{0,0,1\}$. We rescale so that the equations read as follows :

$$- \varepsilon^2 \Delta u_\varepsilon + k \times u_\varepsilon = f - \nabla p_\varepsilon$$
$$\operatorname{div} u_\varepsilon = 0$$
$$u_\varepsilon = 0 \quad \text{on} \quad \partial\Omega_\varepsilon \, . \tag{3.27}$$

We introduce a **"fast" geometrical variable**

$$y = x/\varepsilon \tag{3.28}$$

and we look for u_ε under the following ansatz

$$u_\varepsilon = u_0(x,y) + \varepsilon u_1(x,y) + \ldots \tag{3.29}$$

where in (3.29) at the end of the computation y is replaced by x/ε. In (3.29) $u_j(x,y)$ is defined in $\Omega \times \mathcal{Y}$ and satisfies

$$u_j(x,y) = 0 \quad \text{for} \quad y \in S (= \partial\mathcal{O})$$
$$u_j(x,y) \qquad \text{is periodic in} \quad y. \tag{3.30}$$

[Ansatz of this type have been introduced in the book A. BENSOUSSAN, J.L. LIONS, G. PAPANICOLAOU [1]].

We observe that the above ansatz, with the boundary conditions (3.30) does **not** take care of $u_\varepsilon = 0$ on $\partial\Omega \backslash$ holes.

We now use (3.29) in (3.27) where we observe that $\dfrac{\partial}{\partial x_j}$ becomes $\varepsilon^{-1} \dfrac{\partial}{\partial y_j} + \dfrac{\partial}{\partial x_j}$. Therefore

$$-\varepsilon^2 \Delta = -\Delta_y - 2\varepsilon\Delta_{xy} - \varepsilon^2 \Delta_x \, ,$$

$$\Delta_{xy} = \sum_{j=1}^{3} \frac{\partial^2}{\partial x_j \, \partial y_j}.$$

Since ∇ becomes $\varepsilon^{-1}\nabla_y + \nabla_x$, if we use (3.29) and

$$p_\varepsilon = p_0(x,y) + \varepsilon p_1(x,y) + \dots \tag{3.31}$$

with similar properties on the p_j's, we obtain

$$\nabla_y p_0 = 0 , \tag{3.32}$$

$$- \Delta_y u_0 + k \times u_0 = f - \nabla_y p_1 - \nabla_x p_0$$
$$\operatorname{div}_y u_0 = 0. \tag{3.33}$$

We notice that since $\operatorname{div} u_\varepsilon = 0$, we have

$$\operatorname{div}_y u_1 + \operatorname{div}_x u_0 = 0 . \tag{3.34}$$

But $\displaystyle\int_{\mathcal{Y}} \operatorname{div}_y u_1 dy = 0$ so that

$$\operatorname{div}_x \int_{\mathcal{Y}} u_0 \, dy = 0. \tag{3.35}$$

Moreover u_o **should satisfy** (3.30)
These conditions uniquely define u_0 as a function of $\nabla_x \, p_0$.
Indeed let us set

$$g = f - \nabla_x \, p_0. \tag{3.36}$$

With these notations we rewrite (3.33) in the form

$$- \Delta_y \varphi + k \times \varphi = g - \nabla_y \pi$$
$$\operatorname{div}_y \varphi = 0 \qquad \text{in } \mathcal{Y}$$
$$\varphi = 0 \quad \text{in } S, \quad \varphi \quad \text{periodic} . \tag{3.37}$$

Then, if one can solve (3.37), one has $u_0(x,y) = \varphi(x,y)$ where $g = g(x)$ is given by (3.26) (x being here a parameter).
But (3.37) **admits a unique solution** which is given by

$$\varphi = \Psi(y)g, \quad \Psi(y) = \|\Psi_{i,j}(y)\|, \; i,j = 1,2,3 , \tag{3.38}$$

where $\Psi(y)$ is such that $\varphi \in H^1(\mathcal{Y})^3$ (as a function of y).
Therefore

$$u_0(x,y) = \Psi(y).(f - \nabla_x p_0) \tag{3.39}$$

But u_0 should satisfy to (3.35), hence

$$\text{div}_x \, \mathcal{D}(f - \nabla_x \, p_0) = 0 \qquad (3.40)$$

where

$$\mathcal{D} = \int_y \Psi(y) dy \quad \text{is defined as the Darcy's matrix.} \qquad (3.41)$$

The best boundary conditions we can impose on p_0 correspond to

$$(\int_y u_0 \, dy) \, n = 0 \qquad \text{on } \Gamma = \partial\Omega \qquad (3.42)$$

Consequently

$$\mathcal{D}(f - \nabla_x \, p_0). \, n = 0 \quad \text{on } \Gamma \qquad (3.43)$$

Summing up :

1) **The Darcy's matrix $\mathcal{D}$ is given by (3.41) where ψ is given by** (3.38), **φ being the solution of** (3.37).

2) **The pressure $p_0(x)$ is obtained by solving the Neumann problem for the second order elliptic problem** (3.40), (3.43) (one can verify that $\mathcal{D}$ is coercive).

3) **The average velocity is given by the Darcy's law**

$$\int_y u_0(x,y) dy = \mathcal{D}(f - \nabla_x p_0).$$

We emphasize that the matrix $\mathcal{D}$ **depends on the rotation**, since the rotation $k \times \varphi$ is part of (3.37). $\qquad\qquad\qquad\qquad\qquad\qquad\square$

3.5   Bibliography

A. BENSOUSSAN, J.L. LIONS, and G. PAPANICOLAOU [1] *Asymptotic Analysis for Periodic Structures*, North Holland, (1978).

H.P. GREENSPAN [1] *The theory of rotating fluids.* Cambridge Monographs on Mechanics and Applied Mathematics, (1969).

J. LERAY [1] Essai sur le mouvement plan d'un liquide visqueux que limitent des parois. J.M.P.A. t. XIII (1934), p. 331-418.

[2] Sur le mouvement d'un liquide visqueux emplissant l'espace. Acta Math. 63 (1934), p. 193-248.

J.L. LIONS [1] Quelques résultats d'existence dans les équations aux dérivées partielles non linéaires. Bull S.M.F. 87 (1959), p. 245-273.

[2] *Some methods in the Mathematical Analysis of Systems and their control.* Gordon Breach, New York and Science Press (Beijing), (1981).

[3] Lectures at the College de France (1995, 1996).

J.L. LIONS and G. PRODI [1] Un théorème d'existence et d'unicité dans les équations de Navier Stokes en dimension 2. C.R.A.S. Paris, t. 248, (1959), p. 3519-3521..

J.L. LIONS, R. TEMAM and S. WANG [1] Geostrophic asymptotics of the primitive questions of the atmosphere. Top. Methods Non linear Analysis, 24, (1994), p. 253-287.

[2] A simple global model for the general circulation of the atmosphere. C.P.A.M., Vol L. (1997), p. 707-752.

P.L. LIONS, and N. MASMOUDI [1] Incompressible limit for a viscous compressible fluid. To appear.

N. MASMOUDI [1] Ekman layers of rotating fluids, the case of general initial data. To appear.

4 Complexity of the geometry and domain decomposition method

4.1 Exterior virtual control.

Let Ω be a bounded open set of $\mathbf{R}^d$ ($d = 2$ or 3 in (most of) the applications -but not all of them !). Let A be a second order linear elliptic operator in Ω, given in the form

$$A\varphi = -\sum_{i,j=1}^{d} \frac{\partial}{\partial x_i}\left(a_{ij}(x)\frac{\partial\varphi}{\partial x_j}\right) \tag{4.1}$$

where

$$a_{ij} \in L^\infty(\Omega), \{a_{ij}\} \text{ **not necessarily** a symmetric matrix,}$$

$$\sum_{i,j} a_{ij}(x)\xi_i\,\xi_j \geq \alpha \sum_i \xi_1^2 \quad \forall\,\xi_i \in \mathbf{R},\ \alpha > 0, \quad \text{a.e. in } \Omega. \tag{4.2}$$

Let f be given in $L^2(\Omega)$. The problem

$$Au = f \text{ in } \Omega, \quad u \in H_0^1(\Omega) \tag{4.3}$$

admits a unique solution.

We want to present a systematic method to **approximate** the solution u of (4.1) **by the solutions of problems similar to (4.3) but considered in a domain Q "simpler" than Ω, $Q \supset \bar{\Omega}$.** ☐

Remark 4.1.

For $u, v \in H_0^1(\Omega)$, we define

$$a(u,v) = \sum_{i,j=1}^{d} \int_\Omega a_{ij}(x)\,\frac{\partial u}{\partial x_j}\,\frac{\partial v}{\partial x_i}\,dx . \tag{4.4}$$

Then (4.1) is equivalent to

$$a(u,v) = (f,v) \quad \left(= \int_\Omega fv \, dx\right) \quad \forall \, v \in H_0^1(\Omega),$$
$$u \in H_0^1(\Omega). \tag{4.5}$$

Remark 4.2.

We shall need below **a regularity hypothesis on a_{ij}'s** . □

We introduce now

$$Q = \text{"simple" domain such that}$$
$$\bar\Omega \subset Q \quad \text{(cf. Figure 1)}. \tag{4.6}$$

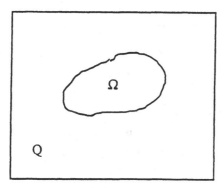

Figure 1

One can take

$$Q = \text{cube}$$

or

$$Q = \text{sphere}$$

or

$$Q = \mathbf{R}^d \text{ as well}.$$

We denote by A **any extension in Q of the operator** A, with the requirement that A is still elliptic in Q.

Remark 4.3.

If the a_{ij}'s are $C^1(\bar\Omega)$, one can always extend the functions a_{ij} in Q in such a way that $a_{ij} \in C^1(\bar Q)$ and that (4.2) holds true in Q.

Said otherwise, we can consider that A **is given in** Q, satisfying (4.2) in Q, and that in (4.4) we deal with the **restriction of A to Ω**. □

We introduce now

$$\mathcal{O} = \quad \text{open set in } Q \,,$$
$$\mathcal{O} \subset \Omega' = \bar\Omega^c \quad \text{(cf. Figure 2)} \tag{4.7}$$

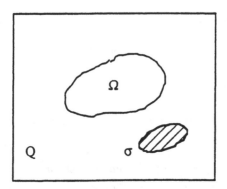

Figure 2

and we consider in Q the **state equation**

$$Au = \tilde{f} + \lambda 1_{\mathcal{O}} \quad \text{in } Q ,$$
$$u \in H_0^1(Q) \tag{4.8}$$

where $\tilde{f}$ is any extension of f in Q (for instance $\tilde{f} = 0$ outside Ω), and where

$$\lambda \in L^2(\mathcal{O}) \quad \textbf{is a virtual control to be defined,}$$
$$1_{\mathcal{O}} = \text{characteristic function of } \mathcal{O}. \tag{4.9}$$

Remark 4.4.
We can consider other types of boundary conditions than Dirichlet in (4.8). For instance we can take u to be **periodic** or we can choose on ∂Q "transparent boundary conditions". □

Of course, given λ (the virtual control) in an arbitrary fashion, there is no reason why the restriction of the solution u of (4.8) should satisfy $u = 0$ on Γ. **But of course if u denotes the restriction of the solution of (4.8) to Ω one has $Au = f$** ! Therefore the only thing which remains is

> **choose, if possible, $\lambda \in L^2(\mathcal{O})$ in such a way that**
> u **is "small" on** Γ. $\tag{4.10}$

We prove first that **it is possible** to achieve (4.10) :

> **the range of the mapping**
> $\lambda \longrightarrow u|_\Gamma = \text{trace of} \quad u \quad \text{on } \Gamma = \partial\Omega$
> **is dense in $L^2(\Gamma)$, when we assume that the coefficients**
> a_{ij} **are smooth enough.** $\tag{4.11}$

Proof.

Let θ and w be the solutions of

$$A\theta = \tilde{f}, \; \theta \in H_0^1(\Omega),$$
$$Aw = \lambda 1_{\mathcal{O}}, \; w \in H_0^1(\Omega). \tag{4.12}$$

Of course $u = \theta + w$ and everything amounts to prove that $\lambda \to w|_\Gamma$ has a range dense in $L^2(\Gamma)$.

We use Hahn Banach theorem (so that this proof is **not constructive**.) We therefore consider $g \in L^2(\Gamma)$ such that

$$\int_\Gamma gw \, d\Gamma = 0 \qquad \forall \lambda \in L^2(\mathcal{O}), \tag{4.13}$$

and we want to prove that $g = 0$. We introduce the adjoint state p in $H_0^1(Q)$ defined by

$$a^*(p, v) = \int_\Gamma gv \, d\Gamma \qquad \forall v \in H_0^1(Q), \tag{4.14}$$

where

$$a^*(\varphi, \psi) = \sum \int_Q a_{ji}(x) \frac{\partial \varphi}{\partial x_i} \frac{\partial \psi}{\partial x_j} \, dx.$$

Then if w is the solution of (4.12) one has

$$\int_\Gamma gw \, d\Gamma = a^*(p, w) = a(w, p) = (\lambda 1_{\mathcal{O}}, p) \tag{4.15}$$

so that (4.13) **is equivalent** to

$$(\lambda 1_{\mathcal{O}}, p) = 0 \qquad \forall \lambda \in L^2(\mathcal{O})$$

i.e.

$$p = 0 \qquad \text{in} \quad \mathcal{O}. \tag{4.16}$$

But it follows from (4.14) that

$$A^*p = 0 \qquad \text{in} \quad \Omega' = \bar{\Omega}^c \quad (\text{in} \quad Q). \tag{4.17}$$

Therefore, **according to the unique continuation theorem,**

$$p \equiv 0 \qquad \text{in} \quad \Omega'. \tag{4.18}$$

Remark 4.5.

It is in order to be able to conclude (4.18) that we have to assume that the coefficients are smooth enough. $\qquad \square$

Remark 4.6.

In the applications, we are going to use the **virtual control method** in a finite elements approximation of the problem. **The unique continuation property does not seem** to have been studied in a **systematic** fashion for finite elements approximations. It is clear that $\mathcal{O}$ should intersect a **"sufficient number"** of elements of the triangulation !   □

We can now conclude. Since $p \in H_0^1(Q)$, the trace of p on Γ is uniquely defined, so that (4.18) implies

$$p = 0 \quad \text{on} \quad \Gamma .$$

It follows from (4.14) that

$$A^* p = 0 \quad \text{in} \quad \Omega$$

which, together with (4.19), implies that $p \equiv 0$ in Ω. Consequently $p \equiv 0$ in Q and therefore $g = 0$.   □

Remark 4.7.

It remains now to choose the virtual control λ in such a way that $\|u\|_{L^2(\Gamma)}$ **is small**, which is possible according to (4.11).

The virtual control method has been introduced in 1998 in several lectures of O. PIRONNEAU and of the A. and published (for much more realistic situations) in the notes J.L. LIONS and O. PIRONNEAU [1],[2],[3].   □

Remark 4.8. Let us assume that A is given by

$$A\varphi = - \sum_{i,j=1}^{d} \frac{\partial}{\partial x_i} \left(a_{i,j}(x) \frac{\partial \varphi}{\partial x_j} \right) + a_0 \varphi \tag{4.19}$$

where the a_{ij}'s are as in (4.1),(4.2) and where

$$a_0(x) \geq \alpha_0 > 0 \quad \text{a.e. in} \quad \Omega, \quad a_0 \in L^\infty(\Omega) .$$

Let us also assume that the a_{ij}'s and a_0's are extended in Q as before. We can then consider the **Neumann** boundary value problem

$$Au = f \quad \text{in} \quad \Omega$$

$$\frac{\partial u}{\partial n_A} = 0 \quad \text{on} \quad \Gamma \tag{4.20}$$

$$\frac{\partial u}{\partial n_A} = \sum_{i,j} a_{ij}(x) \frac{\partial u}{\partial \chi_j} n_i ,$$

$n = \{n_i\}$ = normal to Γ (directed towards the exterior of Ω to fix ideas). Of course (4.20) is taken in the weak sense

$$a(u,v) = \int_\Omega fv \, dx \quad \forall v \in H^1(\Omega), \, u \in H^1(\Omega)$$

but the solution is strong if the a_{ij}'s are, say, in $C^1(\bar{\Omega})$.

We then consider **the same problem** (4.8) **as before**. The solution u is "smooth", in the sense

$$u \in H^2(Q).$$

We can then consider $\dfrac{\partial u}{\partial n_A}|_\Gamma$ (it is in $H^{1/2}(\Gamma)$, cf. J.L. LIONS, E. MA-GENES [1]) and we want to choose (if possible) the virtual control λ in such a way that

$$\left\|\frac{\partial u}{\partial n_A}\right\|_{L^2(\Gamma)} \quad \text{is small} \tag{4.21}$$

This is indeed possible thanks to the following property.

the range of the mapping $\lambda \to \dfrac{\partial u}{\partial n_A} |\Gamma$ **is dense in** $L^2(\Gamma)$. \quad (4.22)

For the proof, which follows the proof of (4.11), we can assume that $\tilde{f} = 0$. Let g be in $L^2(\Gamma)$ such that

$$\int_\Gamma g \frac{\partial u}{\partial n_A} d\Gamma = 0 \quad \forall \lambda \in L^2(\mathcal{O}) . \tag{4.23}$$

We have to prove that $g = 0$. We introduce p in $L^2(Q)$, solution of

$$\int_Q p(A\varphi)dx = \int_\Gamma g \frac{\partial \varphi}{\partial n_A} d\Gamma \quad \forall \varphi \in H^2(Q) \cap H_0^1(Q) \tag{4.24}$$

(this is a very weak solution, defined by transposition, following E. MA-GENES and the A., cf. Bibliography).

Then

$$\int_\Gamma g \frac{\partial u}{\partial n_A} d\Gamma = \int_Q p(Au)dx = \int_{\mathcal{O}} p\lambda \, dx$$

so that (4.23) is equivalent to $\displaystyle\int_{\mathcal{O}} p\lambda \, dx = 0 \quad \forall \lambda \in L^2(\mathcal{O})$, i.e. $p = 0$ in $\mathcal{O}$.

But it follows from (4.24) that

$$A^*p = 0 \text{ in } \Omega'$$

so that, by the unique continuation theorem, $p \equiv 0$ in Ω'.

To conclude that $p \equiv 0$ in Q, one cannot use here traces of u on Γ (as we did for (4.11)). One uses here Pierre-Louis LIONS [1]. Let $p = \{p^+ \text{ in } \Omega, p^- \text{ in } \Omega'\}$. We know that $p^- = 0$.

Let us consider θ solution of

$$A\theta = p^+ \quad \text{in} \quad \Omega ,$$

$$\frac{\partial \theta}{\partial \eta_A} = 0 \quad \text{on} \quad \Gamma$$

and let φ be a smooth continuation of θ in Q, such that

$$\frac{\partial \varphi}{\partial \eta_A} = 0 \quad \text{on} \quad \Gamma, \ \varphi \in H_0^1(Q) \ .$$

If we choose this function φ in (4.24), we have $\displaystyle\int_\Gamma g \frac{\partial \varphi}{\partial \eta_A} \, d\Gamma = 0$ so that

$$\int_Q p \, A\varphi \, dx = 0$$

and since $p^- = 0$, this equality reduces to $\displaystyle\int_\Omega p^+ (A\varphi) dx = 0$ and since on Ω one has $A\varphi = A\theta = p^+$ it follows that

$$\int_\Omega (p^+)^2 \, dx = 0 \quad \text{hence} \quad p^+ = 0 \quad \text{in} \quad \Omega \quad \text{and} \quad p \equiv 0 \quad \text{in} \quad Q \ .$$

□

Remark 4.9.

The **"density lemmas"** (4.11) or (4.22) extend to many other situations (higher order equations, systems of equations). For instance let u be the solution of the Stokes system

$$\begin{aligned}
& -\Delta u = f - \nabla p, \\
& \text{div } u = 0 \quad \text{in } \Omega, \\
& u = 0 \quad \text{on } \Gamma = \partial\Omega, \ \Omega \subset \mathbf{R}^3 \ .
\end{aligned} \qquad (4.25)$$

We introduce the "virtual control" λ and the equations

$$\begin{aligned}
& -\Delta u = f - \nabla p + \lambda 1_{\mathcal{O}}, \\
& \text{div } u = 0 \quad \text{in } Q, \\
& u = 0 \quad \text{on } \partial Q \ .
\end{aligned} \qquad (4.26)$$

If we assume that λ spans a 2-dimensional subspace of $(L^2(\mathcal{O}))^3$, then the range of $u|_\Gamma$ is dense in E,
where

$$E = \{g | g \in (L^2(\Gamma))^3, \ \int_\Gamma gn \, d\Gamma = 0\}.$$

If λ spans a 1-dimensional subspace of $(L^2(\mathcal{O})^3$, then the density property is **only generally true with respect to** Ω (or $\mathcal{O}$). (Cf. E. ZUAZUA and the A. in the Bibliography). □

Remark 4.10.
Let Ω be a domain with a hole C (cf. Figure 3).

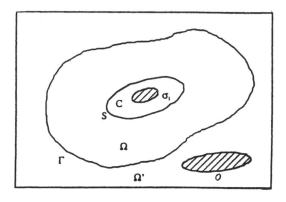

Figure 3

The boundary of Ω is given by

$$\partial\Omega = \Gamma \cup S .$$

We introduce

$$Q = \bar{\Omega} \cup C \cup \Omega' .$$

We want to solve (for instance)

$$Au = f \quad \text{in} \quad \Omega \quad , \quad u = 0 \quad \text{on} \quad \Gamma \cup S \tag{4.27}$$

We introduce two virtual controls located in $\mathcal{O}$ and in $\mathcal{O}_1$ (cf. Figure 3) and we consider in Q the problem

$$Au = f + \lambda 1_{\mathcal{O}} + \lambda_1\, 1_{\mathcal{O}_1} , \quad u \in H_0^1(Q)$$
$$\lambda \in L^2(\mathcal{O}) , \; \lambda_1 \in L^2(\mathcal{O}_1) . \tag{4.28}$$

For λ, λ_1 given arbitrarily in $L^2(\mathcal{O}) \times L^2(\mathcal{O}_1)$, the restriction of u to Γ satisfies (4.27), so that it "only" remains to choose λ **and** λ_1 in such a way that

$$\|u\|_{L^2(\Gamma)} + \|u\|_{L^2(S)} \qquad \text{is small} .$$

This is indeed possible (same proof as before) with the **two** virtual controls λ, λ_1, but would **not** be true with only one virtual control. $\quad\square$

Remark 4.11.
The analogous of properties (4.11) or (4.22) is in general **an open question for non linear problems**.

For instance it is an open question **for Navier Stokes Equations** (we **conjecture** it is true).

We return to this question in Section 4.3 below.   □

Remark 4.12.
Density properties of the type of (4.11) have been studied in A. OSSES and J.P. PUEL [1] for **effective control** problems.   □

Remark 4.13.
The set $\mathcal{O}$ where the virtual control is located has not to be an **open** set of Ω', it suffices for $\mathcal{O}$ to be a **"uniqueness"** set.   □

Remark 4.14.
The same properties hold true for **parabolic** (evolution) **problems.** cf. the A. [2]. The situation is different for **hyperbolic** problems.   □

It remains to show **how** one can choose λ so that $\|u\|_{L^2(\Gamma)}$ is small. We address this question in the following Section.   □

4.2   Approximate controllability method.

We consider problem (4.3) and we introduce (4.8). Since the mapping $\lambda \to u \mid_\Gamma$ has a range which is dense in $L^2(\Gamma)$, there exists λ such that

$$\|u\|_{L^2(\Gamma)} \quad \text{is arbitrary small} . \tag{4.29}$$

Remark 4.15.
One can show that, if everything is smooth enough, $\lambda \to u\|_\Gamma$ has a range dense in spaces which are **"smaller"** than $L^2(\Gamma)$, for instance in $H^s(\Gamma)$, $s \geq 0$.

Finding a **(virtual) control** satisfying (4.29) is a problem of **approximate controllability**.

Problems of this type, but for **evolution** problems and for **effective** controls have been studied from a **Numerical point of view** in R. GLOWINSKI and J.L. LIONS [1] [2], C. CARTHEL, R. GLOWINSKI and J.L. LIONS [1] (**exact** controllability for **hyperbolic** equations was studied previously, from the numerical point of view, in R. GLOWINSKI, C. H. LI and J.L. LIONS [1]).

As always in controllability problems (exact or approximate), the control (should it be virtual or effective) which achieves property (4.29)) in a more precise form, say

$$\|u\|_{L^2(\Gamma)} \leq \varepsilon_0 , \; \varepsilon_0 \text{ given} , \tag{4.30}$$

is **not** unique. Therefore the natural thing to do is to consider the problem

$$\inf_{\lambda} . \frac{1}{2}\|\lambda\|^2_{L^2(\mathcal{O})} ,$$

$$u \quad \text{subject to (2.1 bis)} . \tag{4.31}$$

This is not a very convenient control problem because of the **state constraints** (4.30).

Therefore it is standard procedure to introduce the functional

$$J_\beta(\lambda) = \frac{1}{2}\|\lambda\|^2_{L^2(\mathcal{O})} + \frac{\beta}{2}\int_\Gamma u^2 \, d\Gamma \, , \ \beta > 0 \qquad (4.32)$$

and to consider

$$\inf. \, J_\beta(\lambda) \, , \ \lambda \in L^2(\mathcal{O}) \, . \qquad (4.33)$$
□

Remark 4.16.
One can think of $J_\beta(\lambda)$ as a **virtual cost function**. □

Remark 4.17.
We can think of the term $\dfrac{\beta}{2}\displaystyle\int_\Gamma u^2 d\Gamma$ which appears in (4.32) as **a penalty term**. We return to the **choice** of β below. □

Remark 4.18.
Problem (4.33) admits **a unique solution**. It can be approximated by a **gradient algorithms** (for instance !!) as we now explain. □

The first variation of $J_\beta(\lambda)$ is given by

$$\delta J_\beta(\lambda) = (\lambda, \delta\lambda)_\mathcal{O} + \beta \int_\Gamma u \, \delta u \, d\Gamma \, , \qquad (4.34)$$

where $(\lambda, \delta\lambda)_\mathcal{O} = \int_\mathcal{O} \lambda \, \delta\lambda \, dx$.

We introduce p defined by

$$p \in H^1_0(Q) \, ,$$
$$a^*(p, \varphi) = \int_\Gamma u\varphi d\Gamma \quad \forall \, \varphi \in H^1_0(Q) \, . \qquad (4.35)$$

Then $\displaystyle\int_\Gamma u \, \delta u \, d\Gamma = a^*(p, \delta u) = a(\delta u, p) = (\delta\lambda, p)_\mathcal{O}$

so that

$$\delta J_\beta(\lambda) = (\lambda + \beta p, \delta\lambda)_\mathcal{O} \, . \qquad (4.36)$$

The algorithm is then : assume λ^n, p^n is known. Define

$$\lambda^{n+1} = \lambda^n - \rho(\lambda^n + \beta p^n) \quad \text{on } \mathcal{O},$$
$$\rho > 0 \quad \text{small enough} \, , \qquad (4.37)$$

compute u^{n+1} by

$$a(u^{n+1}, \varphi) = (\tilde{f}, \varphi) + (\lambda^{n+1}, \varphi)_\mathcal{O} \quad \forall \, \varphi \in H^1_0(Q),$$
$$u^{n+1} \in H^1_0(Q) \, , \qquad (4.38)$$

then compute p^{n+1} by

$$a^*(p^{n+1}, \varphi) = \int_\Gamma u^{n+1} \varphi \, d\Gamma \qquad \forall \varphi \in H_0^1(Q) \qquad (4.39)$$

and proceed. The algorithm converges for ρ small enough.   □

The choice of β can now be made using duality arguments, as we now show.

We recall that

$$u = \theta + w, \ A\theta = \tilde{f} \ \text{ in } Q, \ \theta \in H_0^1(Q),$$
$$w \ \text{ is given by (1.12) i.e. } \ Aw = \lambda|_O, w \in H_0^1(Q) \ . \qquad (4.40)$$

We define

$$F_1(\lambda) = \frac{1}{2} \int_O \lambda^2 \, dx,$$
$$F_2(g) = \frac{\beta}{2} \int_\Gamma (\theta + g)^2 d\Gamma \ , \ g \in L^2(\Gamma),$$
$$L\lambda = w|_\Gamma \ , \ L \in \mathcal{L}(L^2(O); L^2(\Gamma)) \ . \qquad (4.41)$$

Then

$$\inf J_\beta(\lambda) = \inf[F_1(\lambda) + F_2(L\lambda)] \ . \qquad (4.42)$$

According to the FENCHEL-ROCKAFELLAR duality theorem [1] one has

$$\inf J_\beta(\lambda) = - \inf_{g \in L^2(\Gamma)} [F_1^*(L^*g) + F_2(-g)] \qquad (4.43)$$

where F^* denotes the conjugate of the (proper) convex function F, i.e.

$$F^*(h) = \sup_{\hat{h}} (h, \hat{h}) - F(\hat{h})$$

F functional on $\mathcal{H}, \hat{h} \in \mathcal{H}, h \in \mathcal{H}' = $ dual of $\mathcal{H}$,

and where L^* is the adjoint of L. If $g \in L^2(\Gamma), L^*g$ is defined by

$$L^*g = p|_O \ ,$$
$$a^*(p, \varphi) = \int_\Gamma g\varphi \, d\Gamma \quad \forall \varphi \in H_0^1(\Omega), \ p \in H_0^1(\Omega) \ . \qquad (4.44)$$

One has

$$F_1^*(\lambda) = F_1(\lambda) \ ,$$
$$F_2^*(g) = \frac{1}{2\beta} \int_\Gamma g^2 \, d\Gamma - \int_\Gamma g\theta \, d\Gamma \ . \qquad (4.45)$$

Therefore **the dual problem of** inf $.J_\beta(\lambda)$ **is given by**

$$\inf_g[\frac{1}{2}\int_O p^2 \, dx + \frac{1}{2\beta}\int_\Gamma g^2 \, d\Gamma + \int_\Gamma g\theta \, d\Gamma] \, , \qquad (4.46)$$

where p is given by (4.44).

Let us now compute the dual problem of (4.31), (4.30), which is equivalent to

$$\inf . F_1(\lambda) + F_3(L\lambda) \, , \qquad (4.47)$$

where F_3 is given by

$$F_3(g) = \begin{vmatrix} 0 & \text{if } \theta + g \in \varepsilon_0 \, B \, , \, B = \quad \text{unit ball of } L^2(\Gamma) \, , \\ +\infty & \text{otherwise} \, . \end{vmatrix}$$

Then

$$F_3^*(g) = \sup .(g, \hat{g}) \, , \quad \theta + \hat{g} = \varepsilon_0 \, b \, , \, b \in B \, ,$$

$$= \sup(g, \varepsilon_0 \, b - \theta) = \varepsilon_0 \, \|g\|_{L^2(\Gamma)} - (g, \theta)_{L^2(\Gamma)} \, .$$

Therefore **the dual problem of** (4.31), (4.30) **is given by**

$$\inf_g[\frac{1}{2}\int_O p^2 \, dx + \varepsilon_0 \, \|g\|_{L^2(\Gamma)} + \int_\Gamma g\theta \, d\Gamma] \, , \qquad (4.48)$$

where p is given by (4.44).

Let g_0 (resp. g_1) be the solution of (4.46) (resp. (4.48)).

At the minimum of (4.46) we have (by the Euler equation)

$$\int_O p_0^2 \, dx + \frac{1}{\beta}\int_\Gamma g_0^2 \, d\Gamma + \int_\Gamma g_0\theta \, d\Gamma \qquad (4.49)$$

and at the minimum of (4.48) we have (by variational inequalities)

$$\int_O p_1^2 \, dx + \varepsilon_0(\int_\Gamma g_1^2 \, d\Gamma)^{1/2} + \int_\Gamma g_1\theta \, d\Gamma \, , \qquad (4.50)$$

where p_0(resp.p_1) is the solution of (4.44) for $g = g_0$ (resp. g_1).

We want to have $g_0 = g_1$ (or, at least, as close as posible). Then $p_0 = p_1$ so that comparing (4.49) and (4.50) we have

$$\frac{1}{\beta}\int_\Gamma g^2 \, d\Gamma = \varepsilon_0(\int_\Gamma g^2 \, d\Gamma)^{1/2} \quad \text{(where} \quad g_0 = g_1 = g) \, .$$

Consequently

$$\beta = \frac{1}{\varepsilon_0} \, \|g\|_{L^2(\Gamma)}. \qquad (4.51)$$

In the iterative algorithm (since $g = u^n|_\Gamma$ with the notations of (4.37), (4.38), (4.39)) we take

$$\beta_n = \frac{1}{\varepsilon_0} \, \|u^n\|_{L^2(\Gamma)}. \tag{4.52}$$

$\square$

Remark 4.19.

Formula (4.52) is analogous to formula (4.21), given in a very different context by the same type of duality argument, in R. GLOWINSKI and the A. [1], (1994). $\square$

Remark 4.20.

In **non linear** problems, duality arguments of the above type are (in general) **non valid**. One can use them in **a formal** fashion for the "linear tangent operator" in the iterative algorithm. $\square$

4.3   Non linear problems and fictitious domains.

Let us consider again the geometry of Figure 1, Section 4.1. We consider in $\Omega \subset \mathbf{R}^3$ the Navier Stokes equations which are expressed in the variational form

$$\begin{aligned} a(u,v) + b(u,u,v) &= (f,v) \qquad \forall\, v \in V(\Omega), \\ u &\in V(\Omega) \end{aligned} \tag{4.53}$$

where

$$V(\Omega) = \{v|\ v \in (H_0^1(\Omega))^3\, , \operatorname{div} v = 0 \quad \text{in } \Omega\}, \tag{4.54}$$

$$a(u,v) = \sum_{i,j=1}^{3} \int_\Omega \frac{\partial u_i}{\partial x_j} \frac{\partial v_i}{\partial x_j}\, dx\,,$$

$$b(u,\hat{u},v) = \sum_{i,j} \int_\Omega u_j \frac{\partial \hat{u}_i}{\partial x_j}\, v_i\, dx\,,$$

$$(f,v) = \sum_i \int_\Omega f_i\, v_i\, dx\,, \qquad f_i \in L^2(\Omega)\,. \tag{4.55}$$

Problem (4.53) admits always at least **a** solution, in fact it admits in general an infinite number of solutions.

We want to approximate one of the solutions u of (4.53) using an idea of **virtual control type**. $\square$

With the notations of Figure 2, we introduce, with rather obvious notations, the problem **with virtual control** λ

$$\begin{aligned} a(u,v) + b(u,u,v) &= (\tilde{f},v) + (\lambda,v)_\mathcal{O} \qquad \forall\, v \in V(Q), \\ u &\in V(Q)\,. \end{aligned} \tag{4.56}$$

In (4.56) the forms a and b are extended to Q and the virtual control λ spans a 2-dimensional subspace of $L^2(\mathcal{O})^3$. $\qquad\qquad\square$

Remark 4.21.

We conjecture -but it is not proven- that when λ spans the 2-d subspace

of $L^2(\mathcal{O})^3$ the set of all traces on Γ of all solutions u of (4.56) is dense in

$E = \{g|\ g \in L^2(\Gamma)^3,\ \int_\Gamma gn\ d\Gamma = 0\}$.

The situation becomes much simpler if λ **spans the whole space** $L^2(\mathcal{O})^3$ **and if**

$$\mathcal{O} = \Omega' . \tag{4.57}$$

Then let u be a solution of (4.53). Assuming the solution u to be smooth enough, we extend u to Q in a function, still denoted by u, such that u is smooth in Q, div $u = 0$ in Q, $u = 0$ on ∂Q. Then **we define** λ by

$$(\lambda, v)_{\Omega'} = a(u, v) + b(u, u, v) - (\tilde{f}, v).$$

In the other words

when $\mathcal{O} = \Omega'$, the mapping $\lambda \to u|_\Gamma$ maps $L^2(\mathcal{O})^3$

onto E . $\tag{4.58}$

There is exact controllability. $\qquad\qquad\square$

Remark 4.22.

If λ is subject to span a 2-d subspace of $L^2(\mathcal{O})^3$ the simple argument above fails and the analogous of (4.58) is an open question. $\qquad\qquad\square$

We can now apply (4.58) in several ways for computational purposes. The first possibility is to use a **penalty argument**. One considers the system in Q given in variational form by

$$a(u_\varepsilon, v) + b(u_\varepsilon, u_\varepsilon, v) + \frac{1}{\varepsilon} \int_\Omega' u_\varepsilon v\ dx = (\tilde{f}, v) \qquad \forall v \in V(Q),$$

$$u_\varepsilon \in V(Q) . \tag{4.59}$$

For every $\varepsilon > 0$, there exists a solution u_ε of (4.59) (same method that without **the penalty term** $\frac{1}{\varepsilon} \int_{\Omega'} u_\varepsilon v\ dx$; cf. J.L. LIONS [3]).

Moreover if we take $v = u_\varepsilon$ in (4.59) and if we notice that

$$b(u_\varepsilon, u_\varepsilon, u_\varepsilon) = 0$$

it follows that

$$a(u_\varepsilon, u) + \frac{1}{\varepsilon} \int_{\Omega'} u_\varepsilon^2\ dx = (\tilde{f}, u_\varepsilon). \tag{4.60}$$

Therefore, as $\varepsilon \to 0$, one has :

u_ε remains in a bounded subset of $V(Q)$ and

$$\int_{\Omega'} u_\varepsilon^2 \, dx \le c\varepsilon, \quad c = \text{constant} . \tag{4.61}$$

One can then show that one can extract a subsequence of u_ε, still denoted by u_ε, and which is such that

$$u_\varepsilon \to u \quad \text{in } V(Q) \quad \text{weakly,}$$
$$u = 0 \quad \text{on } \Omega' ,$$
$$a(u, v) + b(u, u, v) = (f, v) \quad \forall \, v \in V(\Omega)$$

i.e. u is a solution of (4.53).   □

Remark 4.23.
This penalty argument on artificial domains has been introduced in the A. [1].   □

Remark 4.24.
Another method is to consider λ in (4.56), when $\mathcal{O} = \Omega'$, as a Lagrange multiplier. This approach has been systematically followed by R. GLOWIN-SKI, T.W. PAN and J. PERIAUX (cf. these A. [1] and the Bibliography therein).   □

Remark 4.25.
We refer also to O. PIRONNEAU [1].   □

Remark 4.26.
Cf. the A. [2] for similar ideas applied to parabolic (evolution) problems.

4.4   Domain decomposition. An introduction to the virtual control method.

Let Ω be an open set of $\mathbf{R}^d$ which is decomposed in

$$\Omega = \Omega_1 \cup \Omega_2 \quad \text{(cf. Figure 4)}$$

where

$$\Omega_1 \cap \Omega_2 \ne \phi.$$

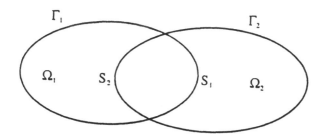

Figure 4

The notations are as follows

$$\partial\Omega_1 = \Gamma_1 \cup S_1 \,,$$
$$\partial\Omega_2 = \Gamma_2 \cup S_2 \,,$$
$$\partial\Omega = \Gamma_1 \cup \Gamma_2 \,, \quad \partial(\Omega_1 \cap \Omega_2) = S_1 \cup S_2 \,.$$

Let A be a second order elliptic operator given in Ω as in Section 4.1. We consider the problem

$$Au = f \quad \text{in } \Omega \,,$$
$$u \in H_0^1(\Omega) \,, \tag{4.62}$$

where f is given in $L^2(\Omega)$.

We now decompose (4.62) using Ω_1, Ω_2 **and the idea of virtual control.**

Let us introduce the decomposition

$$f = f_1 + f_2 \,, \ f_i \in L^2(\Omega_i) \,, \ f_i = 0 \quad \text{outside } \Omega_i \,. \tag{4.63}$$

We introduce the region $\mathcal{O}$ of the virtual control

$$\mathcal{O} \subset \Omega_1 \cap \Omega_2 \tag{4.64}$$

and we define u_1, u_2 as the solutions of

$$Au_1 = f_1 + \lambda 1_{\mathcal{O}} \quad \text{in} \quad \Omega_1 \,,$$
$$u_1 = 0 \quad \text{on} \quad \Gamma_1 \,, \quad \frac{\partial u_1}{\partial n_A} = 0 \quad \text{on} \quad S_1 \,, \tag{4.65}$$

$$Au_2 = f_2 - \lambda 1_{\mathcal{O}} \quad \text{in} \quad \Omega_2 \,,$$
$$u_2 = 0 \quad \text{on} \quad \Gamma_2 \,, \quad \frac{\partial u_2}{\partial n_A} = 0 \quad \text{on} \quad S_2 \,. \tag{4.66}$$

Remark 4.27.
Of course one considers problems (4.65), (4.66) in their weak variational
form.   □

Let us assume for a moment that

one can find a virtual control $\lambda \in L^2(\mathcal{O})$ such that
$$u_1 = 0 \quad \text{on } S_1 \ , \ u_2 = 0 \quad \text{on } S_2 \ . \tag{4.67}$$

Then if $\tilde{u}_i$ denotes the extension of u_i by 0 outside Ω_i, we have (since the
Cauchy data for u_i are 0 on S_i)

$$A\tilde{u}_1 = f_1 + \lambda 1_{\mathcal{O}}$$
$$A\tilde{u}_2 = f_2 - \lambda 1_{\mathcal{O}} \tag{4.68}$$

so that

$$A(\tilde{u}_1 + \tilde{u}_2) = f_1 + f_2 = f \quad \text{in } \ \Omega \ . \tag{4.69}$$

Moreover

$$\tilde{u}_1 + \tilde{u}_2 = 0 \quad \text{on } \ \Gamma_1 \cup \Gamma_2 \tag{4.70}$$

so that

$$\tilde{u}_1 + \tilde{u}_2 = u \ . \tag{4.71}$$

Remark 4.28.
Except in the case where $\mathcal{O} = \Omega_1 \cap \Omega_2$, one **cannot** achieve (4.67).
But we can do it **approximately**, as we show now.   □

We are now going to sketch the proof of

when λ spans $L^2(\mathcal{O})$, then $u_1|_{S_1}$, $u_2|_{S_2}$ **span a set**
which is dense in $L^2(S_1) \times L^2(S_2)$. $\qquad\qquad$ (4.72)

Remark 4.29. It follows from (4.72) that we can choose λ so that
$\|u_1\|_{L^2(S_1)} + \|u_2\|_{L^2(S_2)}$ is arbitrarily small.
In this case (4.71) is in fact an approximation.   □

Proof of (4.72).
As we saw in Section 4.2, we do not restrict the generality if assuming
that $f_1 = f_2 = 0$.
Let $g_i \in L^2(S_i), i = 1, 2$, be such that

$$\int_{S_1} g_1 \, u_1 \, dS_1 + \int_{S_2} g_2 \, u_2 \, dS_2 = 0 \qquad \forall \lambda \in L^2(\mathcal{O}). \tag{4.73}$$

Let us define

$$a_k(u,v) = \sum_{i,j=1}^{d} \int_{\Omega_k} a_{ij}(x) \frac{\partial u}{\partial x_j} \frac{\partial v}{\partial x_i} \, dx \,, \quad k = 1, 2,$$

$$u, v \in H^1(\Omega_k)$$

and let $a_k^*(u,v)$ be the adjoint form of $a_k(u,v)$.

We define $p_i \in H_0^1(\Omega_i)$ by

$$a_i^*(p_i, q_i) = \int_{S_i} g_i \, q_i \, dS_i \qquad \forall \, q_i \in H_0^1(\Omega_i). \tag{4.74}$$

Then, taking $q_i = u_i$ in (4.74) and adding up, we obtain $\int_{S_1} g_1 \, u_1 \, dS_1 + \int_{S_2} g_2 \, u_2 \, dS_2 = a_1^*(p_1, u_1) + a_2^*(p_2, u_2) = a_1(u_1, p_1) + a_2(u_2, p_2) = $ (using (4.65), (4.66) in variational form and with $f_i = 0$) $= (\lambda, 1_{\mathcal{O}} \, p_1 - 1_{\mathcal{O}} \, p_2)$ so that (4.73) is equivalent to

$$p_1 - p_2 = 0 \qquad \text{on} \quad \mathcal{O} \, . \tag{4.75}$$

But according to (4.74)

$$A^* p_1 = 0 \,, \quad A^* p_2 = 0 \qquad \text{in } \Omega_1 \cap \Omega_2 \, . \tag{4.76}$$

If the unique continuation property holds true (i.e. if the coefficeints of A are smooth enough) then it follows from 4(4.75) and (4.76) that

$$p_1 \equiv p_2 \qquad \text{in} \quad \Omega_1 \cap \Omega_2 \, . \tag{4.77}$$

We then define

$$\pi = \{ p_1 \quad \text{in} \quad \Omega_1 \,, \; p_2 \quad \text{in} \quad \Omega_2 \} \, . \tag{4.78}$$

Because of (4.77) there is no ambiguity on $\Omega_1 \cap \Omega_2$ **and we define in this way an element of** $H_0^1(\Omega)$.

In Ω_1 (resp. Ω_2) one has $A^* p_1 = 0$ (resp. $A^* p_2 = 0$) so that (4.78) implies that

$$A^* \pi = 0 \qquad \text{in} \quad \Omega \, .$$

Since $\pi \in H_0^1(\Omega)$, it follows that $\pi \equiv 0$ in Ω so that $g_i = 0$, which proves (4.72). $\qquad \square$

Remark 4.30. Practically all the remarks made in Sections 4.2 and 4.3 apply to the present situation. For the **algorithms** deduced from the above considerations, we refer to O. PIRONNEAU and the A. [1] [2] [3] where much more general cases are presented. $\qquad \square$

4.5   Bibliography

C. CARTHEL, R. GLOWINSKI and J.L. LIONS [1] On exact and approximate boundary controllability for the heat equation : A numerical approach, J.O.T.A. 82 (3), September 1994, p. 429-484.

R. GLOWINSKI and J.L. LIONS [1] Exact and approximate controllability for distributed parameter systems. Acta Numerica (1994), p. 269-378, (1995), p. 159-333.

R. GLOWINSKI, C.H. LI and J.L. LIONS [1] A numerical approach to the exact boundary controllability of the wave equation. Japan J. Appl. Math. 7, (1990), p. 1-76.

R. GLOWINSKI, T.W. PAN and J. PERIAUX [1] A fictitious domain method for flows around moving airfoils : application to store separation. To appear.

J.L. LIONS [1] Sur l'approximation des solutions de certains problèmes aux limites. Rend. Sem. Mat. dell' Univ. di Padova, XXXII (1962), p. 3-54.

[2] Fictitious domains and approximate controllability. Dedicated to S.K. MITTER. To appear.

[3] Quelques méthodes de résolution des problèmes aux limites non linéaires. Dunod. Gauthier Villars, (1969).

J.L. LIONS and E. MAGENES [1] Problèmes aux limites non homogènes et applications.
Vol. 1 Paris. Dunod, (1968).

J.L. LIONS and O. PIRONNEAU [1] Algorithmes parallèles pour la solution de problèmes aux limites. C.R.A.S., 327 (1998), p. 947-952.

[2] Sur le contrôle parallèle des systèmes distribués. C.R.A.S., 327 (1998), p. 993-998.

[3] Domain decomposition methods for CAD. C.R.A.S., 328 (1999), p. 73-80.

J.L. LIONS and E. ZUAZUA [1] A generic result for the Stokes system and its control theoretical consequences, in PDE and its applications, P. MARCELLINI, G. TALENTI and E. VISENTINI, Eds., Dekker Inc., LNPAS 177, (1996), p. 221-235.

P.L. LIONS [1] Personal Communication.

A. OSSES and J.P. PUEL [1] On the controllability of the Laplace equation observed on an interior curve. Revista Mat. Complutense. Vol. II, 2, (1998), p. 403-441.

O. PIRONNEAU [1] Fictitious domains versus boundary fitted meshes. 2^d Conf. on Num. Methods in Eng. La Coruña (June 1993).

5   Decomposition of the energy space

5.1   Setting of the problem.

Let us consider for a moment an abstract formulation. Let V be a (real) Hilbert space and let $a(u, v)$ be a continuous bilinear form on V. We assume a to be coercive, i.e.

$$a(v, v) \geq \alpha\|v\|_V^2 \qquad \forall\, v \in V \,, \alpha > 0 \tag{5.1}$$

where in (5.1) $\|v\|_V$ denotes the norm of v in V (**the energy space**). The form $a(u, v)$ is symmetric **or not**.

Let $v \to L(v)$ be a continuous linear form on V. Then, there exists a unique element $u \in V$ such that

$$a(u, v) = L(v) \qquad \forall\, v \in V \,. \tag{5.2}$$

Let us now consider

$$V_i = \quad \text{closed subspace of } V \,, \ i = 1, 2. \tag{5.3}$$

We assume that in general

$$V_1 \cap V_2 \neq \{0\} \,. \tag{5.4}$$

Remark 5.1.
What we are going to present applies to the case when $V_1 \cap V_2 = \{0\}$, but some of the remarks to follow are trivial in such a case. □

We assume also that

$$V = V_1 + V_2 \tag{5.5}$$

in the sense :
$\forall\, v \in V$, there exists a decomposition

$$v = v_1 + v_2, \quad v_i \in V_i \,,$$

and, in fact, there is an infinite number of decompositions if $V_1 \cap V_2 \neq \{0\}$.
We want to obtain **a decomposition**

$$u = u_1 + u_2 \quad u_i \in V_i \tag{5.6}$$

of the solution u of (5.2), with a "simple" algorithm for the computation of u_i. □

Remark 5.2.

Decompositions of the type (5.5) apply to the geometrical situation considered in Section 3.4. Cf. HELP [1] (F. HECHT, J.L. LIONS, O. PIRONNEAU). But there are very many other decompositions which may be useful in practical applications.   □

We present now a decomposition method based on the idea **of virtual controls**. It can be considered as an Introduction to the paper of R. GLOWINSKI and O. PIRONNEAU and the A. [1].

5.2   Virtual controls.

Let $s_i(u_i, v_i)$ be a continuous bilinear form on V_i. We assume that

$$s_i(u_i, v_i) \quad \text{is a symmetric and coercive on } V_i, \text{ i.e.}$$
$$s_i(v_i, v_i) \geq \sigma_i \|v_i\|_{V_i}^2 \quad \forall v_i \in V_i , \ \sigma_i > 0 \tag{5.7}$$

where $\|v_i\|_{V_i} = \|v_i\|_V$.

Given $u \in V$, the decomposition (5.6) is not unique. We consider the set K

$$K = \{v_i, v_2 \mid v_i \in V_i , \ v_i + v_2 = u\}. \tag{5.8}$$

This set K is closed and convex in $V_1 \times V_2$, therefore **there exists a unique decomposition** $u = u_1 + u_2$, $u_i \in V_i$ **such that**

$$s_1(u_1) + s_2(u_2) = \inf_{v_1, v_2 \in K} s_1(v_1) + s_2(v_2) \tag{5.9}$$

where we have set $s_1(u_1) = s_1(u_1, u_1)$ etc.   □

We introduce now **the virtual controls** $\lambda_1, \lambda_2 \in V_1 \times V_2$.

Given $\lambda = \{\lambda_1, \lambda_2\} \in V_1 \times V_2$, one defines uniquely $u_i = u_i(\lambda) \in V_i$ as the solution of

$$s_1(u_1 - \lambda_1, v_1) + a(\lambda_1 + \lambda_2, v_1) = L(v_1) \quad \forall v_1 \in V_1 ,$$
$$s_2(u_2 - \lambda_2, v_2) + a(\lambda_1 + \lambda_2, v_2) = L(v_2) \quad \forall v_2 \in V_2 . \tag{5.10}$$

Remark 5.3.

In (5.10) u_1 (for instance) is given by the solution of

$$s_1(u_1, v_1) = s_1(\lambda_1, v_1) - a(\lambda_1 + \lambda_2, v_1) + L(v_1) \quad \forall v_1 \in V_1 .$$

□

The problem is now to choose the λ_i's in "**the best possible way**" so that $u_1 + u_2$ is the solution u of (5.2) and the decomposition is the one which achieves (5.9).

To this effect we introduce the **cost function**

$$\mathcal{J}_\varepsilon(\lambda_1, \lambda_2) = \frac{1}{2}(s_1(\lambda_1) + (s_2(\lambda_2)) + \frac{1}{2\varepsilon}(s_1(u_1 - \lambda_1) + (s_2(u_2 - \lambda_2)) ,$$
(5.11)

where $\varepsilon > 0$ is given.

It is straightforward to verify that there exists a unique couple $\lambda_{1\varepsilon}, \lambda_{2\varepsilon} \in V_1 \times V_2$ such that

$$\mathcal{J}_\varepsilon(\lambda_{1\varepsilon}, \lambda_{2\varepsilon}) = \inf_{\lambda_1, \lambda_2} \mathcal{J}_\varepsilon(\lambda_1, \lambda_2) .$$
(5.12)

We denote by $u_{i\varepsilon}$ the solutions of (5.10) for $\lambda_i = \lambda_{i\varepsilon}$.
We are now going to show

$$\text{as } \varepsilon \to u_i \quad \text{in } V_i \quad \textbf{weakly, where}$$

$$u_1 + u_2 = u = \quad \textbf{solution of (5.2) and (5.9) holds true.}$$
(5.13)

In order to prove (5.13), let us begin with the following observation. Let u be the solution of (5.2) and let us consider a decomposition

$$u = v_1 + v_2 , \quad v_i \in V_i .$$

Let us choose $\lambda_i = v_i$. Then the unique solution of (5.10) is given by $u_i = \lambda_i$ so that

$$\mathcal{J}_\varepsilon(v_1, v_2) = \frac{1}{2}(s_1(v_1) + s_2(v_2)) .$$
(5.14)

Therefore

$$\mathcal{J}_\varepsilon(\lambda_{1\varepsilon}, \lambda_{2\varepsilon}) \le \mathcal{J}_\varepsilon(v_1, v_2) \quad \forall \varepsilon$$

gives

$$\overline{\lim} \mathcal{J}_\varepsilon(\lambda_{1\varepsilon}, \lambda_{2\varepsilon}) \le \frac{1}{2}(s_1(v_1) + s_2(v_2)) \quad \forall \{v_1, v_2\} \in K .$$

hence

$$\overline{\lim} \mathcal{J}_\varepsilon(\lambda_{1\varepsilon}, \lambda_{2\varepsilon}) \le \inf_{\{v_1, v_2\} \in K} \frac{1}{2}(s_1(v_1) + s_2(v_2)) .$$
(5.15)

But (5.15) implies that

$$s_i(u_{i\varepsilon} - \lambda_{i\varepsilon}) \le C\varepsilon$$
$$s_i(\lambda_{i\varepsilon}) \le C , \quad \text{where } C \text{ denotes various constants independant of } \varepsilon .$$
(5.16)

Therefore we can extract subsequences, still denoted by $u_{i\varepsilon}, \lambda_{i\varepsilon}$ such that

$$u_{i\varepsilon}, \lambda_{i\varepsilon} \to u_i, \lambda_i \quad \text{with } u_i = \lambda_i \quad \text{in } V_i \text{ weakly} .$$
(5.17)

One can then pass to the limit in (5.10) where $\lambda_i = \lambda_{i\varepsilon}$ and one finds that

$$a(u_1 + u_2, v_1) = L(v_1)$$

$$a(u_1 + u_2, v_2) = L(v_2)$$

i.e. $u_1 + u_2 = u$ solution of (5.2).

It remains to show (5.9). By the definition of $\mathcal{J}_\varepsilon(\lambda_1, \lambda_2)$ we see that

$$\mathcal{J}_\varepsilon(\lambda_{1\varepsilon}, \lambda_{2\varepsilon}) \geq \frac{1}{2}(s_1(\lambda_{1\varepsilon}) + s_2(\lambda_{2\varepsilon}))$$

hence

$$\underline{\lim}\,\mathcal{J}_\varepsilon(\lambda_{1\varepsilon}, \lambda_{2\varepsilon}) \geq \frac{1}{2}(s_1(u_1) + s_2(u_2)) . \tag{5.18}$$

Comparing (5.18) with (5.15) we obtain (5.9) so that (5.13) is proven.

$$\square$$

We now briefly indicate in the next section how to obtain a descent algorithm for the computation of $\lambda_{i\varepsilon}$, $u_{i\varepsilon}$ and we conclude by a number of remarks and problems.

5.3 Algorithms and problems.

Remark 5.4.

Before we proceed, it is clear that one should choose the forms s_i satisfying (5.7) and such that the solution of (5.10) is **"as simple as possible"**.

On this question, cf. R. GLOWINSKI, J.L. LIONS and O. PIRONNEAU, loc. cit.

$$\square$$

We now compute the first variation of $\mathcal{J}_\varepsilon(\lambda_1, \lambda_2)$. We have

$$\delta\mathcal{J}_\varepsilon(\lambda_1, \lambda_2) = s_1(\lambda_1, \delta\lambda_1) + (s_2(\lambda_2, \delta\lambda_2) +$$
$$+ \frac{1}{\varepsilon}[s_1(u_1 - \lambda_1, \delta u_1 - \delta\lambda_1) + (s_2(u_2 - \lambda_2) - \delta u_2 - \delta\lambda_2)] . \tag{5.19}$$

We introduce $p_i \in V_i$ as the solution of

$$s_i(p_i, v_i) = a^*(u_1 - \lambda_1 + u_2 - \lambda_2, v_i) \qquad \forall\, v_i \in V_i , \tag{5.20}$$

where a^* is the adjoint of a, i.e. $a^*(u, v) = a(v, u)$.

But it follows from (5.10) that

$$s_i(\delta u_i - \delta\lambda_i, v_i) + a(\delta\lambda_i + \delta\lambda_2, v_i) = 0 \qquad \forall\, v_i \in V_i \tag{5.21}$$

so that

$$s_1(u_1 - \lambda_1, \delta u_1 - \delta \lambda_1) + s_2(u_2 - \lambda_2, \delta u_2 - \delta \lambda_2)$$
$$= -a(\delta \lambda_1 + \delta \lambda_2, u_1 - \lambda_1 + u_2 - \lambda_2) =$$
$$= -a^*(u_1 - \lambda_1 + u_2 - \lambda_2, \delta \lambda_1 + \delta \lambda_2) = \quad \text{(by using (5.20))}$$
$$- s_1(p_1, \delta \lambda_1) - s_2(p_2, \delta \lambda_2)$$

so that (5.19) gives

$$\varepsilon \delta \mathcal{J}_\varepsilon(\lambda_1, \lambda_2) = s_1(\varepsilon \lambda_1 - p_1, \delta \lambda_1) + s_2(\varepsilon \lambda_2 - p_2, \delta \lambda_2) . \tag{5.22}$$

The following gradient algorithm follows :

$$\lambda_1^{n+1} = \lambda_1^n - \rho(\varepsilon \lambda_1^n - p_1^n),$$
$$\lambda_2^{n+1} = \lambda_2^n - \rho(\varepsilon \lambda_2^n - p_2^n),$$
$$\rho > 0 \quad \text{small enough.} \tag{5.23}$$

Then one computes u_1^{n+1}, u_2^{n+1} by solving

$$s_1(u_1^{n+1} - \lambda_1^{n+1}, v_1) + a(\lambda_1^{n+1} + \lambda_2^{n+1}, v_1) = L(v_1) \quad \forall v_1 \in V_1 ,$$
$$s_2(u_2^{n+1} - \lambda_2^{n+1}, v_2) + a(\lambda_1^{n+1} + \lambda_2^{n+1}, v_2) = L(v_2) \quad \forall v_2 \in V_2 , \tag{5.24}$$

one then defines p_i^{n+1} by solving

$$s_i(p_i^{n+1}, v_i) = a^*(u_1^{n+1} - \lambda_1^{n+1} + u_2^{n+1} - \lambda_2^{n+1}, v_i) \quad \forall v_i \in V_i ,$$
$$i = 1, 2 \tag{5.25}$$

And one proceeds in this way. □

Remark 5.5.
The same techniques apply to the case where

$$V = V_1 + \cdots + V_m \tag{5.26}$$

with $V_i \cap V_j$ not necessarily $\{0\}$ for $i \neq j$. cf. R. GLOWINSKI, J.L. LIONS and O. PIRONNEAU, loc. cit., where conjugate gradient techniques are indicated. □

Remark 5.6.
The method is **completely general**. It applies to all linear boundary value problems. It can also be applied to non linear problems. the case of Navier Stokes equations being given in the paper R. GLOWINSKI, J.L. LIONS, T.W. PAN, O. PIRONNEAU [1]. □

5.4   Bibliography

R. GLOWINSKI, J.L. LIONS, O. PIRONNEAU [1] Decomposition of energy spaces and applications. C.R.A.S. Paris, 1999.

R. GLOWINSKI, J.L. LIONS, T.W. PAN, O. PIRONNEAU [1] Decomposition of energy spaces, virtual control and applications. To appear.

F. HECHT, J.L. LIONS, O. PIRONNEAU [1] Domain decomposition algorithms for C.A.D. Dedicated to I. NECAS, to appear.

Flow And Heat Transfer
In Pressing Of Glass Products

K. Laevksy, B.J. van der Linden, R.M.M. Mattheij

Department of Mathematics and Computer Science,
Eindhoven University of Technology,
PO Box 513, 5600 MB The Netherlands

Abstract

In studying glass morphology often models are used that describe it
as a strongly viscous Newtonian fluid. In this paper we shall study
one of the problems encountered in glass technology. It is dealing with
producing packing glass by a so-called pressing process. The pressing
problem actually deals with the morphology of a bottle or jar. We
first show how to deal with the temperature, by a suitable dimension
analysis. In this analysis we see the dominance of the convection over
the other modes of heat transfer. However, at the end of the pressing
— a stage called *the dwell* — flow is absent and we have to deal with
conduction and radiation in a heat-only problem. A similar analysis is
made for the *re-heating*, where the internal stresses are relaxed before
the final, blowing stage of the forming process. We give a number of
numerical examples to sustain our results.

Contents

1   Introduction

For many years, glass technology has been a craft based on expertise and experimental knowledge, reasonably sufficient to keep the products and production competitive. Over the last twenty years mathematical modelling of the various aspects of production has become increasingly decisive, however. This is induced in part by fierce competition from other materials, notably polymers, which, e.g., have found their way into the food packing industry. For another, this is a consequence of environmental concerns. It is not so much the waste (glass is 100% recyclable, a strong advantage to most competitors) as the energy consumption. One should realize that the melting process of sand to liquid glass makes up the largest cost factor of the product. The relative importance of the current industry is illustrated by the following numbers: In the European Union about 25 megatons of glass is being produced, which represents fifty billion euro worth. The industry employs more than 200,000 people. Two-thirds of the glass production is meant for packing (jars and bottles). Float glass (used for panes) makes up most of the other quarter. The rest is for special products like CRTs and fibers.

Production of container glass products goes more or less along the following lines. First grains of silica (typically available in the form of sand) and additives, like soda, are heated in a tank. This can be an enormous structure with a typical length of several tens of meters and a width of a couple of meters. The height is less impressive and rarely exceeds one meter. Gas burners or electrode heaters provide the necessary heat to heat the material to around 1400°C. At one end, the liquid glass comes out and is either led to a pressing or blowing machine or it ends up on a bed of liquid tin, where it spreads out to become float glass (panes, wind-shields, etc.). In the latter case the major problems are the need for a smooth flow from the oven on the bed and controlling the spreading and flattening. The pressing and blowing process is used in producing packing glass. To obtain a glass form a two-stage process is often used: First a blob of hot glass is pressed into a mould to form a so-called *parison*. It is cooled down (the mould is kept at 500°C) such that a small skin of solid glass is formed. The parison is then blown into its final shape. Such pressing/blowing machinery can produce a number of products at the same time; as a result a more or less steady flow of glass products is coming out on a belt. The products then have to be cooled down in a controlled way such that the remaining stresses are as small as possible (and thus the strength is optimal).

Sometimes only pressing is needed. This is the case in the production of CRTs, where a stamp is pressed into liquid glass and after being lifting, a certain morphology should have been transferred onto the glass screen.

All these processes involve the flow of the (viscous) glass in combination with heat exchange. Although these two are closely intertwined we shall show in this paper that in they can often be decoupled. In cases where convective heat transfer is predominant this effectively leads to isothermal flow problems on one hand and temperature problems on the other. In the stages where flow is absent or negligible we are left with a pure heat problem.

This paper is written as follows. In Section 2 we shall derive the basic flow equations that will play a role in our models. We discuss the pressing of glass in a mould. We describe the model and pay special attention to the heat exchange problem. In Section 3 we discuss the actual pressing phase of the process. In Sections 4 and 5 we look at the post-pressing treatments of the parison: the dwell and the re-heating; these to stages do not involve convection and allow for a thorough treatment of the underlying heat problem.

2   Modelling The Problem

In many cases the process of glass production consists of three main phases. The first one is the pressing phase, the second is the re-heating phase, and the final one the blowing phase. In this section we first describe the process as it is used in industry. The different stages of the process are shown in Figures 2.1 and 2.2. Throughout this section we refer to these figures.

1. A gob of glass leaving the furnace (tank), enters into a configuration consisting of two parts: the mould and the plunger (a). Then, the pressing of the glass takes place in the following way. The plunger moves up inside of the mould (b) and forces the glass to fill the free space in between. At the end of this stage the glass is left in the mould for a second (c), which is called the *dwell*. Stages (a), (b) and (c) form the *pressing phase*.

2. After the dwell the *parison* — as the half-product is called — is taken out of the mould and left outside for a couple of seconds (d) giving it the possibility to *re-heat*, i.e. soften the internal temperature gradients.

3. The parison is then placed into a second mould (e) and is blown in to its final shape (f). The latter two stages form the *blowing phase* of the press-and-blow process.

In order to make a sufficient mathematical model of the process it is necessary to mention the basic characteristics and numerical parameters of the process:

$\eta_0 = \eta(T_g) = 10^4$ kg/m s   –   the dynamic viscosity of the glass
$\kappa = 3.50$ m^{-1} –   absorption coefficient
$\rho = 2500$ kg/m^3 –   the density of glass
$c_p = 1350.0$ J/kg K   –   specific heat
$k^c = 1.71$ W/m K   –   conductivity
$L_0 = 10^{-2}$ m   –   the typical scale for the parison   (2.1)
$n = 1.50$ –   refractive index
$T_g = 1250°$C   –   the temperature of the glass
$T_m = 700°$C   –   the temperature of the mould
$T_p = 1000°$C   –   the temperature of the plunger
$V_0 = 10^{-1}$ m/s   –   the typical velocity of the plunger

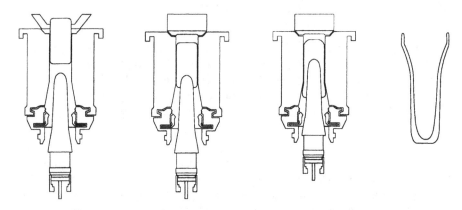

Figure 2.1: The various stages of the pressing phase in a press and blow process: a) The glass enters the mould; b) The plunger presses the glass into form; c) During the dwell the outside of the glass is cooled and solidified; d) The plunger is reheated to reduce temperature gradients.

2.1 Modelling of The Flow

As it was already implicitly assumed above, the glass — at sufficiently high temperatures — can be considered a viscous fluid. Glass may be viewed as a frozen liquid, i.e. it has an amorphous structure. At sufficiently high temperatures (say above 600 °C) it behaves like an incompressible Newtonian fluid, which means that for a given dynamic viscosity η, a velocity $\mathbf{v}$ and a pressure p, the stress tensor σ is given by

$$\sigma = -p\mathbf{I} + \eta(\nabla\mathbf{v} + \nabla\mathbf{v}^T) \tag{2.2}$$

This constitutive relation should be used to close the equations that actually describe the motion of glass gob, the momentum equation (2.3) and the

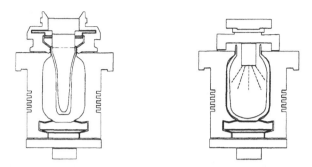

Figure 2.2: The various stages of the blowing phase: e) The plunger is put into the blow mould; f) The glass is blown into its final stage.

continuity equation (2.4):

$$\rho\left(\frac{\partial \mathbf{v}}{\partial t} + (\nabla\mathbf{v})^T\mathbf{v}\right) = \rho\mathbf{f} + \nabla\cdot\boldsymbol{\sigma}, \tag{2.3}$$

where ρ denotes the mass density and $\mathbf{f}$ the volume forces on the blob,

$$\nabla\cdot\mathbf{v} = 0. \tag{2.4}$$

Using (2.2) in (2.3) we obtain

$$\rho\left(\frac{\partial \mathbf{v}}{\partial t} + (\nabla\mathbf{v})^T\mathbf{v}\right) = \rho\mathbf{f} - \nabla p + \nabla\cdot\left[\eta(\nabla\mathbf{v} + \nabla\mathbf{v}^T)\right] \tag{2.5}$$

In the problem we shall study in this paper we anticipate the viscous forces $(\nabla\cdot\boldsymbol{\sigma})$ to dominate in (2.3). To see this we shall reformulate our equations in dimensionless form, for which we need some characteristic quantities.

First we remark that the only acting volume force in the process is gravity, so $\|\mathbf{f}\| = g \approx 10\,\text{m/s}^2$. We define

$$\tilde{\mathbf{f}} := \frac{1}{g}\mathbf{f}. \tag{2.6}$$

For a fixed temperature, the viscosity η is assumed constant, say $\eta_0 \approx 10^4\,\text{kg/m s}$ for a reference temperature of 800°C. Normally, there is no need to introduce a dimensionless viscosity, but we shall nevertheless do this, as will become clear in the subsequent sections. Thus, let

$$\tilde{\eta} := \frac{1}{\eta_0}\eta. \tag{2.7}$$

A typical average velocity V_0 (which is 10^{-1}m/s or much smaller), say $V_0 \approx 10^{-1}\text{m/s}$, can be used as a characteristic velocity. As a characteristic length scale we take $L_0 (\approx 10^{-2}\text{m})$. We now define the dimensionless quantities

$$\tilde{\mathbf{x}} := \frac{\mathbf{x}}{L_0}, \quad \tilde{\mathbf{v}} := \frac{\mathbf{v}}{V_0}, \quad \tilde{p} := \frac{L_0}{\eta_0 V_0} p. \tag{2.8}$$

A proper choice for characteristic time scale is the ratio L_0/V_0 ($\approx 10^{-1}\text{s}$). So, let us finally define

$$\tilde{t} := \frac{V_0}{L_0} t. \tag{2.9}$$

In this problem the Reynolds number (Re) and the Froude number (Fr), defined by

$$\text{Re} := \frac{V_0 L_0 \rho}{\eta_0}, \quad \text{Fr} := \frac{\rho g L_0^2}{\eta_0 V_0}$$

— an important characteristics. The Reynolds number indicates the ratio between inertial forces and viscous forces and the quotient of the Reynolds number and the Froude number indicates the ratio between volume forces (i.e. gravity) and viscous forces. The two numbers are estimated by

$$\text{Re} \approx 10^{-4}, \quad \text{Fr} \approx 10^{-3}. \tag{2.10}$$

Substituting all dimensionless quantities into (2.4), (2.5) yields

$$\text{Re}\left(\frac{\partial \tilde{\mathbf{v}}}{\partial \tilde{t}} + (\nabla \tilde{\mathbf{v}})^T \tilde{\mathbf{v}}\right) = \text{Fr}\tilde{\mathbf{f}} - \nabla \tilde{p} + \nabla \cdot \left[\tilde{\eta}\left(\nabla \tilde{\mathbf{v}} + \nabla \tilde{\mathbf{v}}^T\right)\right]$$
$$\tag{2.11}$$

$$\nabla \cdot \tilde{\mathbf{v}} = 0$$

All spatial derivatives in (2.11) have to be taken with respect to the dimensionless variable $\tilde{\mathbf{x}}$. ¿From this we conclude that the viscous forces dominate indeed. Thus, the equations describing the flow are (rewritten in their dimensionless form)

$$\nabla p = \nabla \cdot \left[\eta\left(\nabla \mathbf{v} + \nabla \mathbf{v}^T\right)\right]$$
$$\tag{2.12}$$

$$\nabla \cdot \mathbf{v} = 0$$

These equations are of course the *Stokes creeping flow equations*. They require further boundary conditions in order to be able to solve for the vector $\mathbf{v}$. Actually, these will be kinematic constraints, changing with time t, describing the evolution of the gob. They have in common that at least one part of the boundary is free. Hence, besides finding the velocity $\mathbf{v}(t)$ we then need to find this free boundary. The actual displacements $\mathbf{x}$ satisfy the ordinary differential equation:

$$\frac{d\mathbf{x}}{dt} = \mathbf{v}(\mathbf{x}). \tag{2.13}$$

Numerically we shall deal with these problems in a two stage sweep: Suppose we have a domain $\mathcal{G}(t)$, describing the glass gob. Then solve (2.12) (approximately) and use the velocity field on the boundary to compute a new domain at time $t + \Delta t$, using (2.13) and the boundary conditions.

The results of particular simulation, velocity magnitude and pressure field, are depicted on Figure 3.1.

2.2 Modelling of The Heat

The energy equation for an incompressible fluid is given by

$$\rho c_p \frac{DT}{Dt} = -\nabla \cdot \mathbf{q} + \Phi, \tag{2.14}$$

where the heat flux $\mathbf{q}$ is due to the heat transfer mechanisms of conduction and radiation, and the source term Φ comes from the internal heat generation by action of viscous and volume forces. Because of the elevated temperatures in this process and the *semi-transparency* of glass — it absorbs, emits and transmits radiative energy — knowledge of radiation is necessary. Because of the high temperatures and the importance of radiation the heat flux, differently from the usual formulations of the energy equation, consists of a conductive heat flux $\mathbf{q}^c$ and a radiative heat flux $\mathbf{q}^r$, or

$$\mathbf{q} = \mathbf{q}^c + \mathbf{q}^r. \tag{2.15}$$

During the rest of the analysis we assume that all material properties are constant throughout the medium and time. Furthermore, we assume that the conduction obeys Fourier's law, which states that

$$\mathbf{q}^c := -k^c \nabla T, \tag{2.16}$$

where k^c is the *conductivity*, a material property.

The radiative heat flux cannot be expressed as simply. An approach generally made in industry is to make the often not sustainable assumption that the problem is *optically thick*. This means that the optical thickness τ — defined by $\tau := \kappa L$, where κ is the *absorption coefficient*, which denotes the amount of radiative energy absorbed by the medium per meter, and L is a characteristic length — is much greater than one. For typical values of $\kappa \sim 100\mathrm{m}^{-1}$ for dark glasses, and $L \sim 0.01$ (the thickness of the parison), we see that this assumption is violated. The results later in this section show the lamentable effect this has on the accuracy of the solution.

The reason to make this assumption in practice is made anyway, is that radiation can be accounted for in a computationally cheap way. Especially in higher dimensions, the only other options are:

- Not to account for radiation — and as we have seen in the previous chapter, radiation sometimes plays only a very minor role, even at elevated temperatures;

- To use higher order approximations — in general these are computationally very intensive, e.g. Monte Carlo methods, Modified Diffusion, Ray Tracing.

However, if we restrict ourselves to one dimension we can use both the optically thick method and the exact solution to the heat problem (2.14), and see if there are significant shortcomings with the use of the optically thick approximation.

The derivation of the optically thick approximation — also called the *Rosseland* or *diffusion approximation* — can be found in many text books on radiative heat transfer such as [5] and [1]. Here, we only state the result. The radiative heat flux under this approximation can be written as

$$\mathbf{q}^r := -k^r(T)\nabla T, \tag{2.17}$$

which resembles the expression for the conductive heat flux (2.16). In this equation $k^r(T)$ is called the *R*osseland parameter and is given by

$$k^r(T) := \frac{4}{3}\frac{n^2\bar{\sigma}T^3}{\kappa},$$

where n is the refractive index (a material property) and $\bar{\sigma}$ is the Stefan-Boltzmann constant. We see that the only difference with the conductive heat flux is the non-linearity of the diffusion coefficient. In many applications this can be implemented without a drastic change of the existing code.

For an exact solution we need to go deeper into the theory behind radiative heat transfer. The emitted radiative energy by a medium is proportional to the *total blackbody intensity* $B(T)$, which by definition is the amount of energy per unit area, per unit spherical angle of a perfect emitter-absorber. Because it is already implicitly integrated over all wavelengths, the total blackbody intensity is only a function of the temperature T. Planck found this upper limit to emission to be

$$B(T) = n^2\frac{\bar{\sigma}T^4}{\pi}. \tag{2.18}$$

Because of the T^4 term, radiation becomes progressively more important than conduction. We will also use the notation $B(\mathbf{x})$, defined as $B(\mathbf{x}) := B(T(\mathbf{x}))$, where the blackbody is a function of the position $\mathbf{x}$.

In practice most of the other radiative quantities are dependent on the wavelength λ of the radiation, but here we consider so-called *gray radiation*, where this dependence is absent. So, we can define the *radiative intensity* $I(\mathbf{x}, \mathbf{s})$ which is defined as the amount of radiative energy per unit area, per unit solid angle, at a certain position $\mathbf{x}$, travelling in a certain direction $\mathbf{s}$. If the radiation is unpolarised and a local thermal equilibrium can be assumed (see [3]), this quantity solely and totally describes electromagnetic radiation. The behaviour of the radiative intensity is determined by the *radiative transfer equation*:

$$\mathbf{s} \cdot \nabla I(\mathbf{x}, \mathbf{s}) = \kappa(\mathbf{x})B(\mathbf{x}) - \kappa(\mathbf{x})I(\mathbf{x}, \mathbf{s}), \tag{2.19}$$

where κ is the *absorption coefficient* denoting the amount of radiative energy absorbed by the medium per unit length. The boundary conditions for this equation for diffusely emitting/reflecting boundaries are given by

$$I(\mathbf{x}_w, \mathbf{s}) = \epsilon(\mathbf{x}_w)B(\mathbf{x}_w) + \frac{\rho(\mathbf{x}_w)}{\pi} \int_{\mathbf{n} \cdot \mathbf{s}' < 0} I(\mathbf{x}_w, \mathbf{s}')\mathbf{n} \cdot \mathbf{s}' \, d\Omega', \quad \forall \mathbf{s} \cdot \mathbf{n} > 0. \tag{2.20}$$

Here, we used $\epsilon(\mathbf{x}_w)$ for the *emissivity* of the boundary, where $\mathbf{x}_w$ is an arbitrary point on the boundary. Further, $\rho(\mathbf{x}_w)$ is the *(diffuse) reflectivity*. Both are material properties. Finally, $\mathbf{n}$ is the outward pointing normal at the boundary.

Then, if we know the intensity, we can determine the radiative heat flux $\mathbf{q}^r$ by

$$\mathbf{q}^r(\mathbf{x}) = \int_{4\pi} \mathbf{s}I(\mathbf{x}, \mathbf{s}) \, d\Omega. \tag{2.21}$$

In higher dimensions, it is very elaborate to work with these equations. However, a lot can be said about the behaviour of radiative heat transport if we restrict ourselves to one dimension only. The radiative transfer equation (2.19) can be simplified for the quasi-one-dimensional case into the following equations. Here, we mean by quasi-one-dimensional that we do not restrict the *directions* to one dimension. Doing so would result in Rosseland-like solutions. So, let

$$\tilde{\mu}\frac{\partial I^+}{\partial \tau}(\tau, \tilde{\mu}) + I^+(\tau, \tilde{\mu}) = B(\tau), \tag{2.22}$$

$$-\tilde{\mu}\frac{\partial I^-}{\partial \tau}(\tau, \tilde{\mu}) + I^-(\tau, \tilde{\mu}) = B(\tau), \tag{2.23}$$

where I^+ and I^- denote the intensity, $\tau = \int_x \kappa(x)dx$ is the dimensionless optical co-ordinate, and $\tilde{\mu} = |\cos \vartheta|$, where ϑ is the angle of the direction vector with the x-axis. Here we wrote the intensity in two different components I^+ and I^- with a $\tilde{\mu} \in (0, 1]$, because it simplifies relations that follow.

If we look at the problem on the open interval $x \in (x_o, x_1)$, with corresponding dimensionless optical co-ordinate $\tau \in (0, \tau_1)$, the formal solution for the above equations is:

$$I^+(\tau, \tilde{\mu}) = c_o e^{-\tau/\tilde{\mu}} + \frac{1}{\tilde{\mu}} \int_0^\tau e^{(s-\tau)/\tilde{\mu}} B(s) \, ds; \tag{2.24}$$

$$I^-(\tau, \tilde{\mu}) = c_1 e^{-\tau/\tilde{\mu}} + \frac{1}{\tilde{\mu}} \int_\tau^{\tau_1} e^{(\tau-s)/\tilde{\mu}} B(s) \, ds. \tag{2.25}$$

Integrated over all directions these give the two *hemispherical* fluxes q^+ and q^-, defined by

$$q^+(\tau) := 2\pi \int_0^1 \tilde{\mu} I^+(\tau, \tilde{\mu})\, d\tilde{\mu} = 2\pi c_0 E_3(\tau) + 2\pi \int_0^\tau E_2(\tau - s) B(s)\, ds;$$

(2.26)

$$q^-(\tau) := 2\pi \int_0^1 \tilde{\mu} I^-(\tau, \tilde{\mu})\, d\tilde{\mu} = 2\pi c_1 E_3(\tau_1 - \tau) + 2\pi \int_\tau^{\tau_1} E_2(\tau - s) B(s)\, ds,$$

(2.27)

where E_i is the i-th exponential integral. Then, we can construct the radiative heat flux with

$$q^r := 2\pi \int_0^1 \tilde{\mu} \left[I^+(\tau, \tilde{\mu}) - I^-(\tau, \tilde{\mu})\right] d\tilde{\mu} = q^+ - q^-. \qquad (2.28)$$

Combining (2.26–2.28) then gives

$$\begin{aligned} q^r &= 2\pi c_0 E_3(\tau) + 2\pi \int_0^\tau E_2(\tau - s) B(s)\, ds \\ &\quad - 2\pi c_1 E_3(\tau_1 - \tau) - 2\pi \int_\tau^{\tau_1} E_2(s - \tau) B(s)\, ds, \qquad (2.29) \end{aligned}$$

Since in the heat equations we use the gradient of the heat flux, we differentiate (2.29) immediately to find

$$\frac{dq^r}{d\tau}(\tau) = 4\pi B(\tau) - 2\pi c_0 E_2(\tau) - 2\pi c_1 E_2(\tau_1 - \tau) - 2\pi \int_0^{\tau_1} E_1(|\tau - s|) B(s)\, ds.$$

(2.30)

In this equation c_0 and c_1 are still undetermined; they are depending on the radiative properties of the boundaries, which we have silently assumed to be diffusely emitting and reflecting. The case with no reflection — $\rho = 0$ in (2.20) — is the simplest. Because of Kirchhoff's law:

$$\epsilon + \rho = 1, \qquad (2.31)$$

we find that $\epsilon = 1$; in other words we are dealing with perfectly black walls and the emitted intensities at the boundaries are equal to the blackbody intensity. As we will see in the Section 5, this situation applies in the reheating phase when the parison is situated in an infinite atmosphere.

However, for the dwell this is not valid. During the dwell the glass is enclosed by the metal mould on both ends and metal typically has a reflectivity of $\rho \approx 0.9$. Therefore we must have a closer look of (2.20), which rephrased for one dimension and integrated over the right directions reads

$$2\pi c_0 := q^+(0) = 2\pi \epsilon_0 B_0 + \rho_0 q^-(0); \qquad (2.32)$$

$$2\pi c_1 := q^-(\tau_1) = 2\pi \epsilon_1 B_1 + \rho_1 q^+(\tau_1). \qquad (2.33)$$

Here B_0 and B_1 are the blackbody intensities due to the temperature of the respective walls, and not due to the temperature of the glass at those walls. If conduction is neglected, it should be noted that there usually is a temperature jump at the boundary. The hemispherical fluxes at the wall have special names: $q_0^+ := q^+(0)$ and $q_1^- := q^-(\tau_1)$ are called the *radiosities* of the respective boundaries, while $q_0^- := q^-(0)$ and $q^+ := q^+(\tau_1)$ are called the *irradition* at those walls. Applying (2.26,2.27) to (2.32,2.33), and then applying the latter two again, gives

$$c_0 = \frac{1}{1 - \rho_0 \rho_1 E_2^2(\tau_1)} \left(\epsilon_0 B_0 + \rho_0 \epsilon_1 B_1 E_2(\tau_1) + \rho_0 \int_0^{\tau_1} E_1(s) B(s) \, ds \right.$$
$$\left. + \rho_0 \rho_1 E_2(\tau_1) \int_0^{\tau_1} E_1(\tau_1 - s) B(s) \, ds \right) ;$$
(2.34)

$$c_1 = \frac{1}{1 - \rho_0 \rho_1 E_2^2(\tau_1)} \left(\epsilon_1 B_1 + \rho_1 \epsilon_0 B_0 E_2(\tau_1) + \rho_1 \int_0^{\tau_1} E_1(s) B(\tau_1 - s) \, ds \right.$$
$$\left. + \rho_0 \rho_1 E_2(\tau_1) \int_0^{\tau_1} E_1(s) B(s) \, ds \right) .$$
(2.35)

If we know the temperature distribution and the temperature of the walls we can use (2.18) directly to compute c_0 and c_1 from these. Although in most heat problems we do not know the temperature distribution *a priori*, the use of (2.30) together with (2.34,2.35) is next to trivial in explicit or iterative methods.

For higher dimensions a similar but more complex derivation can be done. The method where the Discrete Ordinate Method — described in [4] — and ray tracing are combined to approximate the formal solution. A discussion of this method is beyond the scope of this article. The theory behind it can be found in [7].

3   Pressing phase

As usual in viscous fluid flow the energy equation is ignored because in an incompressible Newtonian fluid with constant viscosity it is not coupled to the equations of motion. In the present case the high viscous forces might generate heat by friction, such that the temperature rises and the viscosity decreases. In order to investigate this possibility consider the energy equation (2.14) for incompressible flow. Using (2.15), (2.16) and (2.17) we can rewrite it as follows:

$$\rho c_p \left(\frac{\partial T}{\partial t} + \mathbf{v} \cdot \nabla T \right) = k^c \nabla^2 T + \nabla \cdot (k^r(T) \nabla T) + \eta \left((\nabla \mathbf{v} + \nabla \mathbf{v}^T) : \nabla \mathbf{v} \right),$$
(3.1)

where c_p is the heat capacity, T is the absolute temperature, η is the viscosity, k^c is the heat conductivity, and $k^r(T)$ is the Rosseland parameter as it was defined before.

Let us introduce a dimensionless temperature variable $\tilde{T}$:

$$T = T_m + \Delta T \tilde{T}, \tag{3.2}$$

where $\Delta T = T_g - T_m$ (T_g, T_m are the temperatures of the glass and the mould, as defined in (2.1)). Using dimensionless variables (2.7), (2.8) and (3.2) the equation above reads as follows

$$\frac{\partial \tilde{T}}{\partial \tilde{t}} + \tilde{\mathbf{v}} \cdot \nabla \tilde{T} = \frac{1}{\text{Pe}} \nabla^2 \tilde{T} + \nabla \cdot \left(\frac{k^r(T)}{k^c} \frac{1}{\text{Pe}} \nabla \tilde{T} \right) + \frac{\text{Ec}}{\text{Re}} \tilde{\eta} \left((\nabla \tilde{\mathbf{v}} + \nabla \tilde{\mathbf{v}}^T) : \nabla \tilde{\mathbf{v}} \right).$$
$$\tag{3.3}$$

Both the dimensionless numbers $1/\text{Pe}$ and Ec/Re, defined by

$$\text{Pe} := \frac{\rho c_p L_0 V_0}{k^c}, \quad \text{Re} := \frac{V_0 L_0 \rho}{\eta_0}, \quad \text{Ec} := \frac{V_0^2}{c_p \Delta T}$$

are of order 10^{-4}, so the energy equation (3.1) simplifies to:

$$\frac{dT}{dt} = \frac{\partial T}{\partial t} + \mathbf{v} \cdot \nabla T = 0, \tag{3.4}$$

so the temperature remains constant. Thus, assuming uniform temperature distribution in the glass gob, we can compute the flow using correspondent constant viscosity η. As a result velocity and pressure fields can be obtained.

4 Dwell

During the dwell — the stage when the glass is kept in the mould after all the air has been forced out — there is no flow. If the radius and the height of the bottle is much bigger than its thickness, as is usually the case, we can locally approximate the behaviour of the temperature as being one dimensional. In this case, the two walls of our approximation in Section 2.2 are the mould on one side and the plunger on the other. It is assumed that the glass makes perfect contact with both the mould and the plunger, so we can assume Dirichlet boundary conditions on either side. The thickness of the glass layer is denoted with L. In this case, (2.14) simplifies to

$$\rho c_p \frac{\partial T}{\partial t} = -\frac{\partial q^c}{\partial x} - \frac{\partial q^r}{\partial x}, \quad t > 0, \quad 0 < x < L; \tag{4.1}$$

to which the following boundary and initial conditions are added:

$$T(t,0) = T_{\text{mould}}, \quad T(t,L) = T_{\text{parison}}, \quad \text{and} \quad T(0,x) = T_0(x). \tag{4.2}$$

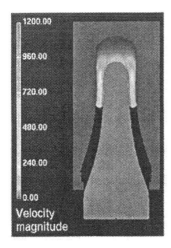 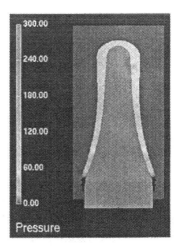

Figure 3.1: Velocity magnitude and pressure field.

Now, we define the *Rosseland number* Ro to be

$$\text{Ro} := \frac{k^c \kappa}{4n^2 \tilde{\sigma} T_w^3},\tag{4.3}$$

where T_w is some characteristic temperature (e.g. the temperature of the walls or the average thereof). If we assume optical thickness, the Rosseland approximation (2.17) holds, the ratio between the radiative and conductive diffusion parameters is

$$\frac{k^r}{k^c} = \frac{\vartheta^3}{3\text{Ro}},$$

in which $\vartheta := T/T_w$ is the dimensionless temperature. Remembering we assumed the material properties to be constant, and defining the *thermal diffusivity* to be $\alpha := k^c/\rho c_p$, (4.1) becomes, after dividing left and righthand side by T_w and k^c:

$$\frac{1}{\alpha}\frac{\partial T}{\partial t} = \frac{\partial}{\partial x}\left[\left(1 + \frac{\vartheta^3}{3\text{Ro}}\right)\frac{\partial T}{\partial x}\right].$$

Introducing the *Fourier number* (dimensionless time) $\varphi := \alpha t/L^2$ and the dimensionless coordinate $\xi := x/L$ this simplifies to, after division by T_w,

$$\frac{\partial \vartheta}{\partial \varphi} = \frac{\partial}{\partial \xi}\left[\left(1 + \frac{\vartheta^3}{3\text{Ro}}\right)\frac{\partial \vartheta}{\partial \xi}\right].\tag{4.4}$$

Expressed in the optical coordinate $\tau := \kappa L$ and the *optical Fourier number* $\varphi_r := \alpha \kappa^2 t$, this can be written as:

$$\frac{\partial \vartheta}{\partial \varphi_r} = \frac{\partial}{\partial \tau}\left[\left(1 + \frac{\vartheta^3}{3\text{Ro}}\right)\frac{\partial \vartheta}{\partial \tau}\right].\tag{4.5}$$

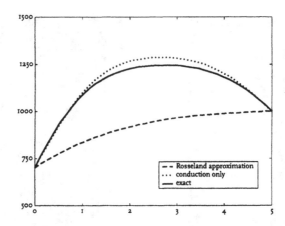

Figure 4.1: Temperature profile after one second dwell (various methods).

Finally, if we do not use the Rosseland approximation, but some exact method to obtain the radiative flux, we can perform the same coordinate transformations. If we furthermore define the *dimensionless radiative heat flux* to be $Q^r := q^r/4n^2\tilde{\sigma}T^4$, we find that

$$\frac{\partial \vartheta}{\partial \varphi_r} = \frac{\partial^2 \vartheta}{\partial \tau^2} - \frac{1}{\text{Ro}} \frac{\partial Q^r}{\partial \tau}. \tag{4.6}$$

Here we can use the method outlined in Section 2.2 to obtain values for the heat flux and its divergence.

In Figure 4.1, we see the results of the one-dimensional problem for a computation simulating a dwell time of one second. The most eye-catching is the erroneous result, the Rosseland approximation gives us in this case. We should have been warned by the small optical thickness (of order one in the production of jars and bottles). This result is worrying as in industry it is widely used 'just to take care of radiation'. The results here show that far better results are achieved by simply neglecting the radiation. Still, one has to take care, since after longer periods the exact method and the radtion neglecting method will deviate severely, too.

The reason the Rosseland approximation fails here, is that it overestimates the radiative energy transport. The glass is not thick enough to achieve diffusion-like behaviour. The Dirichlet boundary-conditions applied to both boundaries enforce a large gradient of the thin glass sample. It is this gradient that makes the problem conduction driven, aggravating the results of the Rosseland approximation, which is basically enlarging the effective conductivity.

We see from these calculations that for short times (i.e. very small Fourier numbers), the conduction-driven problem can be approached by omission of the radiation. This apparently is still valid for the cases in which the optical thickness τ_1 and Rosseland number do not directly indicate the radiation is not crucially unimportant. Yet, be aware that still significant errors are made by such a simple approximation (maximum 50°C in this example). So, depending on the importance of the temperature and its gradient one can choose between accuracy and speed. In this case, for example, a quite big difference of 50 degrees during the dwell does not give rise to an erroneous prediction of the remainder of the process. Simply omitting the radiation, would therefore be most likely candidate for simulations in more dimensions, where the implementation of the exact method brings severe performance penalties.

5 Reheating phase

The heat problem in the reheating phase is basically the same as during the dwell: only the boundary conditions differ. During reheating the parison is standing outside the mould and without plunger in an open atmosphere. This open atmosphere from an radiative point of view lets itself be accurately modelled as a blackbody, with the ambient temperature as driving force. The contact with the surrounding atmosphere is, unlike during the dwell, not a perfect contact. Rather we have to apply a Robin boundary conditions describing the exchange of heat with this surrounding atmosphere. Since, the *ambient temperature* T_∞ — say standard room temperature (20°C) — is much cooler than the parison, we expect it to act as a heat sink. The infinite surroundings of the parison can be seen as a blackbody boundary as shown in Figure 5.1. On the outside we have both convective and radiative cooling of the parison. The convective heat q_0^{conv} entering the parison on the outside is due to free convection of the surrounding air. It can be calculated by:

$$q_0^{\mathrm{conv}}(t) = h(T_\infty - T(t,0)), \tag{5.1}$$

where h is the *convective heat transfer coefficient*. Methods to compute this can be found in books like [6]. Typically its value is $1\text{--}2\,\mathrm{W/m}^2$.

How the radiative properties are taken into account, depends on the model of radiation that is chosen. For the exact method we calculate the intensity of the atmosphere with (2.18), then (2.34) leads to $c_0 = n^2 \bar{\sigma} T_\infty^4 / \pi$. This coefficient then is used in (2.30), which we use to calculate the radiative heat flux gradient throughout the glass. If the Rosseland method is used, however, the radiative exchange of the glass with its surroundings can only be applied at the boundary. We find a similar expression to (5.1) that states:

$$q_0^{\mathrm{rad}}(t) = \varepsilon \bar{\sigma}(T_\infty^4 - T(t,0)^4). \tag{5.2}$$

The emissivity ε, being equal to the absorpsivity, can be determined as follows. Given the intensity B_∞ entering the glass at certain direction $\tilde{\mu}$, it will travel through the glass, then through the inside and then through the glass

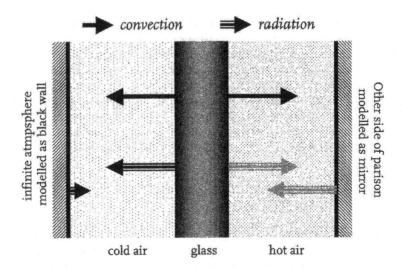

Figure 5.1: The model for the reheating phase. The inner side of the parison 'sees' the other side, and thus, the irradiance at this boundary is equal to its radiosity.

again. Because the hot air does not absorb, we can see from (2.24) that an intensity of $B_\infty e^{-2\tau_1/\tilde\mu}$ caused by the entering radiation, is leaving the glass again. We can now calculate ε using

$$\varepsilon = \frac{2\pi B_\infty - 2\pi \int_0^1 \tilde\mu B_\infty e^{-2\tau_1/\tilde\mu}\, d\mu}{2\pi B_\infty} = 1 - 2E_2(2\tau_1).$$

On the inner side of the parison, the convective heat exchange can be modelled as before:

$$q_1^{\mathrm{conv}}(t) = h(T(t,L) - T_{\mathrm{hot}}), \tag{5.3}$$

where T_{hot} is the temperature of the hot air inside the parison. As for the radiative exchange, we do not 'see' the surrounding atmosphere (directly), but the other inner side of the parison instead. The radiative boundary condition for this side of the glass can be modelled as specularly reflective. Depending on the radiation model this can be simplified further. If we are using a diffusion approximation all the radiative heat has to leave at boundary. However, since there is a specular wall between the glass and the infinite surroundings of the parison, and since air in absence of high vapour and CO_2 concentrations does not absorb radiative heat, the radiative heat exchange at the surface equals zero.

If we use the exact radiation model, treating specular reflections is even more involved than the diffuse reflection we have seen in Section 2. As shown

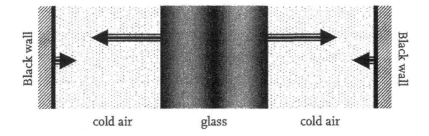

cold air   glass   cold air

Figure 5.2: Additional model for the reheating phase. The specularly reflective boundary is tackled by reflecting the domain.

in Figure 5.2, in one dimension we can evade this, by simply doing what the reflective boundary does: make a reflection. So, rather than using (5.1) over the physical interval $(0, \tau_1)$, we use this equation over the interval $(0, 2\tau_1)$, where we extend the blackbody intensity by using:

$$B(\tau) = B(2\tau_1 - \tau), \quad \forall \ \tau \in (\tau_1, 2\tau 1). \tag{5.4}$$

The problem is written now almost identical to (4.1); only the boundary conditions are different, as can be seen in

$$\rho c_p \frac{\partial T}{\partial t} = -\frac{\partial q^c}{\partial x} - \frac{\partial q^r}{\partial x}, \quad t > 0, \quad 0 < x < L \tag{5.5}$$

to which the following initial condition is added:

$$T(0, x) = T_0(x). \tag{5.6}$$

The boundary conditions are different for different models. For diffusion models (like the Rosseland model or simple neglect of radiation), we have to include the term for radiative heat loss *at* the boundary. Then, the boundary equation becomes the non-linear Robin-condition

$$\frac{\partial T}{\partial x} = \frac{h}{k^{\text{eff}}} \left(T_\infty - T(t, 0)\right) + \frac{\varepsilon \bar{\sigma}}{k^{\text{eff}}} \left(T_\infty^4 - T(t, 0)^4\right), \tag{5.7}$$

where k^{eff} is the *effective conductivity*. In the Rosseland approximation this is equal to $(k^c + k^r(T))$, whereas it is simply k^c if we neglect radiation.

If we use the exact method for the radiation, any loss of heat through radiation to its surroundings is already taken care of by the q^r term. The boundary conditions therefore only have to prescribe the convection at the boundaries:

$$\frac{\partial T}{\partial x} = \frac{h}{k^{\text{eff}}} \left(T_\infty - T(t, 0)\right) \tag{5.8}$$

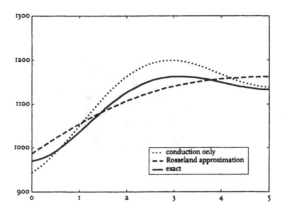

Figure 5.3: Temperature profile after one second dwell (exact method) and one second reheating (various methods).

As we can see in Figure 5.3, more so than during the dwell, the three methods give very different results. Unlike during the dwell, however, the Rosseland approximation now gives a good estimate of the energy being extracted during the reheat. In this case, omission of the radiative heat transfer also leads to large errors, especially concerning the cooling down of the glass. Both simplifications under-perform in approximating the temperature gradients, which are so important during the reheating.

The conclusion is clear. During reheating, in a case where the neither the conduction (as in the optical thing case) nor the radiation (as in the optical thick case) is predominant, only an exact approach gives trustworthy results. If it has been identified either the temperature itself or its gradient is critical to the functioning of the process, effort has to be made to get the radiative heat transfer right. Two numbers, the optical thickness τ_1 and the Rosseland number Ro, can assist in determining whether this effort has to be made. From comparison of the results of the dwell and the reheat, however, we see that these two numbers by themselves are not conclusive. The Dirichlet-conditions applied in the dwell, and thus applying a large temperature gradient over a small distance, forced the conduction to be dominant. Natural boundary conditions as during the reheat, however, give value to the two afore mentioned numbers.

The results presented were derived for the simple one-dimensional case. In higher dimensions the simplicity of the Rosseland equation becomes even more tempting as the computational complexity of the exact solution raises. For the best approximations of three-dimensional problems methods as the Monte Carlo Method and the Ray Trace Method are utilised. However, the large computational times associated with these methods are usually considered to be prohibitive for implementation in simulations of processes in the production of glass. Newer methods include the algebraic for of the Ray

Trace Method, presented in [7], and the *Modified Diffusion Approximation* as discussed in [2]. Given the results presented it here it seems wise to implement a method other than the (standard) Rosseland approximation for thin to medium optical thicknesses, or to accept that the temperature and temperature gradients of the simulation are plainly inaccurate.

References

[1] S. Chandrasekhar. *Radiative Transfer.* Dover Publications, Inc., Toronto, 1960.

[2] F.T. Lentes and N. Siedow. Three-dimensional radiative heat transfer in glass cooling processes. Technical Report Berichte Nr. 4, ITWM, Kaiserslautern, 1998.

[3] D. Mihalas and B. Weibel-Mihalas. *Foundations of Radiation Hydrodynamics.* Oxford University Press, Oxford, 1984.

[4] M.F. Modest. *Radiative Heat Transfer.* McGraw-Hill, Inc., Singapore, 1993.

[5] M.N. Özişik. *Radiative Transfer and Interactions with Conduction and Convection.* John Wiley and Sons, Inc., Toronto, 1973.

[6] B.D. Tapley and Th.R. Poston. *Eschbach's Handbook Of Engineering Fundamentals, Fourth Edition.* John Wiley and Sons, Inc., Toronto, 1989.

[7] B.J. van der Linden and R.M.M. Mattheij. A new method for solving radiative heat problems in glass. *International Journal of Forming Processes,* 2(2–3):41–61, 1999.

Drag Reduction by Active Control
for Flow past Cylinders

J.-W. He[*], M. Chevalier[*], R. Glowinski[*], R. Metcalfe[†],
A. Nordlander[†], and J. Periaux[‡]

May 4, 2000

[*]Department of Mathematics, University of Houston,
Houston, TX 77204-3476;
[†]Department of Mechanical Engineering, University of Houston,
Houston, TX 77204-4792;
[‡]Dassault Aviation, 78 quai Marcel Dassault,
92214 Saint Cloud, France

Abstract

The main objective of this article is to investigate computational methods
for the active control and drag optimization of incompressible viscous flow
past cylinders, using the two-dimensional Navier-Stokes equations as the
flow model. The computational methodology relies on the following ingredients: space discretization of the Navier-Stokes equations by finite element
approximations, time discretization by a second order accurate two step im-
plicit/explicit finite difference scheme, calculation of the cost function gradient by the adjoint equation approach, minimization of the cost function by a
quasi-Newton method à la BFGS. The above methods have been applied to
predict the optimal forcing-control strategies in reducing drag for flow around
a circular cylinder using either an oscillatory rotation or blowing and suction.
In the case of oscillatory forcing, a drag reduction of 31% at Reynolds number 200 and 61% at Reynolds number 1000 was demonstrated. Using only
three blowing-suction slots, we have been able to completely suppress the
formation of the Von-Karman vortex street up to Reynolds number 200 with
a significant *net* drag reduction. We conclude this article by an appendix
describing a bisection method which allows very substantial storage memory savings at reasonable extra computational time when applying adjoint
equation based methodologies to the solution of control problems modeled
by time dependent equations.

Contents

1   Introduction

Engineers have not waited for mathematicians to successfully address flow control problems (see, e.g., refs [Hak89] and [BH90] for a review on flow control from the engineering point of view); indeed, Prandtl as early as 1915 was concerned with flow control and was designing ingenious systems to suppress or delay boundary layer separation [Pra25]. The last decade has seen an explosive growth of investigations and publications of a mathematical nature concerning various aspects of the control of viscous flow, a good example being provided by [Gun95] and [Sri98]. Actually these two volumes also contain some articles related to the computational aspect of the optimal control of viscous flow, but usually, the geometries are fairly simple and the Reynolds numbers fairly low. It is interesting to observe that the *Journal of Computational Physics* has published recently articles on the above topics (see refs. [HR96], [GB97]), [IR98]); however in those articles, once again, the geometry is simple and/or the Reynolds number is low.

The main goal of this article is to investigate computational methods for the active control and drag optimization of incompressible viscous flow past cylinders, using the two-dimensional Navier-Stokes equations as the flow model. The computational methodology relies on the following ingredients: space discretization of the Navier-Stokes equations by finite element approximations, time discretization by a second order accurate two step implicit/explicit finite difference scheme, calculation of the cost function gradient by the adjoint equation approach, and minimization of the cost function by a quasi-Newton method à la BFGS. Motivated in part by the experimental work of Tokumaru and Dimotakis, [TD91], the above methods have been applied to the boundary control by rotation of the flow around a circular cylinder and show 30% to 60% drag reduction, compared to the fixed cylinder configuration, for Reynolds numbers in the range of 200 to 1000. More recently, we have been able to predict the drag reduction optimal forcing-control strategies for flow around a circular cylinder using blowing and suction. Using only three blowing-suction slots, we have been able to completely suppress the formation of the Von-Karman vortex street up to Reynolds number 200 with a further *net* drag reduction compared to control by rotation.

From a methodological point of view, some of the methods used here are clearly related to those employed by our former collaborator M. Berggren in [Ber98] for the boundary control by blowing and suction of incompressible viscous flow in bounded cavities.

The organization of the remainder of the paper is as follows: In Section 2 we formulate the flow control problem, and address its time discretization in Section 3. The important problem of the space discretization by finite element method is discussed in Section 4; a special attention is given there to velocity spaces which are discretely divergence-free in order to reduce the number of algebraic constraints in the control problem. The full discretization of the control problem is addressed in Section 5. Since we intend to use solution methods based on a quasi-Newton algorithm à la BFGS (see Section 6) attention is focused in Section 5 on the derivation of the gradient of the fully discrete cost function via the classical adjoint equation method. The flow simulator (actually a Navier-Stokes equations solver) is further discussed in Section 7 where it is validated on well documented flow around cylinder test problems for various values of the Reynolds number. Finally the results of various numerical experiments for flow control past a cylinder are discussed in Sections 8 and 9; they definitely show that a substantial drag reduction can be obtained using either an oscillatory rotation or blowing and suction.

A superficial inspection may give the feeling that *adjoint equation* based methodologies for the solution of *control* (or *inverse*) *problems* modeled by time dependent *nonlinear partial differential equations* are extremely *storage memory* demanding. In the appendix concluding this article we will show that indeed this is not true and that substantial memory savings can be achieved through a *bisection method* generalizing a related method described in, e.g., [HG98]; we will also show that the additional computational time associated to the method is very reasonable.

2 Formulation of the Flow Control Problem

2.1 Fluid flow formulation

Let Ω be a region of $\mathbb{R}^d$ ($d = 2, 3$ in practice); we denote by Γ the boundary $\partial\Omega$ of Ω. We suppose that Ω is filled with a Newtonian incompressible viscous fluid of *density* ρ and *viscosity* μ; we suppose that the temperature is constant. Under these circumstances the flow of such a fluid is modeled by the following system of Navier Stokes equations:

$$\rho[\partial_t \mathbf{y} + (\mathbf{y} \cdot \nabla)\mathbf{y}] = \nabla \cdot \sigma + \rho \mathbf{f} \quad \text{in } \Omega \times (0, T), \tag{1}$$
$$\nabla \cdot \mathbf{y} = 0 \quad \text{in } \Omega \times (0, T) \quad \text{(incompressibility condition)} . \tag{2}$$

In (1), (2), $\mathbf{y} = \{y_i\}_{i=1}^d$ denotes the *velocity* field, π the *pressure*, $\mathbf{f}$ a density of external forces per mass unit, and σ ($= \sigma(\mathbf{y}, \pi)$) the stress tensor defined by

$$\sigma = 2\mu \mathbf{D}(\mathbf{y}) - \pi \mathbf{I}$$

with the *rate of deformation tensor* $\mathbf{D}(\mathbf{y})$ defined by

$$\mathbf{D}(\mathbf{y}) = \frac{1}{2}(\nabla \mathbf{y} + \nabla \mathbf{y}^t).$$

We also have

$$\partial_t = \frac{\partial}{\partial t}, \quad \nabla^2 = \sum_{i=1}^d \frac{\partial^2}{\partial x_i^2}, \quad \nabla \cdot \mathbf{y} = \sum_{i=1}^d \frac{\partial y_i}{\partial x_i}, \quad (\mathbf{y} \cdot \nabla)\mathbf{z} = \left\{ \sum_{j=1}^d y_j \frac{\partial z_i}{\partial x_j} \right\}_{i=1}^d .$$

In the above equations $(0, T)$ is the *time interval* during which the flow is considered.

Equations (1), (2) have to be completed by further conditions, such as the following *initial condition*

$$\mathbf{y}(0) = \mathbf{y}_0 \quad (\text{with } \nabla \cdot \mathbf{y}_0 = 0) \tag{3}$$

and boundary conditions. Let us consider the typical situation, of interest to us, described in Figure 1, corresponding to an external flow around a cylinder of cross-section B; we assume that the classical two-dimensional reduction holds.

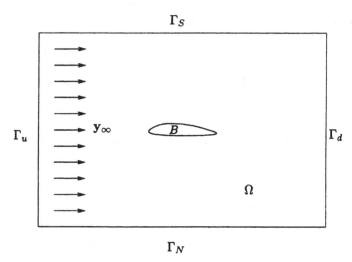

Figure 1: External flow around a cylinder of cross-section B.

In fact $\Gamma_u \cup \Gamma_d \cup \Gamma_N \cup \Gamma_S$ is an *artificial* boundary which has to be taken sufficiently far from B, so that the corresponding flow is a good approximation of the unbounded external flow around B. Typical boundary conditions are

$$\mathbf{y} = \mathbf{y}_\infty \quad \text{on } (\Gamma_u \cup \Gamma_N \cup \Gamma_S) \times (0, T), \tag{4}$$

$$\sigma \mathbf{n} = 0 \quad \cdot \text{on } \Gamma_d \times (0, T) \quad (\text{downstream boundary condition}), \tag{5}$$

with $\mathbf{n}$ the *unit vector* of the outward normal on Γ. We are voluntarily vague, concerning the boundary conditions on ∂B, since they will be part of the control process. Let us conclude this paragraph by recalling that the *Reynolds number* Re is classically defined by

$$Re = \rho U L / \mu, \tag{6}$$

with U a *characteristic velocity* ($|\mathbf{y}_\infty|$, here) and L a *characteristic* length (the thickness of B, for example).

Our goal in this article is to prescribe on ∂B boundary conditions of the *Dirichlet type* (i.e., velocity imposed on ∂B) so that some flow related performance criterion (the cost function) will be minimized under reasonable constraints on the control variables.

2.2 Formulation of the control problem

The flow control problem to be discussed in this article consists of minimizing a *drag* related cost function via controls acting on ∂B; this problem can be formulated as follows (using classical control formalism):

$$\begin{cases} \mathbf{u} \in \mathcal{U}, \\ J(\mathbf{u}) \leq J(\mathbf{v}), \quad \forall \mathbf{v} \in \mathcal{U}, \end{cases} \tag{7}$$

where, in (7), the *control space* $\mathcal{U}$ is a vector space of vector valued functions $\mathbf{v}$ defined on $\partial B \times (0, T)$ and satisfying

$$\int_{\partial B} \mathbf{v}(t) \cdot \mathbf{n} \, ds = 0 \quad \text{for } t \in (0, T), \tag{8}$$

and where the *cost function* J is defined by

$$J(\mathbf{v}) = \frac{\epsilon}{2} \left(\int_0^T \|\mathbf{v}(t)\|_\alpha^2 \, dt + \int_0^T \int_{\partial B} |\partial_t \mathbf{v}(x, t)|^2 \, ds \, dt \right) + \int_0^T P_d(t) \, dt, \tag{9}$$

with $\epsilon \, (\geq 0)$ a *regularization* parameter.

In (8), (9), we have used the following notation

- $\varphi(t)$ for the function $x \to \varphi(x, t)$.

- ds for the superficial measure on ∂B.

- $\|\mathbf{v}\|_\alpha$ for a norm of $\mathbf{v}$ defined on ∂B, involving space derivatives of order α, with α possibly non integer (the readers afraid of these mathematical complications do not have to worry, since in this article we shall consider boundary controls functions of t only).

- $P_d(t)$ is the power needed to overcome, at time t, the drag exerted on B in the direction opposite to that of the motion; $P_d(t)$ is defined by

$$P_d(t) = \int_{\partial B} \sigma \, \mathbf{n} \cdot (\mathbf{v} - \mathbf{y}_\infty) \, ds. \tag{10}$$

- Finally, $\mathbf{y}$ is the solution to the following Navier-Stokes system

$$\rho \left[\partial_t \mathbf{y} + (\mathbf{y} \cdot \nabla) \mathbf{y} \right] = \nabla \cdot \sigma + \rho \mathbf{f} \quad \text{in } \Omega \times (0, T), \tag{11}$$

$$\nabla \cdot \mathbf{y} = 0 \quad \text{in } \Omega \times (0, T), \tag{12}$$

$$\mathbf{y}(0) = \mathbf{y}_0 \quad (\text{with } \nabla \cdot \mathbf{y}_0 = 0), \tag{13}$$

$$\sigma \, \mathbf{n} = 0 \quad \text{on } \Gamma_d \times (0, T), \tag{14}$$

$$\mathbf{y} = \mathbf{y}_\infty \quad \text{on } (\Gamma_u \cup \Gamma_N \cup \Gamma_S) \times (0, T), \tag{15}$$

$$\mathbf{y} = \mathbf{v} \quad \text{on } \partial B \times (0, T). \tag{16}$$

Remark 2.1   The flux condition (8) is not essential and can be easily relaxed if, for example, the downstream boundary conditions are of the Neumann type (like those in (14)).

Remark 2.2   The Navier-Stokes equation (11) can also be written:

$$\rho \left[\partial_t \mathbf{y} + (\mathbf{y} \cdot \nabla) \mathbf{y} \right] - \mu \nabla^2 \mathbf{y} + \nabla \pi = \rho \mathbf{f} \quad \text{in } \Omega \times (0, T),$$

however form (11) is better suited to the drag reduction problem (7), since, as (10), it involves the stress tensor σ.

Remark 2.3   In the particular case of incompressible viscous flow we have

$$\int_0^T P_d(t) \, dt = \int_0^T E_d(t) \, dt + (K(T) - K(0))$$

$$+ \int_0^T P_c(t) \, dt + \int_0^T P_\infty(t) \, dt - \int_0^T P_f(t) \, dt \qquad (17)$$

where, in (17),

- $E_d(t) = 2\mu \int_\Omega |\mathbf{D}(\mathbf{y} - \mathbf{y}_\infty)|^2 \, dx$ is the *viscous dissipation energy*,

- $K(t) = \frac{\rho}{2} \int_\Omega |\mathbf{y}(t) - \mathbf{y}_\infty|^2 \, dx$ is a *kinetic energy*,

- $P_c(t) = \frac{\rho}{2} \int_{\partial B} |\mathbf{v} - \mathbf{y}_\infty|^2 \mathbf{v} \cdot \mathbf{n} \, ds$ is a *control related power*,

- $P_\infty(t) = \frac{\rho}{2} \int_{\Gamma_\infty^d} |\mathbf{y} - \mathbf{y}_\infty|^2 \mathbf{y} \cdot \mathbf{n} \, ds$ is a *downstream boundary related power*,

- $P_f(t) = \rho \int_\Omega \mathbf{f}(t) \cdot (\mathbf{y}(t) - \mathbf{y}_\infty) \, dx$ is the *external forcing power*.

Some observations are in order, such as:

1. If Γ_d "goes to infinity" in the Ox_1 direction, then $\mathbf{y} \to \mathbf{y}_\infty$ which implies in turn that $P_\infty \to 0$.

2. Whenever the control is absent, i.e., $\mathbf{v} = \mathbf{0}$, we have $P_c(t) = 0$.

We can summarize relation (17) by noting that: " the drag work is equal to the energy dissipated by viscosity + the kinetic energy variation between 0 and T + the control associated work + the downstream boundary associated work − the external forcing work." The above observation has the following consequence: instead of minimizing $J(\cdot)$ defined by (9) we can minimize the cost function $J(\cdot)$ defined by

$$J(\mathbf{v}) = \frac{\epsilon}{2} \left(\int_0^T \|\mathbf{v}(t)\|_\alpha^2 \, dt + \int_0^T \int_{\partial B} |\partial_t \mathbf{v}(x, t)|^2 \, ds \, dt \right) + \int_0^T E_d(t) \, dt$$

$$+ (K(T) - K(0)) + \int_0^T P_c(t) \, dt + \int_0^T P_\infty(t) \, dt - \int_0^T P_f(t) \, dt;$$

this function is simpler than the one defined by (9) since it does not involve boundary integrals of stress-tensor related quantities. However we kept working with the cost function defined by (9) since it seems to lead to more accurate results.

In order to apply *quasi-Newton type methods* à la BFGS to the solution of the control problem (7) it is instrumental to know how to compute the gradient of the cost function $J_h^{\Delta t}(\cdot)$, obtained from the full space-time discretization of the control problem (7), since we shall solve the discrete variant of problem (7) via the (necessary) *optimality condition*

$$\nabla J_h^{\Delta t}(\mathbf{u}_h^{\Delta t}) = \mathbf{0},$$

where $\mathbf{u}_h^{\Delta t}$ is a solution of the fully discrete control problem. The calculation of $\nabla J_h^{\Delta t}$ will be discussed in Section 5.

3 Time discretization of the control problem

3.1 Generalities

In order to facilitate the task of the reader unfamiliar with control methodology (not to say philosophy) we are going to discuss first the *time discretization* issue. The space and consequently full space/time discretization issues will be addressed in Section 4. This approach of fractioning the computational difficulties has the definite advantage that some practitioners will be able to use the material in this article for other types of space approximations than the finite element ones discussed in Section 4 (one may prefer spectral methods, for example).

3.2 Formulation of the time discrete control problem

We define first a time discretization step Δt by $\Delta t = T/N$, with N a (large) positive integer. Concentrating on problem (7), we approximate it by

$$\begin{cases} \mathbf{u}^{\Delta t} \in \mathcal{U}^{\Delta t}, \\ J^{\Delta t}(\mathbf{u}^{\Delta t}) \leq J^{\Delta t}(\mathbf{v}), \quad \forall \mathbf{v} \in \mathcal{U}^{\Delta t}, \end{cases} \tag{18}$$

with

$$\mathcal{U}^{\Delta t} = \Lambda^N, \tag{19}$$

$$\Lambda = \{ \lambda \mid \lambda \in (H^\alpha(\partial\Omega))^d, \int_{\partial B} \lambda \cdot \mathbf{n}\, ds = 0 \}, \tag{20}$$

and

$$J^{\Delta t}(\mathbf{v}) = \frac{\epsilon}{2} \Delta t \sum_{n=1}^{N} \left(\|\mathbf{v}^n\|_\alpha^2 + \left\| \frac{\mathbf{v}^n - \mathbf{v}^{n-1}}{\Delta t} \right\|_0^2 \right) + \Delta t \sum_{n=1}^{N} P_d^n, \tag{21}$$

where, in (21), P_d^n is the discrete drag power defined (with obvious notation) by

$$P_d^n = \int_{\partial B} \sigma(\mathbf{y}^n, \pi^n)\,\mathbf{n} \cdot (\mathbf{v}^n - \mathbf{y}_\infty)\,ds$$

with $\{(\mathbf{y}^n, \pi^n)\}_{n=1}^N$ obtained from $\mathbf{v}$ via the solution of the following semi-discrete Navier-Stokes equations

$$\mathbf{y}^0 = \mathbf{y}_0, \tag{22}$$

then

$$\rho\left(\frac{\mathbf{y}^1 - \mathbf{y}^0}{\Delta t} + (\mathbf{y}^0 \cdot \nabla)\mathbf{y}^0\right) = \nabla \cdot \sigma(\tfrac{2}{3}\mathbf{y}^1 + \tfrac{1}{3}\mathbf{y}^0, \pi^1) + \rho\mathbf{f}^1 \quad \text{in } \Omega, \tag{23}$$

$$\nabla \cdot \mathbf{y}^1 = 0 \quad \text{in } \Omega, \tag{24}$$

$$\sigma(\tfrac{2}{3}\mathbf{y}^1 + \tfrac{1}{3}\mathbf{y}^0, \pi^1)\,\mathbf{n} = 0 \quad \text{on } \Gamma_d, \tag{25}$$

$$\mathbf{y}^1 = \mathbf{y}_\infty \quad \text{on } \Gamma_u \cup \Gamma_N \cup \Gamma_S, \tag{26}$$

$$\mathbf{y}^1 = \mathbf{v}^1 \quad \text{on } \partial B, \tag{27}$$

and for $n = 2, \dots N$

$$\rho\left(\frac{\tfrac{3}{2}\mathbf{y}^n - 2\mathbf{y}^{n-1} + \tfrac{1}{2}\mathbf{y}^{n-2}}{\Delta t} + ((2\mathbf{y}^{n-1} - \mathbf{y}^{n-2}) \cdot \nabla)(2\mathbf{y}^{n-1} - \mathbf{y}^{n-2})\right)$$

$$= \nabla \cdot \sigma(\mathbf{y}^n, \pi^n) + \rho\mathbf{f}^n \quad \text{in } \Omega, \tag{28}$$

$$\nabla \cdot \mathbf{y}^n = 0 \quad \text{in } \Omega, \tag{29}$$

$$\sigma(\mathbf{y}^n, \pi^n)\,\mathbf{n} = 0 \quad \text{on } \Gamma_d, \tag{30}$$

$$\mathbf{y}^n = \mathbf{y}_\infty \quad \text{on } \Gamma_u \cup \Gamma_N \cup \Gamma_S, \tag{31}$$

$$\mathbf{y}^n = \mathbf{v}^n \quad \text{on } \partial B. \tag{32}$$

The above scheme is a semi-implicit, second order accurate two-step scheme.

Anticipating the finite element approximation to take place in the following section, we can rewrite (23)-(32) in *variational form*. We obtain thus

$$\rho\int_\Omega \frac{\mathbf{y}^1 - \mathbf{y}^0}{\Delta t} \cdot \mathbf{z}\,dx + 2\mu\int_\Omega \mathbf{D}(\tfrac{2}{3}\mathbf{y}^1 + \tfrac{1}{3}\mathbf{y}^0) : \mathbf{D}(\mathbf{z})\,dx$$

$$+ \rho\int_\Omega (\mathbf{y}^0 \cdot \nabla)\mathbf{y}^0 \cdot \mathbf{z}\,dx$$

$$- \int_\Omega \pi^1 \nabla \cdot \mathbf{z}\,dx = \rho\int_\Omega \mathbf{f}^1 \cdot \mathbf{z}\,dx, \quad \forall \mathbf{z} \in \mathbf{V}_0, \tag{33}$$

$$\int_\Omega \nabla \cdot \mathbf{y}^1 q\,dx = 0, \quad \forall q \in L^2(\Omega), \tag{34}$$

$$\mathbf{y}^1 = \mathbf{y}_\infty \quad \text{on } \Gamma_u \cup \Gamma_N \cup \Gamma_S, \tag{35}$$

$$\mathbf{y}^1 = \mathbf{v}^1 \quad \text{on } \partial B; \tag{36}$$

and for $n = 2, \ldots N$

$$\rho \int_\Omega \frac{\frac{3}{2}\mathbf{y}^n - 2\mathbf{y}^{n-1} + \frac{1}{2}\mathbf{y}^{n-2}}{\Delta t} \cdot \mathbf{z} \, dx + 2\mu \int_\Omega \mathbf{D}(\mathbf{y}^n) : \mathbf{D}(\mathbf{z}) \, dx$$

$$+ \rho \int_\Omega ((2\mathbf{y}^{n-1} - \mathbf{y}^{n-2}) \cdot \nabla)(2\mathbf{y}^{n-1} - \mathbf{y}^{n-2}) \cdot \mathbf{z} \, dx$$

$$- \int_\Omega \pi^n \nabla \cdot \mathbf{z} \, dx = \rho \int_\Omega \mathbf{f}^n \cdot \mathbf{z} \, dx, \quad \forall \mathbf{z} \in \mathbf{V}_0, \tag{37}$$

$$\int_\Omega \nabla \cdot \mathbf{y}^n q \, dx = 0, \quad \forall q \in L^2(\Omega), \tag{38}$$

$$\mathbf{y}^n = \mathbf{y}_\infty \quad \text{on } \Gamma_u \cup \Gamma_N \cup \Gamma_S, \tag{39}$$

$$\mathbf{y}^n = \mathbf{v}^n \quad \text{on } \partial B. \tag{40}$$

In (33), (37), we have used the notation

$$\mathbf{T} : \mathbf{S} = \sum_{i=1}^d \sum_{j=1}^d t_{ij} s_{ij}$$

to denote the scalar-product in $\mathbb{R}^{d^2}$ of the two tensors $\mathbf{T} = \{t_{ij}\}_{i,j=1}^d$ and $\mathbf{S} = \{s_{ij}\}_{i,j=1}^d$, and the space $\mathbf{V}_0$ is defined by

$$\mathbf{V}_0 = \{ \mathbf{z} \,|\, \mathbf{z} \in (H^1(\Omega))^d, \, \mathbf{z} = \mathbf{0} \text{ on } \Gamma_u \cup \Gamma_N \cup \Gamma_S \cup \partial B \}. \tag{41}$$

3.3 Comments on the time discretization of the control problem (7)

Since the time discretization step used in Section 3.2 is a two step one, a *starting procedure* is required; the one we have used, namely, (23)-(27), leads to a generalized Stokes problem to obtain $\{\mathbf{y}^1, \pi^1\}$ which has the same coefficients as the ones used to obtain $\{\mathbf{y}^n, \pi^n\}$ from $\mathbf{v}^n$ and $\mathbf{y}^{n-1}, \mathbf{y}^{n-2}$. As we shall see later on in this article, scheme (22)-(32) albeit partly explicit has shown very good robustness properties when applied to the solution of drag reduction problems.

4 Full discretization of the control problem

4.1 Synopsis

In order to *spatially discretize* the control problem (7), we are going to use a *finite element* approximation, since this kind of approximation is well-suited to handle complicated boundaries and boundary conditions. The discretization to be used combines a continuous piecewise Q_2-approximation for the velocity and a discontinuous P_1-approximation for the pressure. This approximation satisfies a *discrete inf-sup condition*, implying that the discrete problems are well-posed and the approximation is convergent (see, e.g., [GR86], [Pir89], [Gun89], [BF91] for the finite element approximation of the Navier-Stokes equations).

4.2   Discrete flow model

In order to fully discretize the semi-discrete model (22)-(32) we are going to mimic its equivalent formulation. Doing so we obtain

$$\mathbf{y}_h^0 = \mathbf{y}_{0h} \quad \text{(a convenient approximation of } \mathbf{y}_0\text{);}} \tag{42}$$

and

$$\rho \int_{\Omega_h} \frac{\mathbf{y}_h^1 - \mathbf{y}_h^0}{\Delta t} \cdot \mathbf{z} \, dx + 2\mu \int_{\Omega_h} \mathbf{D}(\frac{2}{3}\mathbf{y}_h^1 + \frac{1}{3}\mathbf{y}_h^0) : \mathbf{D}(\mathbf{z}) \, dx$$

$$+ \rho \int_{\Omega_h} (\mathbf{y}_h^0 \cdot \nabla)\mathbf{y}_h^0 \cdot \mathbf{z} \, dx$$

$$- \int_{\Omega_h} \pi_h^1 \nabla \cdot \mathbf{z} \, dx = \rho \int_{\Omega_h} \mathbf{f}^1 \cdot \mathbf{z} \, dx, \quad \forall \mathbf{z} \in \mathbf{V}_{0h}, \tag{43}$$

$$\int_{\Omega_h} \nabla \cdot \mathbf{y}_h^1 q \, dx = 0, \quad \forall q \in \mathbf{P}_h, \tag{44}$$

$$\mathbf{y}_h^1 = \mathbf{y}_\infty \quad \text{on } \Gamma_{u,h} \cup \Gamma_{N,h} \cup \Gamma_{S,h}, \tag{45}$$

$$\mathbf{y}_h^1 = \mathbf{v}^1 \quad \text{on } \partial B_h; \tag{46}$$

then for $n = 2, \ldots N$

$$\rho \int_{\Omega_h} \frac{\frac{3}{2}\mathbf{y}_h^n - 2\mathbf{y}_h^{n-1} + \frac{1}{2}\mathbf{y}_h^{n-2}}{\Delta t} \cdot \mathbf{z} \, dx + 2\mu \int_{\Omega_h} \mathbf{D}(\mathbf{y}_h^n) : \mathbf{D}(\mathbf{z}) \, dx$$

$$+ \rho \int_{\Omega_h} ((2\mathbf{y}_h^{n-1} - \mathbf{y}_h^{n-2}) \cdot \nabla)(2\mathbf{y}_h^{n-1} - \mathbf{y}_h^{n-2}) \cdot \mathbf{z} \, dx$$

$$- \int_{\Omega_h} \pi_h^n \nabla \cdot \mathbf{z} \, dx = \rho \int_{\Omega_h} \mathbf{f}^n \cdot \mathbf{z} \, dx, \quad \forall \mathbf{z} \in \mathbf{V}_{0h}, \tag{47}$$

$$\int_{\Omega_h} \nabla \cdot \mathbf{y}_h^n q \, dx = 0, \quad \forall q \in \mathbf{P}_h, \tag{48}$$

$$\mathbf{y}_h^n = \mathbf{y}_\infty \quad \text{on } \Gamma_{u,h} \cup \Gamma_{N,h} \cup \Gamma_{S,h}, \tag{49}$$

$$\mathbf{y}_h^n = \mathbf{v}^n \quad \text{on } \partial B_h. \tag{50}$$

In formulation (42)-(50) we require

$$y_h^n \in V_h, \quad \forall n = 0, 1, \cdots, N, \tag{51}$$
$$\pi_h^n \in P_h, \quad \forall n = 1, \cdots, N. \tag{52}$$

The spaces V_h and P_h are defined as follows:

$$V_h = \{ z \,|\, z \in (C^0(\bar{\Omega}))^2, \, z|_K \in Q_{2K}^2, \, \forall K \in \mathcal{Q}_h \,\}, \tag{53}$$
$$P_h = \{ q \,|\, q \in L^2(\Omega_h), \, q|_K \in P_1, \, \forall K \in \mathcal{Q}_h \,\}, \tag{54}$$

where $\mathcal{Q}_h$ is a *"quadrangulation"* of Ω_h (Ω_h: finite element approximation of Ω) and

$$Q_{2K} = \{ \varphi \,|\, \varphi \circ F_K \in Q_2 \,\}, \tag{55}$$

with F_K a well-chosen one-to-one mapping from $[0,1]^2$ into K, such that $F_K \in Q_2^2$. We also need to introduce the following subspace V_{0h} of V_h

$$V_{0h} = \{ z \,|\, z \in V_h, \, z = 0 \text{ on } \Gamma_{u,h} \cup \Gamma_{N,h} \cup \Gamma_{S,h} \cup \partial B_h \,\}. \tag{56}$$

Let us introduce now the following subspaces of V_h and V_{0h}

$$W_h = \{ z \,|\, z \in V_h, \int_K q \nabla \cdot z \, dx = 0, \, \forall q \in P_1, \forall K \in \mathcal{Q}_h \,\}, \tag{57}$$
$$W_{0h} = W_h \cap V_{0h}. \tag{58}$$

An equivalent formulation to (42)-(50) is provided then by

$$y_h^0 = y_{0h} \quad \text{in } W_h; \tag{59}$$

and

$$\rho \int_{\Omega_h} \frac{y_h^1 - y_h^0}{\Delta t} \cdot z \, dx + 2\mu \int_{\Omega_h} D(\tfrac{2}{3}y_h^1 + \tfrac{1}{3}y_h^0) : D(z) \, dx$$
$$+ \rho \int_{\Omega_h} (y_h^0 \cdot \nabla) y_h^0 \cdot z \, dx$$
$$= \rho \int_{\Omega_h} f^1 \cdot z \, dx, \quad \forall z \in W_{0h}, \quad y_h^1 \in W_h, \tag{60}$$

$$y_h^1 = y_\infty \quad \text{on } \Gamma_{u,h} \cup \Gamma_{N,h} \cup \Gamma_{S,h}, \tag{61}$$
$$y_h^1 = v^1 \quad \text{on } \partial B_h; \tag{62}$$

then for $n = 2, \ldots N$

$$\rho \int_{\Omega_h} \frac{\frac{3}{2}\mathbf{y}_h^n - 2\mathbf{y}_h^{n-1} + \frac{1}{2}\mathbf{y}_h^{n-2}}{\Delta t} \cdot \mathbf{z} \, dx + 2\mu \int_{\Omega_h} \mathbf{D}(\mathbf{y}_h^n) : \mathbf{D}(\mathbf{z}) \, dx$$

$$+ \rho \int_{\Omega_h} ((2\mathbf{y}_h^{n-1} - \mathbf{y}_h^{n-2}) \cdot \nabla)(2\mathbf{y}_h^{n-1} - \mathbf{y}_h^{n-2}) \cdot \mathbf{z} \, dx$$

$$= \rho \int_{\Omega_h} \mathbf{f}^n \cdot \mathbf{z} \, dx, \quad \forall \mathbf{z} \in \mathbf{W}_{0h}, \quad \mathbf{y}_h^n \in \mathbf{W}_h, \tag{63}$$

$$\mathbf{y}_h^n = \mathbf{y}_\infty \quad \text{on } \Gamma_{u,h} \cup \Gamma_{N,h} \cup \Gamma_{S,h}, \tag{64}$$

$$\mathbf{y}_h^n = \mathbf{v}^n \quad \text{on } \partial B_h. \tag{65}$$

The pressure unknown has been eliminated, at the price of having to use finite element spaces defined by non trivial linear constraints and also the necessity to construct vector bases of $\mathbf{W}_h$ and $\mathbf{W}_{0h}$, satisfying thus these constraints. This issue will be addressed in Section 7.3.

4.3 Formulation of the fully-discrete control problem

Using the above approximation of the Navier-Stokes equations yields the following discrete control problem

$$\begin{cases} \mathbf{u}_h^{\Delta t} \in \mathcal{U}_h^{\Delta t}, \\ J_h^{\Delta t}(\mathbf{u}_h^{\Delta t}) \le J_h^{\Delta t}(\mathbf{v}), \quad \forall \mathbf{v} \in \mathcal{U}_h^{\Delta t}, \end{cases} \tag{66}$$

with

$$\mathcal{U}_h^{\Delta t} = \Lambda_h^N, \tag{67}$$

$$\Lambda_h = \{\lambda | \int_{\partial B} \lambda \cdot \mathbf{n} \, ds = 0, \lambda = \tilde{\lambda}|_{\partial B}, \tilde{\lambda} \in \mathbf{W}_h\}, \tag{68}$$

and

$$J_h^{\Delta t}(\mathbf{v}) = \frac{\epsilon}{2} \Delta t \sum_{n=1}^{N} \left(\|\mathbf{v}^n\|_\alpha^2 + \left\| \frac{\mathbf{v}^n - \mathbf{v}^{n-1}}{\Delta t} \right\|_0^2 \right) + \Delta t \sum_{n=1}^{N} P_{d,h}^n, \tag{69}$$

where the discrete drag power $P_{d,h}^n$ is defined by

$$P_{d,h}^1 = \rho \int_{\Omega_h} \frac{\mathbf{y}_h^1 - \mathbf{y}_h^0}{\Delta t} \cdot \mathbf{y}_b^1 \, dx + 2\mu \int_{\Omega_h} \mathbf{D}(\frac{2}{3}\mathbf{y}_h^1 + \frac{1}{3}\mathbf{y}_h^0) : \mathbf{D}(\mathbf{y}_b^1) \, dx$$

$$+ \rho \int_{\Omega_h} (\mathbf{y}_h^0 \cdot \nabla)\mathbf{y}_h^0 \cdot \mathbf{y}_b^1 \, dx - \rho \int_{\Omega_h} \mathbf{f}^1 \cdot \mathbf{y}_b^1 \, dx, \tag{70}$$

and for $n = 2, \ldots N$

$$P_{d,h}^n = \rho \int_{\Omega_h} \frac{\frac{3}{2}\mathbf{y}_h^n - 2\mathbf{y}_h^{n-1} + \frac{1}{2}\mathbf{y}_h^{n-2}}{\Delta t} \cdot \mathbf{y}_b^n \, dx + 2\mu \int_{\Omega_h} \mathbf{D}(\mathbf{y}_h^n) : \mathbf{D}(\mathbf{y}_b^n) \, dx$$

$$+ \rho \int_{\Omega_h} ((2\mathbf{y}_h^{n-1} - \mathbf{y}_h^{n-2}) \cdot \nabla)(2\mathbf{y}_h^{n-1} - \mathbf{y}_h^{n-2}) \cdot \mathbf{y}_b^n \, dx$$

$$- \rho \int_{\Omega_h} \mathbf{f}^n \cdot \mathbf{y}_b^n \, dx, \tag{71}$$

with $\mathbf{y}_b^n$ a "lifting" (i.e., an extension) of $\mathbf{v}^n - \mathbf{y}_\infty$, contained in $\mathbf{W}_h$ and vanishing on $\Gamma_{u,h} \cup \Gamma_{N,h} \cup \Gamma_{S,h}$. The above quantities approximate P_d^n defined in Section 3.2. In (70) and (71), $\{\mathbf{y}_h^n\}_{n=0}^N$ is obtained from $\mathbf{v}^n$ via the solution of (59)-(65).

5   Gradient calculation

Computing the gradient $\nabla J_h^{\Delta t}$ of functional $J_h^{\Delta t}$ is at the same time straightforward and complicated; let us comment on this apparently paradoxical statement: computing the gradient is straightforward in the sense that it relies on a well-established and systematical methodology which has been discussed in many articles (see, e.g., [GL94], [GL95], [Ber98], [HG98], [HGMP98]); on the other hand the relative complication of the discrete state equations (59)-(65) makes the calculation of $\nabla J_h^{\Delta t}$ a bit tedious as we can guess from the relations just below. Indeed, the gradient $\nabla J_h^{\Delta t}(\mathbf{v})$ of functional $J_h^{\Delta t}$ at $\mathbf{v}$ is given by

$$< \nabla J_h^{\Delta t}(\mathbf{v}), \mathbf{w} > = \Delta t \sum_{n=1}^N < \mathbf{g}_h^n - \frac{\mathbf{n} \int_{\partial B} \mathbf{g}_h^n \cdot \mathbf{n} \, ds}{\int_{\partial B} ds}, \mathbf{w}^n >, \quad \forall \mathbf{w} \in \mathcal{U}_h^{\Delta t}, \tag{72}$$

with

$$< \mathbf{g}_h^N, \mathbf{w}^N > = \epsilon(\mathbf{v}^N, \mathbf{w}^N)_\alpha + \epsilon \left(\frac{\mathbf{v}^N - \mathbf{v}^{N-1}}{(\Delta t)^2}, \mathbf{w}^N \right)$$

$$+ \rho \int_{\Omega_h} \frac{\frac{3}{2}\mathbf{p}_h^N}{\Delta t} \cdot \tilde{\mathbf{w}}^N \, dx + 2\mu \int_{\Omega_h} \mathbf{D}(\mathbf{p}_h^N) : \mathbf{D}(\tilde{\mathbf{w}}^N) \, dx$$

$$+ \rho \int_{\Omega_h} \frac{\frac{3}{2}\mathbf{y}_h^N - 2\mathbf{y}_h^{N-1} + \frac{1}{2}\mathbf{y}_h^{N-2}}{\Delta t} \cdot \tilde{\mathbf{w}}^N \, dx$$

$$+ \rho \int_{\Omega_h} [(2\mathbf{y}_h^{N-1} - \mathbf{y}_h^{N-2}) \cdot \nabla](2\mathbf{y}_h^{N-1} - \mathbf{y}_h^{N-2}) \cdot \tilde{\mathbf{w}}^N \, dx$$

$$+ 2\mu \int_{\Omega_h} \mathbf{D}(\mathbf{y}_h^N) : \mathbf{D}(\tilde{\mathbf{w}}^N) \, dx - \rho \int_{\Omega_h} \mathbf{f}_h^N \cdot \tilde{\mathbf{w}}^N \, dx, \tag{73}$$

and

$$
<\mathbf{g}_h^{N-1}, \mathbf{w}^{N-1}>= \epsilon\,(\mathbf{v}^{N-1}, \mathbf{w}^{N-1})_\alpha + \epsilon\,\Big(\frac{2\mathbf{v}^{N-1} - \mathbf{v}^{N-2} - \mathbf{v}^N}{(\Delta t)^2}, \mathbf{w}^{N-1}\Big)
$$

$$
+ \ \rho\int_{\Omega_h}\frac{\frac{3}{2}\mathbf{p}_h^{N-1} - 2\mathbf{p}_h^N}{\Delta t}\cdot\tilde{\mathbf{w}}^{N-1}\,dx + 2\mu\int_{\Omega_h}\mathbf{D}(\mathbf{p}_h^{N-1}):\mathbf{D}(\tilde{\mathbf{w}}^{N-1})\,dx
$$

$$
+ \ \rho\int_{\Omega_h}[(2\mathbf{y}_h^{N-1} - \mathbf{y}_h^{N-2})\cdot\nabla](\tilde{\mathbf{w}}^{N-1})\cdot(2\mathbf{p}_h^N)\,dx
$$

$$
+ \ \rho\int_{\Omega_h}(\tilde{\mathbf{w}}^{N-1}\cdot\nabla)(2\mathbf{y}_h^{N-1} - \mathbf{y}_h^{N-2})\cdot(2\mathbf{p}_h^N)\,dx
$$

$$
+ \ \rho\int_{\Omega_h}\frac{\frac{3}{2}\mathbf{y}_h^{N-1} - 2\mathbf{y}_h^{N-2} + \frac{1}{2}\mathbf{y}_h^{N-3}}{\Delta t}\cdot\tilde{\mathbf{w}}^{N-1}\,dx
$$

$$
+ \ \rho\int_{\Omega_h}[(2\mathbf{y}_h^{N-2} - \mathbf{y}_h^{N-3})\cdot\nabla](2\mathbf{y}_h^{N-2} - \mathbf{y}_h^{N-3})\cdot(\tilde{\mathbf{w}}^{N-1})\,dx
$$

$$
+ \ 2\mu\int_{\Omega_h}\mathbf{D}(\mathbf{y}_h^{N-1}):\mathbf{D}(\tilde{\mathbf{w}}^{N-1})\,dx - \rho\int_{\Omega_h}\mathbf{f}_h^{N-1}\cdot\tilde{\mathbf{w}}^{N-1}\,dx, \qquad (74)
$$

then for $n = N - 2, \cdots, 2$

$$
<\mathbf{g}_h^n, \mathbf{w}^n >= \epsilon\,(\mathbf{v}^n, \mathbf{w}^n)_\alpha + \epsilon\,\Big(\frac{2\mathbf{v}^n - \mathbf{v}^{n-1} - \mathbf{v}^{n+1}}{(\Delta t)^2}, \mathbf{w}^n\Big)
$$

$$
+ \ \rho\int_{\Omega_h}\frac{\frac{3}{2}\mathbf{p}_h^n - 2\mathbf{p}_h^{n+1} + \frac{1}{2}\mathbf{p}_h^{n+2}}{\Delta t}\cdot\tilde{\mathbf{w}}^n\,dx + 2\mu\int_{\Omega_h}\mathbf{D}(\mathbf{p}_h^n):\mathbf{D}(\tilde{\mathbf{w}}^n)\,dx
$$

$$
+ \ \rho\int_{\Omega_h}[(2\mathbf{y}_h^n - \mathbf{y}_h^{n-1})\cdot\nabla](\tilde{\mathbf{w}}^n)\cdot(2\mathbf{p}_h^{n+1})\,dx
$$

$$
+ \ \rho\int_{\Omega_h}[(2\mathbf{y}_h^{n+1} - \mathbf{y}_h^n)\cdot\nabla](\tilde{\mathbf{w}}^n)\cdot(-\mathbf{p}_h^{n+2})\,dx
$$

$$
+ \ \rho\int_{\Omega_h}(\tilde{\mathbf{w}}^n\cdot\nabla)(2\mathbf{y}_h^n - \mathbf{y}_h^{n-1})\cdot(2\mathbf{p}_h^{n+1})\,dx
$$

$$
+ \ \rho\int_{\Omega_h}(\tilde{\mathbf{w}}^n\cdot\nabla)(2\mathbf{y}_h^{n+1} - \mathbf{y}_h^n)\cdot(-\mathbf{p}_h^{n+2})\,dx
$$

$$
+ \ \rho\int_{\Omega_h}\frac{\frac{3}{2}\mathbf{y}_h^n - 2\mathbf{y}_h^{n-1} + \frac{1}{2}\mathbf{y}_h^{n-2}}{\Delta t}\cdot\tilde{\mathbf{w}}^n\,dx
$$

$$
+ \ \rho\int_{\Omega_h}[(2\mathbf{y}_h^{n-1} - \mathbf{y}_h^{n-2})\cdot\nabla](2\mathbf{y}_h^{n-1} - \mathbf{y}_h^{n-2})\cdot(\tilde{\mathbf{w}}^n)\,dx
$$

$$
+ \ 2\mu\int_{\Omega_h}\mathbf{D}(\mathbf{y}_h^n):\mathbf{D}(\tilde{\mathbf{w}}^n)\,dx - \rho\int_{\Omega_h}\mathbf{f}_h^n\cdot\tilde{\mathbf{w}}^n\,dx, \qquad (75)
$$

and, finally, for $n = 1$

$$< \mathbf{g}_h^1, \mathbf{w}^1 > = \epsilon \, (\mathbf{v}^1, \mathbf{w}^1)_\alpha + \left(\frac{2\mathbf{v}^1 - \mathbf{v}^0 - \mathbf{v}^2}{(\Delta t)^2}, \mathbf{w}^1 \right)$$

$$+ \rho \int_{\Omega_h} \frac{\mathbf{p}_h^1 - 2\mathbf{p}_h^2 + \frac{1}{2}\mathbf{p}_h^3}{\Delta t} \cdot \tilde{\mathbf{w}}^1 \, dx + 2\mu \int_{\Omega_h} \mathbf{D}(\frac{2}{3}\mathbf{p}_h^1) : \mathbf{D}(\tilde{\mathbf{w}}^1) \, dx$$

$$+ \rho \int_{\Omega_h} [(2\mathbf{y}_h^1 - \mathbf{y}_h^0) \cdot \nabla](\tilde{\mathbf{w}}^1) \cdot (2\mathbf{p}_h^2) \, dx$$

$$+ \rho \int_{\Omega_h} [(2\mathbf{y}_h^2 - \mathbf{y}_h^1) \cdot \nabla](\tilde{\mathbf{w}}^1) \cdot (-\mathbf{p}_h^3) \, dx$$

$$+ \rho \int_{\Omega_h} (\tilde{\mathbf{w}}^1 \cdot \nabla)(2\mathbf{y}_h^1 - \mathbf{y}_h^0) \cdot (2\mathbf{p}_h^2) \, dx$$

$$+ \rho \int_{\Omega_h} (\tilde{\mathbf{w}}^1 \cdot \nabla)(2\mathbf{y}_h^2 - \mathbf{y}_h^1) \cdot (-\mathbf{p}_h^3) \, dx$$

$$+ \rho \int_{\Omega_h} \frac{\mathbf{y}_h^1 - \mathbf{y}_h^0}{\Delta t} \cdot \tilde{\mathbf{w}}^1 \, dx + \rho \int_{\Omega_h} (\mathbf{y}_h^0 \cdot \nabla)(\mathbf{y}_h^0) \cdot (\tilde{\mathbf{w}}^1) \, dx$$

$$+ 2\mu \int_{\Omega_h} \mathbf{D}(\frac{2}{3}\mathbf{y}_h^1 + \frac{1}{3}\mathbf{y}_h^0) : \mathbf{D}(\tilde{\mathbf{w}}^1) \, dx - \rho \int_{\Omega_h} \mathbf{f}_h^1 \cdot \tilde{\mathbf{w}}^1 \, dx. \qquad (76)$$

In (73)-(76) $\tilde{\mathbf{w}}^n$ is a lifting of $\mathbf{w}^n$ contained in $\mathbf{W}_h$ and vanishing on $\Gamma_{u,h} \cup \Gamma_{N,h} \cup \Gamma_{S,h}$, and $\mathbf{p}_h^n \in \mathbf{W}_h$ is the solution of the following discrete adjoint system :

$$\rho \int_{\Omega_h} \frac{\frac{3}{2}\mathbf{p}_h^N}{\Delta t} \cdot \mathbf{z} \, dx + 2\mu \int_{\Omega_h} \mathbf{D}(\mathbf{p}_h^N) : \mathbf{D}(\mathbf{z}) \, dx = 0, \quad \forall \mathbf{z} \in \mathbf{W}_{0h}, \qquad (77)$$

$$\mathbf{p}_h^N = 0 \qquad \text{on } \Gamma_\infty^u, \qquad (78)$$

$$\mathbf{p}_h^N = \mathbf{v}_h^N - \mathbf{y}_\infty \qquad \text{on } \partial B; \qquad (79)$$

and

$$\rho \int_{\Omega_h} \frac{\frac{3}{2}\mathbf{p}_h^{N-1} - 2\mathbf{p}_h^N}{\Delta t} \cdot \mathbf{z} \, dx + 2\mu \int_{\Omega_h} \mathbf{D}(\mathbf{p}_h^{N-1}) : \mathbf{D}(\mathbf{z}) \, dx \qquad (80)$$

$$+ \rho \int_{\Omega_h} [(2\mathbf{y}_h^{N-1} - \mathbf{y}_h^{N-2}) \cdot \nabla]\mathbf{z} \cdot (2\mathbf{p}_h^N) \, dx$$

$$+ \rho \int_{\Omega_h} (\mathbf{z} \cdot \nabla)(2\mathbf{y}_h^{N-1} - \mathbf{y}_h^{N-2}) \cdot (2\mathbf{p}_h^N) \, dx = 0, \quad \forall \mathbf{z} \in \mathbf{W}_{0h}, \qquad (81)$$

$$\mathbf{p}_h^{N-1} = 0 \qquad \text{on } \Gamma_\infty^u, \qquad (82)$$

$$\mathbf{p}_h^{N-1} = \mathbf{v}_h^{N-1} - \mathbf{y}_\infty \qquad \text{on } \partial B; \qquad (83)$$

then for $n = N - 2, \cdots, 2$

$$\rho \int_{\Omega_h} \frac{\frac{3}{2}\mathbf{p}_h^n - 2\mathbf{p}_h^{n+1} + \frac{1}{2}\mathbf{p}_h^{n+2}}{\Delta t} \cdot \mathbf{z} \, dx + 2\mu \int_{\Omega_h} \mathbf{D}(\mathbf{p}_h^n) : \mathbf{D}(\mathbf{z}) \, dx$$

$$+ \rho \int_{\Omega_h} [(2\mathbf{y}_h^n - \mathbf{y}_h^{n-1}) \cdot \nabla]\mathbf{z} \cdot (2\mathbf{p}_h^{n+1}) \, dx$$

$$+ \rho \int_{\Omega_h} [(2\mathbf{y}_h^{n+1} - \mathbf{y}_h^n) \cdot \nabla]\mathbf{z} \cdot (-\mathbf{p}_h^{n+2}) \, dx$$

$$+ \rho \int_{\Omega_h} (\mathbf{z} \cdot \nabla)(2\mathbf{y}_h^n - \mathbf{y}_h^{n-1}) \cdot (2\mathbf{p}_h^{n+1}) \, dx$$

$$+ \rho \int_{\Omega_h} (\mathbf{z} \cdot \nabla)(2\mathbf{y}_h^{n+1} - \mathbf{y}_h^n) \cdot (-\mathbf{p}_h^{n+2}) \, dx = 0, \quad \forall \mathbf{z} \in \mathbf{W}_{0h}, \quad (84)$$

$$\mathbf{p}_h^n = 0 \qquad\qquad \text{on } \Gamma_\infty^u, \tag{85}$$

$$\mathbf{p}_h^n = \mathbf{v}_h^n - \mathbf{y}_\infty \quad \text{on } \partial B; \tag{86}$$

and, finally, for $n = 1$

$$\rho \int_{\Omega_h} \frac{\mathbf{p}_h^1 - 2\mathbf{p}_h^2 + \frac{1}{2}\mathbf{p}_h^3}{\Delta t} \cdot \mathbf{z} \, dx + 2\mu \int_{\Omega_h} \mathbf{D}(\tfrac{2}{3}\mathbf{p}_h^1) : \mathbf{D}(\mathbf{z}) \, dx$$

$$+ \rho \int_{\Omega_h} [(2\mathbf{y}_h^1 - \mathbf{y}_h^0) \cdot \nabla]\mathbf{z} \cdot (2\mathbf{p}_h^2) \, dx + \tag{87}$$

$$+ \rho \int_{\Omega_h} [(2\mathbf{y}_h^2 - \mathbf{y}_h^1) \cdot \nabla]\mathbf{z} \cdot (-\mathbf{p}_h^3) \, dx \tag{88}$$

$$+ \rho \int_{\Omega_h} (\mathbf{z} \cdot \nabla)(2\mathbf{y}_h^1 - \mathbf{y}_h^0) \cdot (2\mathbf{p}_h^2) \, dx \tag{89}$$

$$+ \rho \int_{\Omega_h} (\mathbf{z} \cdot \nabla)(2\mathbf{y}_h^2 - \mathbf{y}_h^1) \cdot (-\mathbf{p}_h^3) \, dx = 0, \quad \forall \mathbf{z} \in \mathbf{W}_{0h}, \quad (90)$$

$$\mathbf{p}_h^1 = 0 \qquad\qquad \text{on } \Gamma_\infty^u, \tag{91}$$

$$\mathbf{p}_h^1 = \mathbf{v}_h^1 - \mathbf{y}_\infty \quad \text{on } \partial B. \tag{92}$$

Once $\nabla J_h^{\Delta t}$ is known, via the solution of the above adjoint system, we can derive optimality conditions which enable us to solve the discrete control problem (66) by various kinds of descent methods such as conjugate gradient, BFGS, etc. The BFGS solution of (66) will be discussed in Section 6 hereafter.

6   A BFGS algorithm for the discrete control problem

In order to solve the discrete control problem (66) we shall employ a *quasi-Newton method* à la BFGS (see, e.g., [LN89] for BFGS algorithms and their implementation); such an algorithm reads as follows when applied to the solution of a generic optimization problem such as

$$\begin{cases} x \in \mathbb{R}^l, \\ f(x) \le f(y), \ \forall y \in \mathbb{R}^l. \end{cases} \tag{93}$$

If f is smooth enough the solution of problem (93) satisfies also

$$\nabla f(x) = 0. \tag{94}$$

The BFGS algorithm applied to the solution of (93), (94) takes the following form:

$$x^0 \in \mathbb{R}^l, \ H^0 \in \mathcal{L}(\mathbb{R}^l, \mathbb{R}^l) \quad \text{are given}, \tag{95}$$

$$g^0 = \nabla f(x^0). \tag{96}$$

For $k \ge 0$, assuming that x^k, H^k and g^k are known, we proceed as follows

$$d^k = -H^k g^k, \tag{97}$$

$$\begin{cases} \rho_k \in \mathbb{R}, \\ f(x^k + \rho_k d^k) \le f(x^k + \rho d^k), \quad \forall \rho \in \mathbb{R}, \end{cases} \tag{98}$$

$$x^{k+1} = x^k + \rho_k d^k, \tag{99}$$

$$g^{k+1} = \nabla f(x^{k+1}), \tag{100}$$

$$s^k = x^{k+1} - x^k, \tag{101}$$

$$y^k = g^{k+1} - g^k, \tag{102}$$

$$H^{k+1} = H^k + \frac{(s^k - H^k y^k) \otimes s^k + s^k \otimes (s^k - H^k y^k)}{(y^k, s^k)} \tag{103}$$

$$- \frac{(s^k - H^k y^k, y^k)}{(y^k, s^k)^2} s^k \otimes s^k.$$

Set $k = k + 1$ and return to (97).

The tensor product of two vectors u and v in $\mathbb{R}^l$ is defined, as usual, as the linear mapping from $\mathbb{R}^l \times \mathbb{R}^l$ into $\mathbb{R}^l$ such that

$$(u \otimes v)w = (v, w)u, \quad \forall w \in \mathbb{R}^l. \tag{104}$$

In (103), (104), we have denoted by $(\cdot, \cdot)$ the Euclidean scalar product used on space $\mathbb{R}^l$; $(\cdot, \cdot)$ is not necessarily the dot product on $\mathbb{R}^l$ (we may have

$(v, w) = Sv \cdot w, \forall v, w \in \mathbb{R}^l$, with S a $l \times l$ matrix, symmetric and positive definite).

Applying the above algorithm to the solution of the discrete control problem (66) is straightforward.

7   Validation of the flow simulator

7.1   Motivation

An important issue for the flow control problem discussed in this article is the quality of the flow simulator, i.e., of the methodology which will be used to solve the Navier-Stokes equations modeling the flow (and also the adjoint equations) in order to compute $\nabla J_h^{\Delta t}$. For the validation of our flow simulator we have chosen as test problem the flow past a circular cylinder at various Reynolds numbers. This test problem has the advantage of combining a simple geometry with a rich flow dynamics and it has always been a traditional benchmarking problem for incompressible viscous flow simulators (see, e.g., [BCM86], [BH96], [BCDM90], [IT90] and the many references therein). Also, this particular geometry has motivated the work of several flow investigators from the experimental points of view (See, e.g., [Ros55], [TD91], [Wil89]).

7.2   Description of the mesh and other parameters

In order to validate our incompressible viscous flow simulator we have chosen as computational domain the two-dimensional region Ω so that

$$\Omega = \Pi \backslash \bar{B}$$

where Π is the rectangle $(-15, 45) \times (-15, 15)$ and B is the disk of center $(0, 0)$ and of radius $a = .5$. The diameter of B will be taken as characteristic length, implying that the Reynolds number is defined as

$$\mathrm{Re} = \frac{2a\rho|\mathbf{y}_\infty|}{\mu}.$$

The simulations will be done with $\mu = 1/200$ and $1/1000$ implying that $\mathrm{Re} = 200$ and 1000 for $\rho = 1$. The finite element mesh used for the calculations at $\mathrm{Re} = 1000$ has been shown in Figure 2, where we have also visualized the mesh in the neighborhood of B; we observe the refinement used close to ∂B in order to better capture the boundary layer. Actually further information for both $\mathrm{Re} = 200$ and $\mathrm{Re} = 1000$ calculations is provided by Table 1.

7.3   Numerical results and comparisons

The goal of these computational experiments is to simulate the development of the vortex street in an unforced laminar wake behind the circular cylinder in Figure 2, for $Re = 200$ and 1000. Although the simulations at $Re = 1000$ are two-dimensional and they do not include the effects of 3D instabilities and turbulence, the high Re 2D simulations are still of interest in comparing with other 2D results and in capturing the key dynamics of the large, 2D vortices, that clearly dominate the high Re flow experiments such as those of Tokumaru and Dimotakis.

Actually, for Re values below $40 - 50$ a stable steady flow is observed with formation of a bubble in the wake. The length of the recirculating zone increases with Re, and beyond a certain critical value the flow becomes unstable. Alternating eddies are formed in an asymmetrical pattern which generates an alternating separation of vortices. These vortices are advected and diffused downstream forming the well-known Karman vortex street. In "actual life" the symmetry breaking is triggered by various causes such as disturbances in the initial and/or boundary conditions. In our simulation the computational mesh and the boundary conditions are perfectly symmetric. As an initial condition we have taken the symmetric solution obtained from a Navier-Stokes calculation where symmetry is systematically enforced at each time step by averaging. This symmetric solution (unstable for Re sufficiently large) is itself used as initial condition for a simulation where the symmetry constraint has been relaxed. The symmetry breaking taking place for Re sufficiently large can be explained by the various truncation and rounding errors taking place in the calculations.

At the initial stage of the symmetry breaking, the growth of the perturbation is linear and the drag coefficient grows first very fast up to a point where the growth becomes oscillating and a saturation is observed. In Figures 3 and 4 we have represented the variations of the drag and the lift versus t, for $Re = 200$ and 1000, respectively.

The periodic regime which is asymptotically reached is characterized by the frequency at which the vortices are shed. For comparison purposes it was found convenient to introduce the *Strouhal* number

$$S_n = \frac{2a}{|\mathbf{y}_\infty|} f_n,$$

which is a nondimensional representation of the shedding frequency. In Table 2, a comparison is given at various Re between the Strouhal numbers from our simulation and those obtained experimentally and computationally by various authors ([Hen97], [Wil89], [Ros55], [BDY89]). The agreement with Henderson's computational data and Williamson's experimental data is very good for Re between 60 and 1000. For more details on these comparisons see [Nor98]. Similarly, in Table 7.3, the time averaged drag coefficient is seen to be in very good agreement with Henderson's results for the steady and periodic state [Hen97]. However the results of Braza et al. [BDY89] are inconsistent for Reynolds number 1000 and do not match other two-dimensional simulations. A well-known effect of having just two dimensions in numerical simulations as opposed to three is that the drag tends to be over-predicted for higher Reynolds numbers, where three-dimensional instabilities would occur. For more details on these drag comparisons see, again, [Nor98].

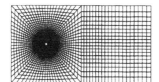 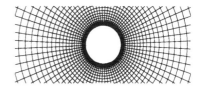

Figure 2: Mesh used for Reynolds number 1000.

Re	Δt	elements	points	nodes	unknowns
200	0.005	2226	2297	9046	11415
1000	0.001	4104	4200	16608	20905

Table 1: Discretization parameters

			S_n		
Re	Present	Henderson	Williamson	Roshko	Braza
60	0.1353	0.1379	0.1356	0.1370	-
80	0.1526	0.1547	0.1521	0.1557	-
100	0.1670	0.1664	0.1640	0.1670	0.16
200	0.1978	0.1971	-	-	0.20
400	0.2207	0.2198	-	-	-
600	0.2306	0.2294	-	-	-
800	0.2353	0.2343	-	-	-
1000	0.2392	0.2372	-	-	0.21

Table 2: Strouhal number for different Reynolds numbers. Comparison with [Hen97], [Wil89], [Ros54], and [BCM86].

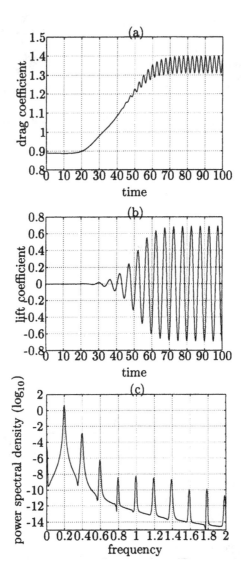

Figure 3: Case of a fixed cylinder in uniform free-stream flow, Re = 200. (a) Drag and (b) lift coefficient. (c) Power spectrum density of the lift coefficient history. The Strouhal number is 0.1978.

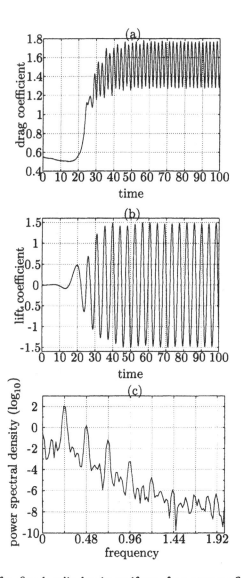

Figure 4: Case of a fixed cylinder in uniform free-stream flow, Re = 1000. (a) Drag and (b) lift coefficient. (c) Power spectrum of the lift coefficient history. The Strouhal number is 0.2392.

	C_D			
Re	Present work	Henderson	Braza et al.	Fornberg
20	2.0064	2.0587	2.19	2.0001
40	1.5047	1.5445	1.58	1.4980
60	1.3859	1.4151	1.35	-
80	1.3489	1.3727	-	-
100	1.3528	1.3500	1.36	-
200	1.3560	1.3412	1.39	-
400	1.4232	1.4142	-	-
600	1.4641	1.4682	-	-
800	1.4979	1.4966	-	-
1000	1.5191	1.5091	1.198	-

Table 3: Drag coefficient for different Reynolds numbers. Comparison with [Hen97], [BCM86], and [For80].

8 Active Control by Rotation

8.1 Synopsis

In this section we investigate via simulation various strategies for the active control by rotation of the flow around a cylinder. In section 8.2 we consider the dynamical behavior of the flow under the effect of forced sinusoidal rotation of the cylinder. Then in Section 8.3 we present the results obtained when applying the optimal control strategy discussed in Sections 2 to 6.

8.2 Active control by forced sinusoidal rotation

The active control discussed in this section is based on *oscillatory rotation* as in the experiments of Tokumaru and Dimotakis [TD91]. If the forcing is *sinusoidal* there are *two degrees of freedom*, namely the *frequency* f_e and the *amplitude* ω_1 of the angular velocity. The forcing Strouhal number is defined as

$$S_e = 2af_e/|\mathbf{y}_\infty|$$

which yields the following forcing angular velocity

$$\omega(t) = \omega_1 \sin(2\pi S_e t).$$

A series of simulations was performed at $\mathrm{Re} = 200$ with different forcing frequencies S_e varying from .35 to 1.65. The amplitude ω_1 of the forcing angular velocity was held fixed to the value 6 for all simulations in this series. Once the transients had died out a spectral analysis of the (time dependent) drag minus its time averaged value was performed, leading to the results shown in Figures 5a, 5b, and 5c, which correspond to $S_e = 0.75$, 1.25 and 1.65, respectively. Several comments are in order :

1. At $S_e = .75$ a perfect lock-in to the forcing frequency can be observed, in which the forcing frequency dominates the dynamics of the flow (in simple terms: *the flow oscillates at the forcing frequency*).

2. At $S_e = 1.25$, there is a competition between the forcing frequency and the natural shedding fundamental frequency. The dynamics corresponds to a quasi-periodic state.

3. At $S_e = 1.65$ the flow dynamics is dominated by the natural shedding frequency (.2 from Table 2); the forcing frequency has little influence on the flow dynamics.

These results agree with those in [KT89] which discusses the active control of flow around cylinders by sinusoidal transversal motions (a kind of chattering control).

Similar experiments were performed at $\mathrm{Re} = 1000$, with $\omega_1 = 5.5$ and $S_e = .625$, 1.325 and 1.425. The corresponding results have been reported in Figures 6a, 6b, and 6c. The computed results suggest the existence of a threshold amplitude for the forcing; we need to operate beyond this threshold for the flow to "feel" the forcing. It was further observed that this threshold is a function of the forcing frequency: higher frequencies require higher amplitude in order for the control to stay effective.

The above results suggest looking for *optimal pairs* $\{\omega_1, S_e\}$ for *drag minimization*. To be more precise, we consider the drag as a function of $\{\omega_1, S_e\}$ and try to minimize this function for $\{\omega_1, S_e\}$ varying in a "reasonable" subset of $\mathbb{R}^2$. For $Re = 200$ standard minimization techniques in finite dimension (2 here) yield $\omega_1 = 6$ and $S_e = .74$ which corresponds to the lock-in case previously described. In Figure 7 we have visualized the contours of the drag, considered as a function of ω_1, and S_e, in the neighborhood of the optimal solution. In Figure 8a we have represented the variation versus time of the optimal sinusoidal control whose action started at time $t = 0$. The transition to low drag has been visualized in Figure 8b which also shows the shedding frequency transition. The drag reduction was found to be of the order of 30%. Finally the lift coefficient has been represented in Figure 8c; we observe that the amplitude of the lift oscillations is substantially reduced. Finally in Figures 9a and 9b we have shown snapshots of the uncontrolled flow and of the optimally forced flow. The shedding of very large vortices has been supplanted by the shedding of much smaller vortices that do not span the wake. This is qualitatively similar to the effects observed by Tokumaru and Dimotakis [TD91]. Details of the vortex shedding for various values of t have been reported in Figure 10; these figures clearly show the significant change in the vortex shedding due to rotation of the cylinder. The shed vortices are detached much sooner from the boundary layer by reversal of the direction of cylinder rotation.

Similar experiments have been carried out for $Re = 1000$; qualitatively the simulated phenomena are identical to those observed for $Re = 200$, however the drag reduction is this time of the order of 60%. The optimal amplitude and frequency are, this time, $\omega_1 = 5.5$ and $S_e = .625$. The results shown in Figures 11 to 14 are self-explanatory.

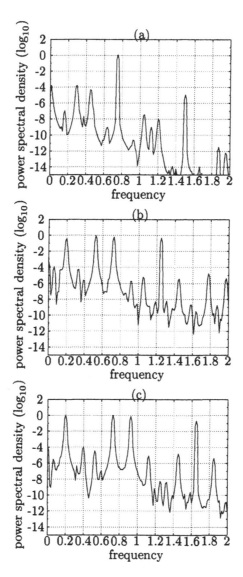

Figure 5: Case of a sinusoidal rotating cylinder in uniform free-stream flow, $Re = 200$. Power spectral density of the lift coefficient history is shown for (a) lock-in ($S_e = 0.75$), (b) quasiperiodic ($S_e = 1.25$), and (c) nonreceptive state ($S_e = 1.65$). The natural Strouhal number (S_n) is 0.1978

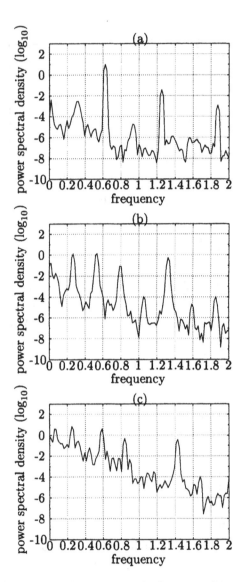

Figure 6: Case of a sinusoidal rotating cylinder in uniform free-stream flow, $Re = 1000$. Power spectral density of the lift coefficient history for (a) lock-in ($S_e = 0.625$), (b) quasiperiodic ($S_e = 1.325$), and (c) nonreceptive state ($S_e = 1.425$). The natural Strouhal number (S_n) is 0.2392

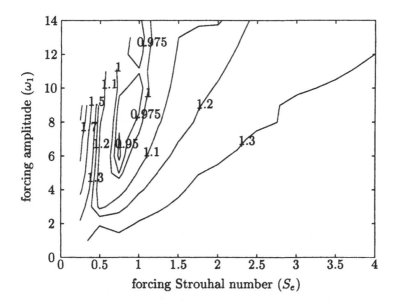

Figure 7: Variation of the drag C_D with S_e and ω_1 at Reynolds number 200.

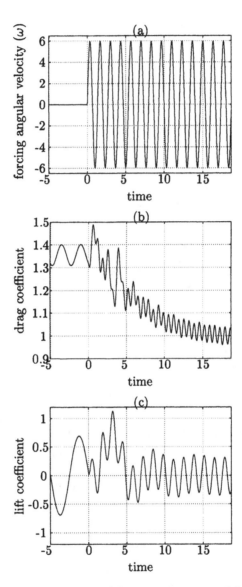

Figure 8: The time evolution of the (a) sinusoidal-optimal forcing, with $S_e = 0.75$ and $\omega_1 = 6.00$, (b) drag C_D, and (c) lift C_L, at Reynolds number 200. Forcing was started at time $t = 0$.

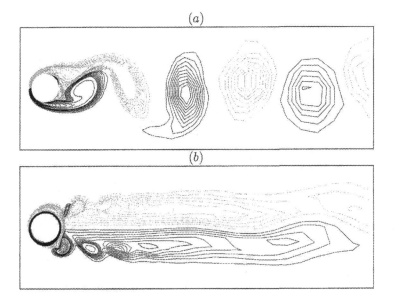

Figure 9: Vorticity contour plot of the wake of the unforced (a) and forced (b) flow at Reynolds number 200.

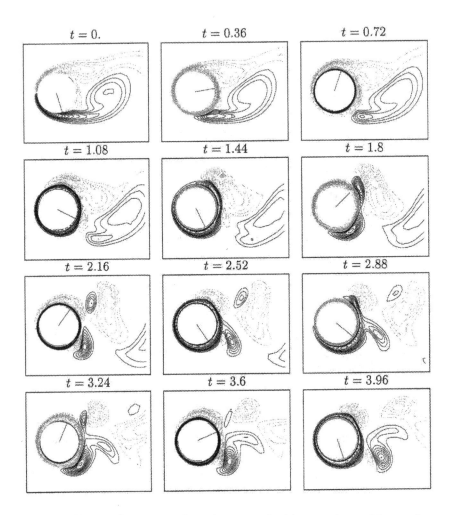

Figure 10: Near-wake region: forced vortex shedding at Reynolds number 200 with $S_e = 0.75$ and $\omega_1 = 6$. The sequence represents the first three forcing period.

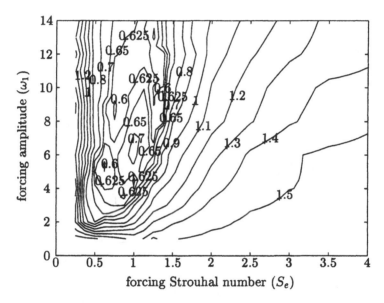

Figure 11: Variation of the drag C_D with S_e and ω_1 at Reynolds number 1000.

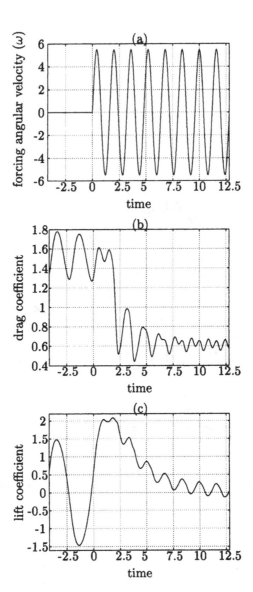

Figure 12: The time evolution of the (a) sinusoidal-optimal forcing, with $S_e = 0.625$ and $\omega_1 = 5.5$, (b) drag C_D and (c) lift C_L, at Reynolds number 1000. Forcing was started at time $t = 0$.

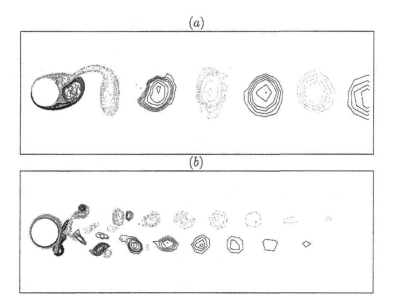

Figure 13: Vorticity contour plot of the wake of the unforced (a) and forced (b) flow at Reynolds number 1000.

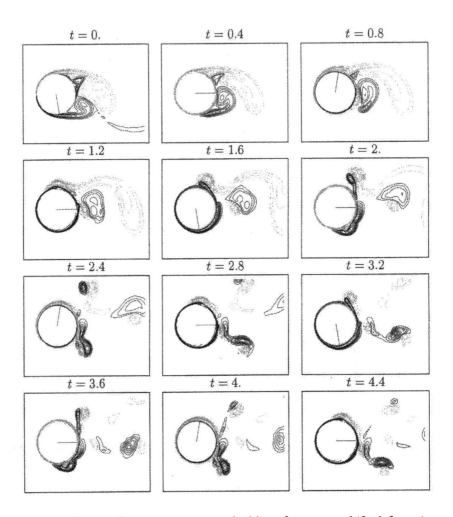

Figure 14: Near-wake region: vortex shedding frequency shifted from its natural shedding frequency ($S_n = 0.2398$) to the forcing frequency ($S_e = 0.625$), Re $= 1000$ and $\omega_1 = 5.5$. The sequence represents the first three forcing period.

8.3 Drag reduction by optimal control

In this section we are going to present the results obtained by applying the methods discussed in Sections 2 to 6 to active flow control by rotation, the cost function being essentially the drag, since the following results have been obtained with $\epsilon = 0$ in (2.9). The values of Re are as in Section 8.2, namely 200 and 1000. As initial guess for the optimal control computation we have used the quasi-optimal forcing obtained in Section 8.2. Typically convergence was obtained in 20 iterations of the BFGS algorithm.

Let us comment first on the result obtained for Re = 200. In Figure 15a we have represented the computed optimal control ($-$) as a function of t and compared it to the optimal sinusoidal control ($--$) obtained in Section 8.2. We observe that the fundamental frequency of the optimal control is very close to the optimal frequency for the sinusoidal control. The power spectral density of the optimal control is shown in Figure 15b.

Similarly we have represented in Figures 16a and 16b the results corresponding to Re = 1000. From these figures we observe that the fundamental frequency of the optimal control and the optimal frequency for the sinusoidal control are even closer than for Re = 200.

From these simulations it follows that:

1. The fundamental frequency of the optimal control is very close to the optimal frequency obtained by the methods of Section 8.2.

2. The optimal control has one fundamental frequency and several harmonics whose frequencies are *odd* multiples of the fundamental frequency.

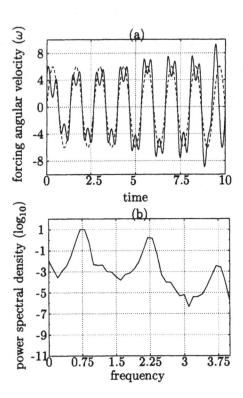

Figure 15: The (a) optimal forcing at Reynolds number 200, and its (b) power spectral density. In (a) the dashed line represents the optimal sinusoidal control.

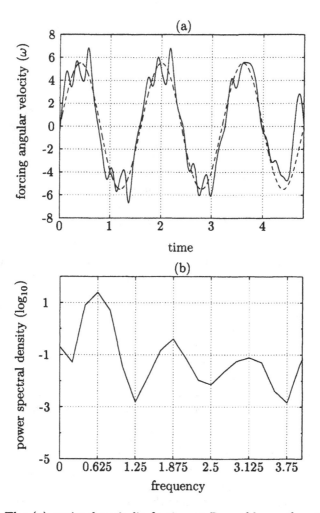

Figure 16: The (a) optimal periodic forcing at Reynolds number 1000, and its (b) power spectral density. In (a) the dashed line represent the optimal sinusoidal control.

8.4   Drag reduction by control in Fourier space

From the results described in Section 8.3, it appears that the optimal controls obtained there were predominantly composed of a sinusoidal mode oscillating at a fundamental frequency superposed with higher harmonic modes. This observation suggests looking for the controls in Fourier space. More precisely the angular velocity $\omega(t)$ will be of the following form:

$$\omega(t) = \sum_{k=1}^{K} \omega_k \sin(2k\pi S_e t - \delta_k) \tag{105}$$

At $\mathrm{Re} = 200$, in order to see what effect additional harmonics may have on the drag reduction, the optimal forcing was sought in the space described by (105) with three different values of K, namely 1, 3 and 10. The time interval for the control $(0, T)$ was chosen such that $T = 3T_f$, with $T_f = 1/S_e$ the forcing period. A piecewise control strategy was used for the solution of the optimal periodic control. Computational results show that the effect of the phase shifts δ_i is small, suggesting taking $\delta_k = 0$ in (105).

The computational experiment reported in Figure 19 corresponds to the following scenario:

- From $t = -T$ to $t = 0$, the cylinder is fixed, there is no control and the flow oscillates at its natural frequency.

- At $t = 0$ control starts by an optimal periodic control in the class given by relation (105).

The optimal periodic control is shown in Figure 19a and its corresponding drag and lift are shown in Figures 19b and 19c.

The optimal periodic control obtained during the 10th piecewise control loop has been used successfully to stabilize the system beyond that loop; the effectiveness of this approach relies on the fact that most transitional effects have been damped out. A deeper analysis of the optimal periodic state reached is in order. From Figure 20, we observe that when the peak rotation speed is reached, a corresponding minimum in the drag occurs, at times $t = 0.6$, $t = 1.25$, and $t = 1.9$. We observe that the effect of the optimal control is to flatten the drag in the neighborhood of its minima and to sharpen it in the neighborhood of the maxima. This can be seen in Figure 20b at times $t = .5$, $t = 1.15$, and $t = 1.85$. Indeed, the sharp peaks in the drag correspond to times when the forcing changes direction, i.e., crosses zero. A very interesting feature can be seen at times $t = 0.2$ and $t = 1.5$, where a zig-zag forcing motion corresponds to a lower peak in the drag. This optimization of the periodic forcing leads to an extra reduction in the drag coefficient from 0.932 to 0.905, or 2.87 %.

From Figure 21, where a vorticity snapshot is presented, it can be seen that qualitatively the structure of the wake remains unchanged from the optimal sinusoidal control forced case to the optimal periodic forced case. This suggests that the effects of the higher frequencies are only felt close to the boundary, but do not significantly affect the wake. The time evolution of the drag, and its pressure and viscous contributions is shown in Figure 22. We observe that the reduction in the pressure drag is slightly higher than for the viscous drag. In Figure 23, a time-averaged profile over the cylinder surface of the viscous drag contribution is shown, in comparison with the unforced and optimal sinusoidal forced cases. The reduction in the viscous drag occurs mainly at the peaks of the profile, namely at $\theta = 60$ and $\theta = 300$.

At $Re = 1000$, we have, from a qualitative point of view, the same behavior as at $Re = 200$ as shown in Figures (24)-(27). Compared to the optimal sinusoidal control, the optimal control brings an additional drag reduction of no more than 2%, suggesting that engineering intuition was right when suggesting to reduce the drag via sinusoidal control.

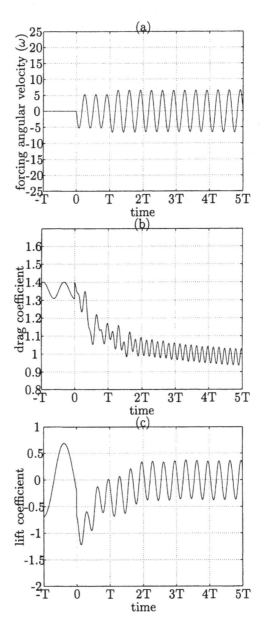

Figure 17: The (a) optimal periodic control (b) the corresponding drag (c) the corresponding lift. $K = 1$

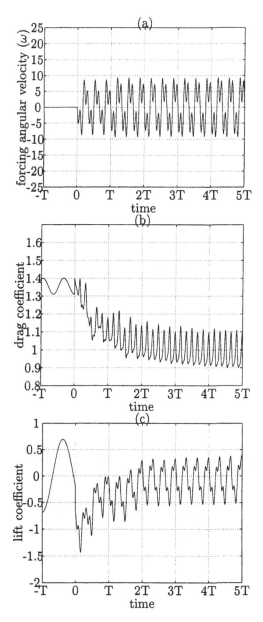

Figure 18: The (a) optimal periodic control (b) the corresponding drag (c) the corresponding lift. $K = 3$

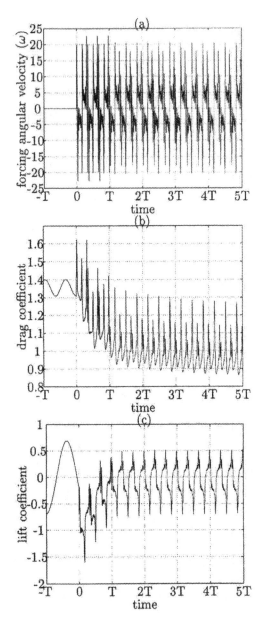

Figure 19: The (a) optimal periodic control (b) the corresponding drag (c) the corresponding lift. $K = 10$

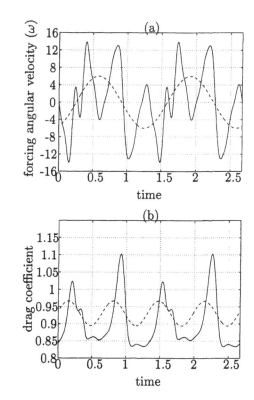

Figure 20: The (a) optimal periodic control (solid) in comparison with the optimal sinusoidal (dashed) control at Reynolds number 200. In (b) the corresponding drag is shown, with an additional reduction of 2.9 % from the optimal sinusoidal forced case.

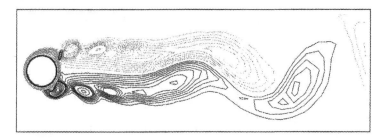

Figure 21: Vorticity contour plot of the wake of the optimally forced flow at Reynolds number 200.

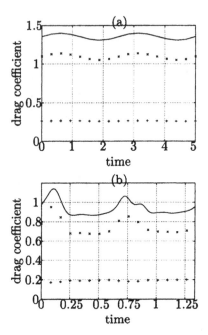

Figure 22: The time evolution of the drag and its viscous and pressure components for (a) the unforced and (b) the optimal periodic forced case, at Reynolds number 200. The total drag is represented by a solid line and the pressure and viscous component by x and + respectively.

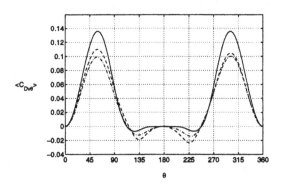

Figure 23: Contribution of the time-averaged viscous drag as a function of the angle, at Reynolds number 200. The solid line represents the unforced case, the dashed line the optimal sinusoidal forced case, and the dash-dotted line the optimal periodic forced case.

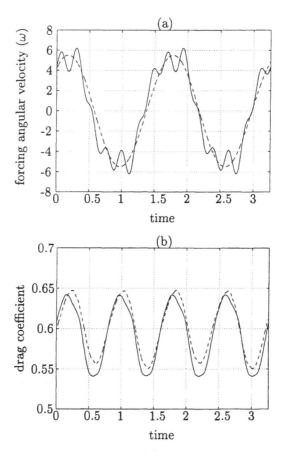

Figure 24: The (a) optimal forcing (solid) in comparison with the quasi-optimal (dashed) forcing at Reynolds number 1000. In (b) the corresponding drag is shown, with an additional reduction of 1.5 % from the quasi-optimally forced case.

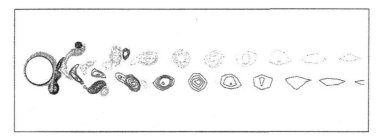

Figure 25: Vorticity contour plot of the wake of the optimally forced flow at Reynolds number 1000.

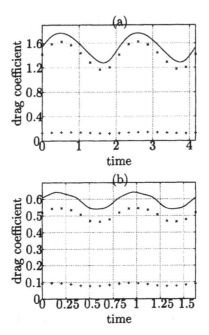

Figure 26: The time evolution of the drag and its viscous and pressure components for (a) the unforced and (b) the optimal periodic forced case, at Reynolds number 1000. The total drag is represented by a solid line and the pressure and viscous component by x and + respectively.

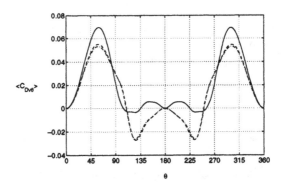

Figure 27: Contribution of the time-averaged viscous drag as a function of the angle, at Reynolds number 1000. The solid line represents the unforced case, the dashed line the forced case, and the dash-dotted line the optimally forced case.

9 Active Control by Blowing and Suction

9.1 Synopsis

A more efficient method, measured in terms of energy expenditure, than controlling the wake through cylinder rotation is to use local blowing and suction. These techniques can also be readily applied to non-cylindrical airfoils. This section will present the simulations performed with two and three slots and compare some of these results to existing results.

9.2 Simulation configuration

We consider the configuration depicted in Figure 28 for the blowing and suction simulations, the flow region being as described in Section 7. At time $t = 0$, with a fully established flow (no transients) corresponding to the prevailing Reynolds number, we start controlling by injection and suction of fluid on the boundary. The fluid is injected at an angle θ_i with the boundary normal (i is the slot number) and several slots may be used simultaneously. The angle θ_i can be either a control parameter itself or be fixed. The angle α_i denotes the angular distance between the leading edge and the ith slot. Each slot has the same parabolic outflow velocity profile, $h_i(\mathbf{x})$, and this profile is scaled by the corresponding slot's own control parameter $c_i(t)$; the corresponding boundary condition reads as follows:

$$\mathbf{y} = c_i(t)\big(\cos\theta_i(t)\mathbf{n} + \sin\theta_i(t)\mathbf{t}\big)h_i(\mathbf{x}), \quad \text{on } \partial B_i \times (0,T),$$

where ∂B_i denotes the part of the boundary where the ith slot is located. The only constraint that we put on the controls $c_i(t)$ is that their sum, at any time t, must be zero. This is the same as enforcing the mass conservation of the physical system. This condition will be relaxed in future simulations. While mass conservation over some reasonable period of time is a reasonable assumption, instantaneous mass conservation is too restrictive.

The slot aperture β_i, is chosen to be $10°$ for all slots throughout the simulation and later will be denoted just as β. A smaller angle would give too few grid points on the slot and larger slots would not have sufficiently local forcing.

Ideally we would like to run the control simulation over a long time interval $(0,T)$ until the flow reaches some asymptotic state, but this is not realistic with the present computer resources. Thus, the control is carried out only over several Strouhal periods.

9.3 Blowing and suction with two slots

9.3.1 Antisymmetrical forcing at $\mathrm{Re} = 470$

To further validate the code and to ensure that it works for the blowing and suction setup, we have simulated some of the experiments conducted

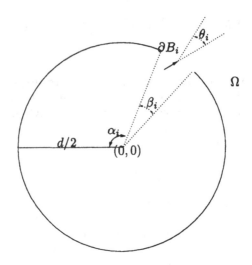

Figure 28: The blowing and suction configuration. The index i denotes the ith slot: each slot has its own configuration. The arrow indicates injection of fluid into the domain where θ_i denotes the angle of incidence of t he velocity profile, α_i the angle from the leading edge, d is the diam eter, β_i is the aperture of the slot, and ∂B_i is the part of the boundary occupied by the slot.

by Williams *et al* [WMA92]. Their experiments were performed in a water tank with a cylinder with small holes drilled along-side in two rows at $\pm 45^\circ$ from the upstream stagnation point. The bleeding from the two rows of holes could be either in phase (symmetrical forcing) or 180° out of phase (antisymmetrical forcing). The Reynolds number in their experiments was set to 470 and we used the same value in our simulations. To measure the forcing they introduced the bleed coefficient, C_b, defined as

$$C_b = \frac{\bar{y}^2 d_j}{y_\infty^2 d},$$

where d_j is the hole diameter which with our notation corresponds to $\beta \approx 8.9^\circ$, $\bar{y}$ is the root-mean-square of the flow velocity at the exit of the unsteady bleed jet, and the other variables are as defined earlier. We take $\beta = 10^\circ$ in our simulations. The diameter and spacing of the holes in the experimental setup were small enough for the effective disturbance to be two-dimensional over the length of the cylinder.

The symmetrical forcing is the more efficient way to tame the wake but we have focused on the antisymmetrical one, mainly due to the fact that the conservation of mass still holds at each instant of time and not just over a forcing cycle as it is the case for symmetrical forcing. As mentioned earlier, this is a constraint in our present code, but it is not intrinsic to the problem at hand. In fact, Williams *et al.* tested both symmetrical and antisymmetrical forcing and concluded that the symmetrical forcing was the more efficient one to symmetrize the wake. We follow the same scenario as Williams *et al.*, namely, we look at the flow for four different forcings: the natural wake ($C_b = 0$), the wake at low amplitude ($C_b = 0.004$), intermediate amplitude ($C_b = 0.019$), and high amplitude mode ($C_b = 0.073$). The excitation frequency was being held fixed at $S_{f_e} = 8.85S$ throughout the simulation where, as earlier, S is the unforced Strouhal number. For the present Reynolds number we have $S = 0.226$.

For these four different cases of antisymmetrical forcing, the main structures in the flow are almost the same as can be seen in Figure 29. By looking at the power spectral density of, for example, the drag, one can clearly observe the various forcing frequencies but they are too weak to shift the phase of the flow if applied out of phase with the unforced flow. Williams *et al.* observed the same behavior.

9.3.2 Flow stabilization at $Re = 60$

It has been shown by, for example, Roussopoulos 1993 [Rou93] and Park *et al.* 1994 [PLH94] that feedback signals from the wake can stabilize the flow and prevent vortex shedding up to certain Reynolds numbers depending on the control method. One of the goals in this work is to find optimal control algorithms, and thus to extend the range of Reynolds numbers at which a stable, low drag flow can be achieved.

For these low Reynolds numbers separation takes place around $\pm120°$ from the leading stagnation point. Park *et al.* showed, via numerical simulations at $Re = 60$, that they could stabilize the flow with a single feedback sensor located in the wake. They used two slots located at $\pm110°$ where the y_2-velocity at a certain point downstream was providing the feedback signal, via the relation

$$f(t) = \gamma \frac{y_2(\mathbf{x} = \mathbf{x}_s, t)}{y_{2_{max}}(t)},$$

where

$$y_{2_{max}}(t) = \max_{\tau \leq t} |y_2(\mathbf{x} = \mathbf{x}_s, \tau)|,$$

γ is a scaling factor and $\mathbf{x}_s$ is the feedback sensor location. They applied the feedback signal, $f(t)$, as a scaling factor of their velocity profile at the slots and they used an antisymmetrical forcing. For higher Reynolds numbers, up to 80, they were able to suppress the primary vortex shedding but at the same time they induced a secondary instability mode which triggered a new vortex generation.

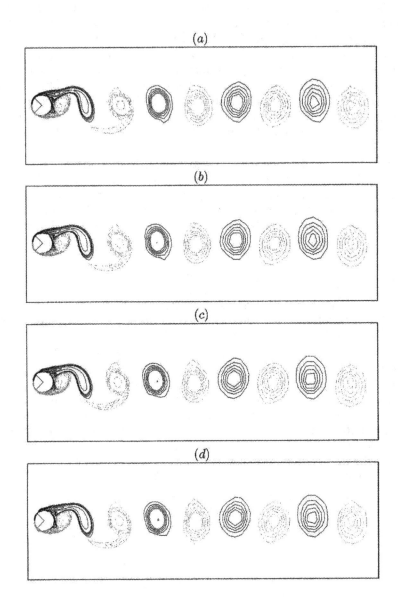

Figure 29: Natural wake (a), $C_b = 0$; low amplitude forcing (b), $C_b = 0.004$; intermediate amplitude forcing (c), $C_b = 0.019$; high amplitude forcing (d), $C_b = 0.073$.

Our approach is somewhat different; instead of using feedback to control the flow we are again looking for an optimal control approach. For these simulations we used B-splines as the control space, mainly due to the fact that the convergence rate goes up with fewer control parameters. Our present computer resources limited the longest run we performed to $t = 32$; we were able to stabilize and symmetrize this flow. The power, P_c, necessary to control the flow, decays quickly with time, as can be seen in Figure 30.a. In Figure 30.b, $c(t)$ is plotted and features the same behavior that Park *et al.* reported for their feedback signal. As soon as the feedback starts acting, the feedback signal amplitude goes down since the wake becomes more symmetrical. Note that our control seems to stabilize the flow faster. After less than fifteen time units the control amplitude is only one percent of the initial value. From Park *et al.* we know that stabilization up to $Re = 60$ is possible which is confirmed by our optimal control simulations. Further simulations are now being performed at larger Reynolds numbers to check if secondary effects can be controlled by the optimal control. In Figure 31 the drag (a) and the corresponding lift (b) are plotted. As the control acts the drag goes down and levels out close to the drag for (symmetrical) steady state flow. Also the lift goes down and becomes almost zero.

In Figure 32 the contours of the vorticity field are depicted for the natural flow (a) and the stabilized (b) flow for $Re = 60$. Both Park *et al.* and Roussopoulos reported that the amount of feedback necessary to maintain the no-vortex shedding is very low. Our results indicate the same behavior, as can be seen in Figure 30.

9.4 Blowing and suction with three slots at $Re = 200$

With the simulations featuring two slots, at $Re = 60$, we were able to stabilize the flow with the optimal control, and as a side effect of that, the wake became symmetrized along the x_1-axis. To improve the possibility of stabilizing higher Reynolds number flows, another slot was added to the cylinder. The additional slot was placed at the trailing edge and the other two slots again were located symmetrically around the x_1-axis.

The introduction of the additional slot brings a completely different structure to the wake compared to the two slot case. This can be seen in Figure 33, where the vorticity contour lines are plotted for different time steps, and where the optimal control is applied. It turns out that the additional slot always injects fluid and the other two always draw in fluid. Note that the additional slot has almost the same effect as a splitter plate placed behind the cylinder; it prevents interaction between the upper and lower part of the flow, see, for example, Kwon and Choi 1996 [KC96]. Thus, this setup requires more active control than the former two slot forcing method. Much more energy will be spent on the control but there is increased control over the flow for a larger range of Reynolds numbers.

The Reynolds number for the three slot configuration was set equal to 200

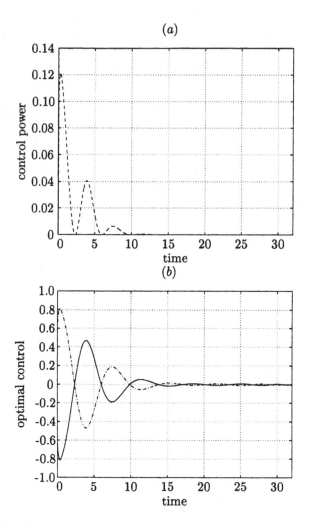

Figure 30: The power, P_c, necessary to control the flow (a). The optimal control, $c(t)$, where dash-dotted and solid lines denote the forcing for the two different slots located at angles $\pm 70°$ respectively (b).

in order to compare qualitatively with the results obtained by rotational oscillations. Two different slot geometries have been investigated: slots located at angles $\{105°, 180°, -105°\}$ and at angles $\{90°, 180°, -90°\}$ respectively. In the former case, the off-axis slots are located slightly before the points of separation based on the experience from the two slot configuration for $Re = 60$. For the first set of slots, two different control spaces have been searched; one with the angles θ_i excluded from the control and one with θ_i included and

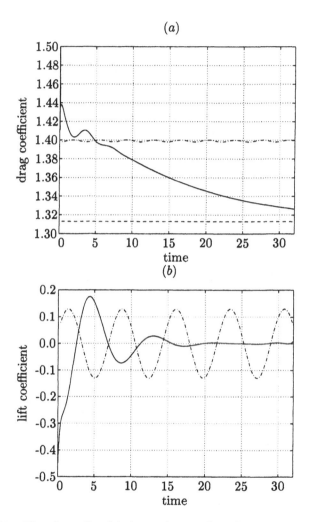

Figure 31: The drag C_D (a) for unforced flow (dash-dotted), forced flow (solid), and steady state flow (dashed). The lift C_L (b) for unforced flow (dash-dotted) and forced flow (solid).

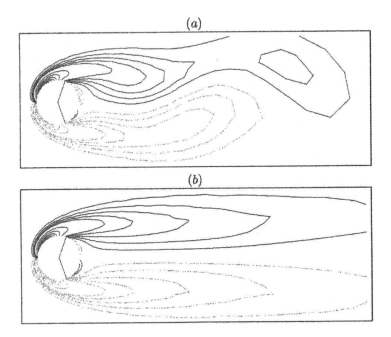

Figure 32: Comparing vorticity contour plot for $\mathbf{R}e = 60$ without control (a) and with the control $c(t)$ (b) where the shaded lines denotes negative vorticity and the solid lines positive vorticity.

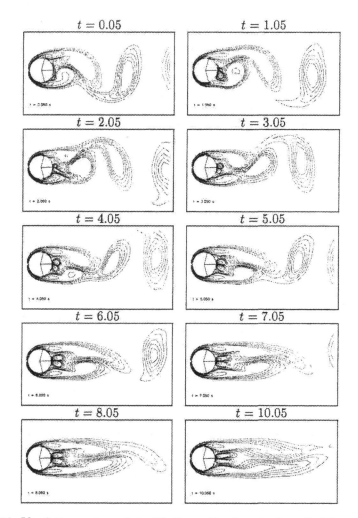

Figure 33: Vorticity contour plot with the optimal control applied from $t = 0$. The angles θ_i are included in the control and $\mathrm{Re} = 200$.

where the control space, $c(t)$ and $\theta_i(t)$, are spanned by a B-spline basis. For the second set of slots we have only run without θ in the control.

In Figure 34, 36, and 39 the optimal control in the first and last piecewise optimal control iteration are plotted for the different configurations (note that the time interval is discontinuous). The same technique is used in Figures 35-40. The optimal control is almost the same with and without the angle θ_i included in the control. The difference between the two geometries is more pronounced though and for the slot locations at $\{90°, 180°, -90°\}$ the optimal control signal is found to be slightly smaller than for the other cases.

In order to compare the different runs in terms of power we have visualized the unforced flow drag power, the steady state flow drag power, and the forced flow drag power. These functions of time are plotted in Figures 35.a, 38.a, and 40.a. We have plotted the lift in Figures 35.b, 38.b, and 40.b. The power necessary to actuate the control and the difference between power saved by applying the control and the actuation power is plotted in Figures 34, 37, and 39. From these figures we can draw the conclusion that by also controlling the angles θ_i control efficiency can be slightly improved. In Figure 36.b the optimal angles θ_i for the off-axis slots are seen to be about 35° inwards towards the x_1-axis. More surprisingly perhaps is that the slot configuration $\{90°, 180°, -90°\}$ gives us higher total power savings. Drag reduction of up to 37% was achieved and by taking into account the power used to drive the control, we could get a net drag reduction of up to 32%.

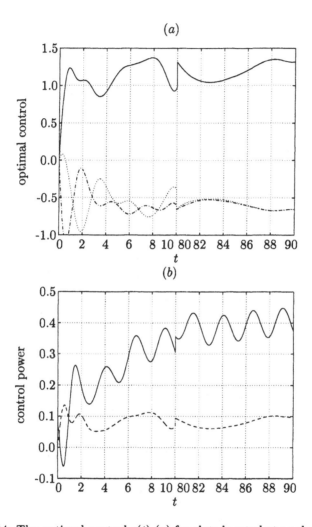

Figure 34: The optimal control $c(t)$ (a) for slots located at angles 105° (dash-dotted), −105° (dotted), and 180° (solid). The corresponding control power P_c (dashed) and the power saving (solid) due to the control (b). The angles θ_i are fixed at 0° and Re = 200. Only the first and the ninth step in the piecewise optimal control strategy are plotted.

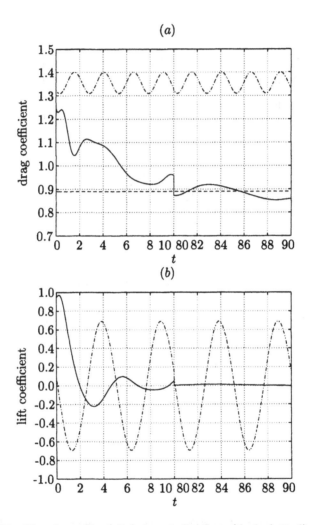

Figure 35: The drag C_D (a) for periodic flow (dash-dotted), steady flow (dashed), and forced flow (solid). The corresponding lift C_L (b) for periodic flow (dash-dotted) and forced flow (solid). The angles θ_i are fixed at 0° and Re = 200.

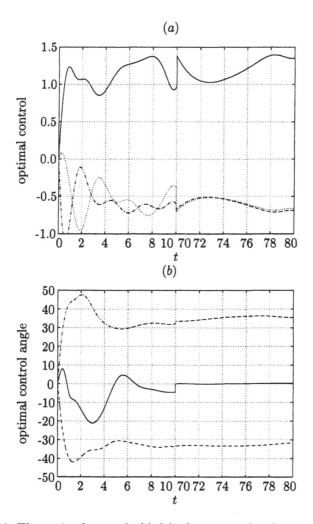

Figure 36: The optimal control $c(t)$ (a) after one and eight piecewise control updates (45 control iterations in total) and the optimal control angle $\theta_i(t)$ (b) for slots located at angles 105° (dash-dotted), −105° (dotted), and 180° (solid) and $Re = 200$.

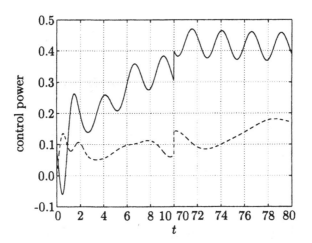

Figure 37: The control power P_c (dashed) and the power saving (solid) due to the control. The angles θ_i are included in the control and $Re = 200$.

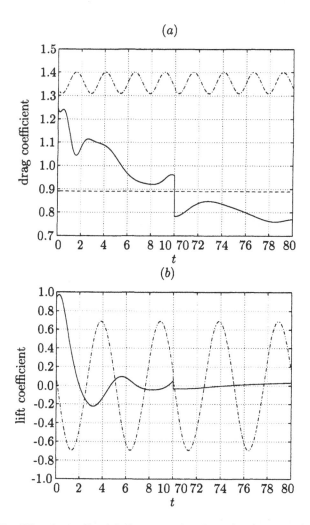

Figure 38: The drag C_D (a) for periodic flow (dash-dotted), steady flow (dashed), and forced flow (solid). The corresponding lift C_L (b) for periodic flow (dash-dotted) and forced flow (solid). The angles θ_i are included in the control and $\text{Re} = 200$.

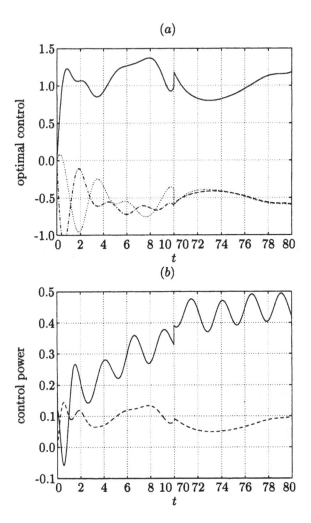

Figure 39: The optimal control $c(t)$ (a) after one and eight piecewise control updates (52 control iterations on total) and the corresponding control power P_c (dashed) and the power saving (solid) due to the control (b) for slots located at angles 90° (dash-dotted), −90° (dotted), and 180° (solid). The angles θ_i are fixed at 0° and $Re = 200$.

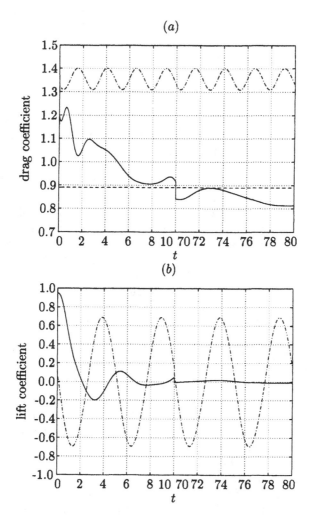

Figure 40: The drag C_D (a) for periodic flow (dash-dotted), steady flow (dashed), and forced flow (solid). The corresponding lift C_L (b) for periodic flow (dash-dotted) and forced flow (solid). The angles θ_i are fixed at $0°$ and $Re = 200$.

10 Conclusions

Through a parametric search in forcing amplitude and frequency, minima were found for the drag reduction coefficient for flow around a spinning cylinder at Reynolds numbers 200 and 1000. These minima corresponded to drag reductions of 31% at Reynolds number 200 and 61% at Reynolds number 1000. These results are qualitatively consistent with the experimental drag reduction of 80% at Reynolds number 15,000 found by Tokumaru and Dimotakis [TD91]. This suggests the potential for significant drag reductions effects, increasing with Reynolds number, at least up to the critical Reynolds number of 300,000.

Under conditions of optimal forcing, it was noted that the wakes were smaller, less energetic, and had smaller spreading angle compared with the unforced case. Also, to generate the flow field necessary for maximum drag reduction, increased amplitude of forcing was required as the oscillation frequencies increased. The quasi-optimal forcing conditions determined by parametric search agreed closely with those found by application of optimal control theory. The theory predicted, and it was confirmed by simulation, that further drag reduction could be achieved by adding higher harmonics to the forcing oscillations. This was achieved by extending the time interval of minimum drag at the expense of slightly higher, narrower peaks of maximum drag; however the improvement is fairly small.

A more efficient forcing technique, measured in terms of energy expenditure, than controlling the wake through cylinder rotation is to use local blowing and suction. We have also been able to predict the optimal forcing-control strategies in reducing drag for flow around a circular cylinder using variable (in space and time) blowing and suction at the walls. Using two slots located at $\pm 110°$, we have been able to stabilize the flow and prevent vortex shedding at Reynolds number 60.

The control power necessary to control the flow decays quickly with time, as can be seen in the left plot of Figure 30. The optimal control forcing features the same behavior that Park *et al.* reported for their feedback signal. As soon as the the control starts having effect, the amplitude of the control goes down since the wake becomes more symmetric. Our control seems to stabilize the flow faster. After less than fifteen time units the control amplitude is only one percent of the initial value and the control power necessary to maintain the non-vortex shedding is very small. Using only 3 blowing-suction slots, we have been able to completely suppress the formation of the Von-Karman vortex street up to Reynolds number 200 with a further *net* drag reduction compared to the control by rotation.

While drag reduction for flow around a circular cylinder using either an oscillatory rotation or blowing and suction provides an excellent demonstration of the application of optimal forcing control theory, it is clearly of little practical significance. However, the application of this theory to more complex shapes like airfoils should lead to some new forcing strategies. We are presently involved in such studies, the results of which will be reported in the near future.

A   A bisection storage memory saving method for the solution of time dependent control problems by adjoint equation based methodologies

A.1   Motivation. Synopsis

A superficial inspection suggests that applying adjoint equation based methods to the solution of control (or inverse) problems modeled by time dependent nonlinear partial differential equations will require huge storage memory savings. One may be lead to believe that the solution of the state equation has to be stored for each time step. If such was the case, it is clear that the adjoint equation approach may not be applicable, for example, for time dependent state equations in three space dimension. Actually very substantial storage memory savings can be achieved through a bisection storage method, the price to be paid being a reasonable additional computational time in the sense that the state equation will have to be integrated more than once (3 to 5 times, typically). In Section A.2 we shall consider a model optimal control problem and use the adjoint equation technique to compute the cost function gradient after time discretization. Then in Section A.3 we will describe the bisection method mentioned above.

A.2   A model optimal control problem

Let us consider the following *optimal control problem*

$$\begin{cases} \mathbf{u} \in \mathcal{U}(= L^2(0,T;\mathbb{R}^c)), \\ J(\mathbf{u}) \le J(\mathbf{v}), \quad \forall \mathbf{v} \in \mathcal{U}, \end{cases} \tag{106}$$

with, in (106), the *cost function* J defined by

$$\begin{aligned} J(\mathbf{v}) = {} & \frac{1}{2}\int_0^T \mathbf{S}\mathbf{v}(t) \cdot \mathbf{v}(t)\, dt + \frac{k_1}{2}\int_0^T \|\mathbf{C}_1\mathbf{y}(t) - \mathbf{z}_1(t)\|_1^2\, dt \\ & + \frac{k_2}{2}\|\mathbf{C}_2\mathbf{y}(T) - \mathbf{z}_2\|_2^2, \end{aligned} \tag{107}$$

$\mathbf{y}$ being a function of $\mathbf{v}$ through the solution of the state equation below:

$$\begin{cases} \dfrac{d\mathbf{y}}{dt} + \mathbf{A}(\mathbf{y},t) = \mathbf{f} + \mathbf{B}\mathbf{v} \quad \text{in } (0,T), \\ \mathbf{y}(0) = \mathbf{y}_0. \end{cases} \tag{108}$$

We suppose that in (106)-(108):

- $T \in (0, +\infty)$

- $\mathbf{S}$ is a time independent $c \times c$ matrix, symmetric and positive definite.

- k_1, k_2 are both nonnegative with $k_1 + k_2 > 0$.

- C_1 is a time independent $m_1 \times d$ matrix, $z_1 \in L^2(0, T; \mathbb{R}^{m_1})$ and $\| \cdot \|_1$ denotes the canonical Euclidean norm of $\mathbb{R}^{m_1}$.

- C_2 is an $m_2 \times d$ matrix, $z_2 \in \mathbb{R}^{m_2}$ and $\| \cdot \|_2$ denotes the canonical Euclidean norm of $\mathbb{R}^{m_2}$.

- $y(t) \in \mathbb{R}^d$, $\forall t \in [0, T]$, $f \in L^2(0, T; \mathbb{R}^d)$, B is a time independent $c \times d$ matrix, $y_0 \in \mathbb{R}^d$, $A : \mathbb{R}^d \times [0, T] \rightarrow \mathbb{R}^d$; we shall assume that A is differentiable with respect to y.

Assuming that problem (106) has a solution u, this solution will verify

$$J'(u) = 0, \tag{109}$$

where J' denotes the differential of J. We can easily show that

$$J'(v) = Sv + B^t p, \quad \forall v \in \mathcal{U}, \tag{110}$$

where, in (106), the *vector-valued function* p is solution of the following *adjoint system*

$$\begin{cases} -\dfrac{dp}{dt} + \dfrac{\partial A}{\partial y}(y, t)^t p = k_1 C_1^t (C_1 y - z_1) & \text{in } (0, T), \\ p(T) = k_2 C_2^t (C_2 y(T) - z_2). \end{cases} \tag{111}$$

Let us briefly discuss now the *time-discretization* of the control problem (106). For simplicity we shall time-discretize (108) by the *forward Euler scheme* with $\Delta t = T/N$, N being a positive integer. We obtain then as discrete control problem:

$$\begin{cases} u^{\Delta t}(= \{u^n\}_{n=1}^N) \in \mathcal{U}^{\Delta t}(= (\mathbb{R}^c)^N), \\ J_{\Delta t}(u^{\Delta t}) \leq J_{\Delta t}(v), \quad \forall v(= \{v^n\}_{n=1}^N) \in \mathcal{U}^{\Delta t}, \end{cases} \tag{112}$$

with

$$J_{\Delta t}(v) = \frac{\Delta t}{2} \sum_{n=1}^N Sv^n \cdot v^n + \frac{k_1 \Delta t}{2} \sum_{n=1}^N \|C_1 y^n - z_1^n\|_1^2 + \frac{k_2}{2} \|C_2 y^N - z_2\|_2^2, \tag{113}$$

with $\{y^n\}_{n=1}^N$ obtained via the solution of the following discrete state equation:

$$\begin{cases} y^0 = y_0; \\ \text{for } n = 1, \cdots, N, \\ \dfrac{y^n - y^{n-1}}{\Delta t} + A(y^{n-1}, (n-1)\Delta t) = f^n + Bv^n. \end{cases} \tag{114}$$

If $\mathbf{u}^{\Delta t}$ is solution of the discrete control problem we have then

$$J'_{\Delta t}(\mathbf{u}^{\Delta t}) = \mathbf{0}, \tag{115}$$

where, in (115), the differential $J'_{\Delta t}$ of $J_{\Delta t}$ is obtained as follows:

$$J'_{\Delta t}(\mathbf{v}) = \{S\mathbf{v}^n + \mathbf{B}^t\mathbf{p}^n\}_{n=1}^N, \quad \forall \mathbf{v} \in \mathcal{U}^{\Delta t}, \tag{116}$$

with $\{\mathbf{p}^n\}_{n=1}^N$ the solution of the following discrete adjoint equation

$$\begin{cases} \mathbf{p}^{N+1} = k_2 \mathbf{C}_2^t (\mathbf{C}_2 \mathbf{y}^N - \mathbf{z}_2), \\ \dfrac{\mathbf{p}^N - \mathbf{p}^{N+1}}{\Delta t} = k_1 \mathbf{C}_1^t (\mathbf{C}_1 \mathbf{y}^N - \mathbf{z}_1^N); \\ \text{for } n = N-1, \cdots, 1, \\ \dfrac{\mathbf{p}^n - \mathbf{p}^{n+1}}{\Delta t} + \dfrac{\partial \mathbf{A}}{\partial \mathbf{y}} (\mathbf{y}^n, n\Delta t)^t \mathbf{p}^{n+1} = k_1 \mathbf{C}_1^t (\mathbf{C}_1 \mathbf{y}^n - \mathbf{z}_1^n); \end{cases} \tag{117}$$

A superficial inspection of relations (116) and (117) suggest that to compute $J'_{\Delta t}(\mathbf{v})$ it is necessary, in general, to store the vector $\{\mathbf{y}^n\}_{n=1}^N$ which for large values of c and N makes the above approach not very practical. We will show in the following section that in fact this difficulty can be easily overcome, the price to pay for the cure being the necessity to solve the discrete state equation more than once (3 to 5 times, typically).

A.3   Description of the bisection storage saving method

Suppose that $N = 2^M$ with $M > 1$ (we have then $\Delta t = 2^{-M}T$). We shall store the components of the discrete state vector $\{\mathbf{y}^n\}_{n=0}^N$ at $t = T_0, T_1, \cdots, T_q, \cdots, T_Q$ with

$$T_q = (1 - 2^{-q})T, \quad 0 \le q \le Q \le M \tag{118}$$

and at the discrete times on interval $(T_Q, T]$; we have thus stored S_Q snap-shots with

$$S_Q = Q + 1 + 2^{M-Q}; \tag{119}$$

see Figure A.1 below for the time location of the stored snapshots.

$$0 \hspace{10em} T$$

Fig A.1. Time location of the stored snapshots ($M = 4$, $Q = 2$).

Consider the function $S : \mathbb{R}_+ \to \mathbb{R}$ defined by

$$S(\xi) = \xi + 1 + 2^{M-\xi}. \tag{120}$$

Function S is minimal at $\xi = \xi^*$ such that

$$S'(\xi^*) = 0, \tag{121}$$

with the derivative S' of S given by

$$S'(\xi) = 1 - \ln 2 \; 2^{M-\xi}. \tag{122}$$

We have then

$$\xi^* = M + \ln \ln 2 / \ln 2. \tag{123}$$

Since $\ln 2 = 0.69 \cdots \approx 1/\sqrt{2}$ we have, from (123), that

$$\xi^* \approx M - 1/2. \tag{124}$$

It follows from (124) that ξ^* is not an integer implying that in terms of memory saving an optimal choice is provided by $Q = M$ ($Q = M - 1$ is another possibility since ξ^* is "almost" the mid-point of interval $[M-1, M]$). we shall suppose from now on that $Q = M$, implying that $T_Q = T_M = T - \Delta t$.

In order to evaluate the cost of computing $J'_{\Delta t}(\mathbf{v})$, from relations (114), (116) and (117), we are going to proceed by induction over Q:

- Suppose that $Q = 0$, i.e., we store the full state vector $\{\mathbf{y}^n\}_{n=0}^N$. To obtain $J'_{\Delta t}(\mathbf{v})$ we "just" have to solve *once* the equation (114) and the adjoint state equation (117).

- Suppose now that $Q = 1$. In order to compute $J'_{\Delta t}(\mathbf{v})$ we have to solve (114) once and store $\mathbf{y}^0$ and $\{\mathbf{y}^n\}_{n=N/2}^N$. Then we solve (117) from $n = N+1$ to $n = N/2$ and compute $\{\mathbf{Sv}^n + \mathbf{B}^t\mathbf{p}^n\}_{n=N/2}^N$. Next we solve (114) from $n = 0$ to $n = N/2 - 1$ and store the corresponding snapshots in the storage areas previously occupied by $\{\mathbf{y}^n\}_{n=N/2+1}^N$. Finally, we solve (117) from $n = N/2$ to $n = 1$ and compute $\{\mathbf{Sv}^n + \mathbf{B}^t\mathbf{p}^n\}_{n=1}^{N/2-1}$. The state equation has been solved 1.5 times and the adjoint equation only once.

- Generalizing the above procedure for $Q > 1$ is straightforward; proceeding by induction we can easily show that we shall have to solve the state equation (114) "$1 + Q/2$ times" and the adjoint equation (117) only once. Thus if $Q = M$ we shall solve the state equation (114) "$1 + M/2$ times" and the adjoint equation (117) only once.

Let us summarize: *Assuming that $Q = M$ we shall store $M + 2 = \log_2 N + 2 = \log_2 4N$ snapshots extracted from $\{\mathbf{y}^n\}_{n=0}^N$ and shall have to solve the discrete state equation (114) $1 + M/2 = \log_2 2\sqrt{N}$ times and the adjoint equation (117) only once.*

To illustrate the above procedure, suppose that $N = 1{,}024 = 2^{10}$. We have then to store 12 snapshots (instead of 1,025 if $Q = 0$) and solve the state equation (114) six times. Since we have to include the cost of integrating the adjoint equation (117) we can say that using $Q = M$ instead of $Q = 0$ implies that

- The required memory is divided by 85.

- We have to solve seven 7 discrete differential equations instead of 2, i.e., a factor of 3.5.

Remark
The above storage saving method is a variant of the one described in [HG98]. Both methods are related to the *automatic differentiation methods* discussed in, e.g., [Gri92].

Acknowledgment

The authors would like to thank M. Berggren, M. Heinkenschloss, J.L. Lions, T.W. Pan, O. Pironneau, B. Stoufflet for helpful comments and suggestions. The support of Dassault Aviation and of the Higher Education Texas Coordinating Board is also acknowledged.

References

[BCDM90] H. M. Badr, M. Coutanceau, S. C. R. Dennis, and C. Ménard. Unsteady flow past a rotating circular cylinder at Reynolds numbers 10^3 and 10^4. *J. Fluid Mech.*, 220:459–484, 1990.

[BCM86] M. Braza, P. Chassaing, and H. H. Minh. Numerical study and physical analysis of the pressure and velocity fields in the near wake of a circular cylinder. *J. Fluid Mech.*, 165:79–130, 1986.

[BDY89] H. M. Badr, S. C. R. Dennis, and P. J. S. Young. Steady and unsteady flow past a rotating circular cylinder at low Reynolds numbers. *Computers and Fluids*, 17(4):579–609, 1989.

[Ber98] M. Berggren. Numerical solution of a flow-control problem: vorticity reduction by dynamic boundary action. *SIAM J. Sci. Comput.*, 19:829–860, 1998.

[BF91] F. Brezzi and M. Fortin. *Mixed and Hybrid Finite Elements Methods*. Springer-Verlag, New York, 1991.

[BH90] D.M. Buschnell and J.N. Hefner. *Viscous Drag Reduction in Boundary Layers*. American Institute of Aeronautics and Astronautics, Washington, DC, 1990.

[BH96] D. Barkley and R. D. Henderson. Three-dimensional Floquet stability analysis of the wake of a circular cylinder. *J. Fluid Mech.*, 322:215–241, 1996.

[For80] B. Fornberg. A numerical study of steady viscous flow past a circular cylinder. *J. Fluid Mech.*, 98:819–855, 1980.

[GB97] O. Ghattas and J.-H. Bark. Optimal control of two- and three-dimensional incompressible Navier-Stokes flows. *J. Comput. Phys.*, 136(2):231–244, 1997.

[GL94] R. Glowinski and J. L. Lions. Exact and approximate controllability for distributed parameter systems (Part I). *Acta Numerica, Cambridge Univ. Press*, pages 269–378, 1994.

[GL95]   R. Glowinski and J. L. Lions. Exact and approximate controlla-
           bility for distributed parameter systems (Part II). *Acta Numer-
           ica, Cambridge Univ. Press*, pages 159–333, 1995.

[GR86]   V. Girault and P. A. Raviart. *Finite Element Methods for Navier-
           Stokes Equations: Theory and Algorithms*. Springer-Verlag, New
           York, 1986.

[Gri92]   A. Griewank. Achieving logarithmic growth of temporal and spa-
           tial complexity in reverse automotic differention. *Optimization
           Methods and Software*, 1:35–54, 1992.

[Gun89]   M. D. Gunzburger. *Finite Element Method for Viscous Incom-
           pressible Flows: a Guide to Theory, Practice, and Algorithms*.
           Academic Press, Boston, 1989.

[Gun95]   M.D. Gunzburger, editor. *Flow Control*, volume 68 of the IMA
           Volumes in Mathematics and its Applications. Springer-Verlag,
           1995.

[Hak89]   M. Gad El Hak.   Flow control.   *Applied Mechanics Reviews*,
           42:261–292, 1989.

[Hen97]   R. D. Henderson. Nonlinear dynamics and patterns in turbulent
           wake transition. *J. Fluid Mech.*, 352:65–112, 1997.

[HG98]   J.W. He and R. Glowinski.   Neumann control of unstable
           parabolic systems: numerical approach. *J. O. T. A.*, 96:1–55,
           1998.

[HGMP98]   J.W. He, R. Glowinski, R. Metcalfe, and J. Periaux. A Numerical
           Approach to the Control and Stabilization of Advection-Diffusion
           Systems: Application to Viscous Drag Reduction. *Int. J. Comp.
           Fluid Mech.*, 11:131–156, 1998.

[HR96]   L. S. Hou and S. S. Ravindran. Computations of boundary op-
           timal control problems for an electrically conducting fluid. *J.
           Comput. Phys.*, 128(2):319–330, 1996.

[IR98]   K. Ito and S. S. Ravindran. A reduced-order method for simula-
           tion and control of fluid flows. *J. Comput. Phys.*, 143(2):403–425,
           1998.

[IT90]   D. B. Ingham and T. Tang. A numerical investigation into the
           steady flow past a rotating circular cylinder at low and interme-
           diate Reynolds numbers. *J. Comput. Phys.*, 87:91–107, 1990.

[KC96]   K. Kwon and H. Choi. Control of laminar vortex shedding behind
           a circular cylinder. *Phys. of Fluids*, 8:479–486, 1996.

[KT89]   G. E. Karniadakis and G. S. Triantafyllou. Frequency selection
          and asymptotic states in laminar wakes. *J. Fluid Mech.*, 199:441–
          469, 1989.

[LN89]   D. C. Liu and J. Nocedal. On the limited memory BFGS method
          for large optimization. *Mathematical Programming*, 45:503–528,
          1989.

[Nor98]   A. Nordlander. Active control and drag optimization for flow past
          a circular cylinder. Research Report UH/MD 248, Department of
          Mathematics, University of Houston, Houston, TX 77204-3476,
          1998.

[Pir89]   O. Pironneau. *Finite Element Method for Fluids*. John Wiley &
          Sons, Chichester, England, 1989.

[PLH94]   D. S. Park, D. M. Ladd, and E. W. Hendricks. Feedback control
          of von Kármán vortex shedding behind a circular cylinder at low
          Reynolds numbers. *Phys. of Fluids*, 6(7):2390–2405, 1994.

[Pra25]   L. Prandtl. The Magnus effect and windpowered ships. *Natur-
          wissenschaften*, 13:93–108, 1925.

[Ros54]   A. Roshko. On the development of turbulent wakes from vortex
          streets. *NACA Rep. 1191*, 1954.

[Ros55]   A. Roshko. On the wake and drag of bluff bodies. *J. Aerosp.
          Sci.*, 22:124–132, 1955.

[Rou93]   K. Roussopoulos. Feedback control of vortex shedding at low
          Reynolds numbers. *J. Fluid Mech.*, 248:267–296, 1993.

[Sri98]   S.S Sritharan, editor. *Optimal Control of Viscous Flows*. SIAM,
          1998.

[TD91]   P. T. Tokumaru and P. E. Dimotakis. Rotary oscillation control
          of a cylinder wake. *J. Fluid Mech.*, 224:77–90, 1991.

[Wil89]   C. H. K. Williamson. Oblique and parallel modes of vortex shed-
          ding in the wake of a circular cylinder at low Reynolds numbers.
          J. Fluid Mech., 206:579–627, 1989.

[WMA92]   D. R. Williams, H. Mansy, and C. Amato. The response and
          symmetry properties of a cylinder wake subjected to localized
          surface excitation. *J. Fluid Mech.*, 234:71–96, 1992.

Signal Processing for Everyone

Gilbert Strang

Department of Mathematics Massachusetts Institute of Technology
Cambridge MA 02139 USA

Contents

1 Introduction

In the past, signal processing was a topic that stayed almost exclusively in electrical engineering. It was only the specialists who applied lowpass filters to remove high frequencies from digital signals. The experts could cancel unwanted noise. They could compress the signal and then reconstruct. It took two-dimensional experts to do the same for images.

The truth is that everyone now deals with digital signals and images (involving large amounts of data). We all need to understand signal processing — *sampling, transforming, and filtering*. These pages are intended to explain these basic operations, using simple examples. We will reach as far as filter banks (in discrete time) and wavelet expansions (in continuous time).

Most signals start their lives in analog form. They become digital by sampling at equal time intervals. If $x_{\text{analog}}(t)$ is a *continuous time signal*, its samples give a discrete time signal:

$$x_{\text{digital}}(n) = x_{\text{analog}}(nT) \quad n = 0, \pm 1, \pm 2, \ldots \tag{1}$$

The *sampling interval* is T. We often normalize to $T = 1$, by a simple rescaling of the time variable.

A device that actually does this sampling is called an *A-to-D converter*. The input is an analog (A) signal, probably from measurements. The output is a digital (D) signal, probably for computer processing. Usually an A-to-D converter loses high frequency information (or mixes it with low frequencies, which is aliasing). Shannon's Theorem will tell us that when there are no high frequencies in the signal, the analog signal can be recovered at all t from its (digital) samples at the discrete times nT.

Notice that the signal is assumed to be infinitely long, with no start and no finish. The time line is $-\infty < t < \infty$. Then the discrete signal $x(n)$ is defined for all integers ($-\infty < n < \infty$). Neither of these assumptions is exactly true for real signals. The realistic assumption (this is often well justified) is that the signal is so long that end effects are not significant. By working with the whole line **R** and all integers **Z**, we can use Fourier methods to the utmost.

And those Fourier methods are very powerful. The chief tool in our analysis will be the Discrete Time Fourier Transform, which turns the samples $x(n) = x_{\text{digital}}(n)$ into the coefficients of a 2π-periodic function $X(\omega)$:

$$X(\omega) = \sum_{n=-\infty}^{\infty} x(n)e^{-in\omega} . \tag{2}$$

All terms are unchanged when ω is increased by 2π. We refer to ω as the *frequency*, and we graph the transform $X(\omega)$ between $\omega = -\pi$ and $\omega = \pi$. Then "low frequencies" refer to frequencies near zero, and "high frequencies" have $|\omega| \approx \pi$.

Two special signals have the lowest and highest frequencies, $\omega = 0$ and $\omega = \pi$. The pure DC signal $x = (\ldots, 1, 1, 1, 1, 1, \ldots)$ has exactly zero frequency. Its transform $X(\omega)$ has a Dirac delta function at $\omega = 0$. More precisely, $X(\omega)$ is a periodic train of delta functions of magnitude 2π. The pure AC signal $x = (\ldots, 1, -1, 1, -1, \ldots)$ has the highest frequency $\omega = \pi$ (and $\omega = -\pi$). Its transform is a train of delta functions at $\omega = \pm\pi, \pm 3\pi, \ldots$.

This alternation between 1 and -1 gives the fastest oscillation of any discrete signal. Between $\omega = 0$ and $\omega = \pm\pi$ is the family of pure sinusoidal signals with frequency $-\pi < \omega < \pi$:

$$x_\omega(n) = e^{in\omega} \qquad \text{for each } n. \tag{3}$$

We are frequently working with systems that respond to these pure inputs with pure outputs. The output has *no change in frequency*. The only change is in amplitude and phase, from multiplication by $H(\omega)$. This is a Linear Time Invariant system:

LTI systems: *The input $x_\omega(n)$ produces the output $H(\omega)x_\omega(n)$.*
$$\tag{4}$$

The amplifying factor $H(\omega)$, also written $H(e^{i\omega})$, is the *frequency response*. It varies from one frequency to another, but separate frequencies stay separate. $H(\omega)$ is an "eigenvalue" of the system, when the eigenvector is the oscillating signal $x_\omega(n)$. A Linear Time Invariant system is often called a *filter*.

We will study filters in detail. First we look again at these special signals — complex exponentials and real sinusoids. Fourier (and Mozart too) assembled all signals out of these pure harmonics.

2 Sinusoidal Signals

The special signal that we call x_ω has the complex exponential $e^{in\omega}$ as its nth sample. Its pure frequency is ω. A more general complex exponential has a positive real amplitude A (not necessarily 1) and a real phase shift θ (not necessarily 0):

$$x(t) = Ae^{i(\omega t + \theta)}. \tag{5}$$

From the great formula $e^{i\theta} = \cos\theta + i\sin\theta$, the real part and imaginary part are a cosine signal and a sine signal:

$$\begin{aligned} \text{Re}\{Ae^{i(\omega t + \theta)}\} &= A\cos(\omega t + \theta) \\ \text{Im}\{Ae^{i(\omega t + \theta)}\} &= A\sin(\omega t + \theta) \end{aligned} \tag{6}$$

Notice how these signals have the same frequency ω and the same amplitude A and a phase difference of $\frac{\pi}{2}$ (or 90°). A cosine function turns into a sine function if we shift by $\frac{\pi}{2}$ radians:

$$\sin(\theta) = \cos\left(\theta - \frac{\pi}{2}\right) .$$

The complex exponential is nice because it is a single function. The sinusoid representation is nice because the cosine and sine are real. Let us emphasize again that one complex function produces two real functions. But two complex functions can also produce *one* real function:

$$\begin{aligned}
e^{i\omega t} + e^{-i\omega t} &= 2\cos\omega \\
e^{i\omega t} - e^{-i\omega t} &= 2\sin\omega .
\end{aligned} \tag{7}$$

So a real sinusoid like $\cos\omega t$ or $\sin\omega t$, apparently with only one pure frequency ω, actually has *two frequencies* ω and $-\omega$!

Note on the relation of frequency to period

The pure sinusoid $x(t) = \cos\omega t$ repeats whenever the time t is increased by $2\pi/\omega$. This is its period T:

$$\text{The } period \text{ of } x(t) = \cos\omega t \text{ is } T = \frac{2\pi}{\omega} . \tag{8}$$

Suppose T is measured in seconds. Then the number of cycles in one second is $f = 1/T$:

$$\text{The } frequency \text{ in Hertz is } f = \frac{1}{T} = \frac{\omega}{2\pi} . \tag{9}$$

Thus 60 cycles per second is the same as 60 Hertz. A radio wave (whose existence was demonstrated by Heinrich Hertz) might have a frequency of 10^5 Hz. Notice the simple relation $f = \omega/2\pi$ between our two measures of frequency. We graph a function $H(\omega)$ between $\omega = -\pi$ and $\omega = \pi$. Using the normalized frequency the graph goes from $f = -0.5$ to $f = 0.5$. When the frequency doubles, the period of the signal is halved.

Note on the relation of phase shift to time shift

The symbol θ represents the phase shift. The total phase is $\omega t + \theta$, the argument of the sine or cosine, and θ clearly produces a shift. But there is another way to obtain the same result. *We could shift the time variable t by an amount θ/ω.* When ω multiplies the shifted time variable $t' = t + (\theta/\omega)$, the result $\omega t' = \omega t + \theta$ accounts for the phase shift.

There is one small point. By worldwide agreement, a "positive" time shift represents a "delay." A function $x(t)$ undergoing a unit delay becomes $x(t-1)$. The event that originally happened at $t = 5$ now waits until $t = 6$ (because then $t - 1 = 5$). The graph shifts one unit *to the right*. This is a fact of life, that replacing the argument by $t' = t - 1$ will delay the event by a

unit time. (And changing the argument to $t'' = t + 1$ will *advance* the event
by a unit time.)

The complication is that *a positive time shift* (a delay) *corresponds to a
negative phase shift* θ. We had one sign in $\omega t + \theta$ and we have the other sign
in $\omega(t - 1)$. So this time shift of 1 corresponds to a phase shift $\theta = -\omega$. In
terms of the period $T = 2\pi/\omega$ this is:

$$\Delta t \text{ corresponds to phase shift } \theta = -\tfrac{2\pi}{T}\Delta t. \tag{10}$$

The phase shift $\theta = \frac{\pi}{2}$ between the sine and cosine corresponds to a time
delay (shift to the right) of a quarter-period $\Delta t = T/4$.

Note on the sampling period and aliasing

To get a useful discrete-time signal from an analog signal, we must sample
often enough. But too many samples will be expensive and possibly redundant (unless we are using the sample to draw a continuous graph, as below).
Of course it is the ratio between the sampling period and the true oscillation
period that is critical. The question is how many samples to take in each
up-down oscillation of a sinusoid $\cos \omega_0 t$.

We start with a small number of samples:

1. *The critical sampling period* T_{Nyquist} *has two samples per oscillation.*
   This can produce the alternating signal $1, -1, 1, \ldots$ with $x(n) = (-1)^n$.
   This has the fastest oscillation and the highest frequency π of any dis-
   crete signal:

   $$\omega_0 T_{\text{Nyquist}} = \pi \quad \text{and} \quad \frac{1}{T_{\text{Nyquist}}} = \frac{\omega_0}{\pi} \quad \text{and} \quad f_{\text{sampling}} = 2 f_{\text{sinusoid}} \tag{11}$$

   This Nyquist rate is the borderline between undersampling and oversampling. If the continuous signal is a combination of many sinusoids,
   the fastest ω_0 in the combination sets its Nyquist rate.

2. Sampling at twice the Nyquist rate is *oversampling*. The samples $1, 0, -1, 0, 1, 0$
   The samples $1, 0, -1, 0, 1, 0, \ldots$ are obtained with $x(n) = \cos \frac{\pi n}{2}$.
   This discrete signal has frequency $\omega = \frac{\pi}{2}$.

3. Sampling at $\frac{2}{3}$ of the Nyquist rate is *undersampling*. This produce the
   same sample values $1, 0, -1, 0, 1, 0, \ldots$ as before. We cannot tell from
   those samples which sinusoid $\cos \omega_0 t$ or $\cos \omega_0 t/3$ is the continuous time
   signal. *This is aliasing!* The slow frequency $\omega_0/3$ is an alias for the
   higher frequency ω_0, because the discrete samples are the same. A
   familiar example is watching a wheel turn in the movies. Often the alias
   frequency is negative and the wheel appears to be rotating backwards.

A very clear case of undersampling is half the Nyquist rate (only one sample per oscillation). That sample will fall at the same point of every oscillation, so all samples are equal. The discrete signal is then DC (direct current with samples $s, s, s, s, \ldots$ of constant value). The alias of $\omega = 2\pi/T$ is $\omega = 0$.

4. Eight samples in a period will give a reasonably good representation. This linear interpolation is often adequate but it is certainly not perfect.

5. Eighty samples in a period will produce a very lifelike graph. A curved sinusoid is actually constructed in MATLAB by linear interpolation.

Shannon Sampling Theorem

Every analog signal $x(t)$ with frequencies not exceeding $\omega_{\max}$ can be perfectly reconstructed from its discrete samples $x(nT)$, provided the sampling rate $1/T$ exceeds $2f_{\max} = \omega_{\max}/\pi$.

The conclusion of the Shannon Sampling Theorem is always amazing to me, that a band-limited analog signal (continuous time and low frequencies only) can be exactly recovered from a *countable number* of samples (discrete time). This fact is fundamental to communications and digital signal processing. Sinusoids can be recovered from samples that are taken faster than the Nyquist rate.

This is not just a statement of good approximation by linear interpolation. It is a case of perfect reconstruction through interpolation by "sinc functions." The interpolation formula (12) will be stated again (with proof). We are sampling the theorem twice in one book, as Shannon would have wished.

Notice that the signal $x(t) = \sin \omega t$ is not correctly reconstructed from its samples when $1/T$ is *exactly* the Nyquist rate ω/π. Each sample $x(nT) = \sin \omega nT = \sin n\pi$ is zero! Shannon's formula (12) will give zero. Reconstruction requires a strict inequality $\omega_{\text{signal}} T_{\text{sampling}} > \pi$, and equality (the Nyquist rate) cuts it too close.

Here is Shannon's interpolation formula. The function $(\sin t)/t$ is known as the *sinc function*. It equals 1 at $t = 0$. By shifting to $t - nT$, we center a sinc function at the sampling point nT. By scaling with π/T, we make it vanish at all other sampling points:

$$x(t) = \sum_{n=-\infty}^{\infty} x(nT) \frac{\sin(t - nT)\pi/T}{(t - nT)\pi/T} \tag{12}$$

This is a case when the D-to-A converter, from samples back to the original function, is an exact inverse of the A-to-D converter. Normally high frequency information is lost in the A-to-D step. It is irretrievably mixed with low frequency information, because of aliasing. Shannon's assumption is that *the signal has no high-frequency information*. It is assumed to be band-limited.

No aliasing occurs, no information is lost, and the transformation from the
A function $x(t)$ to the D signal $x(nT)$ can be reversed.

3 FIR Filters

A filter is the most important operation in signal processing. It acts on a
signal to produce a modified signal. Usually some frequency components of
the input signal are reduced; it is remarkable how simply this can be done.
When the filter is FIR (*finite* impulse response), each output sample $y(n)$ is
just a linear combination of a *finite* number of input samples.

The simplest example is a moving average

$$y(n) = \frac{1}{2}x(n) + \frac{1}{2}x(n-1). \tag{13}$$

This filter combines each sample $x(n)$ with the previous sample $x(n-1)$.
The weights in the linear combination are the filter coefficients $\frac{1}{2}$ and $\frac{1}{2}$. The
filter is *time-invariant* because those coefficients are constant for all time.
The filter is *causal* because it involves no future samples like $x(n+1)$. The
effect $y(n)$ never comes earlier than its causes $x(n)$ and $x(n-1)$. Thus we
have a causal linear time-invariant FIR system.

A noncausal system cannot operate on a real-time signal, because the
input would not be available when the output is required. A pure delay
$y(n) = x(n-1)$ is causal, and acceptable in real-time. A pure advance
$y(n) = x(n+1)$ is anticausal and not acceptable. This certainly applies to
audio signals. For an image the situation is different, because n refers to
position not time. The complete image may be available and the filters in
image processing need not be causal.

What does this particular "running average" filter do to the input signal
$x(n)$? Consider first three special inputs, an impulse and a constant (DC)
signal and an alternating (AC) signal. Here are the outputs from those inputs:

I.   Impulse $x(n)$ = $(\dots, 0, 0, 1, 0, 0, \dots)$

   Impulse Response $y(n)$ = $(\dots, 0, 0, \frac{1}{2}, \frac{1}{2}, 0, \dots)$

   The impulse response contains the filter coefficients!

II.   Constant $x(n)$ = $(\dots, 1, 1, 1, 1, 1, \dots)$

   Averaged Output $y(n)$ = $(\dots, 1, 1, 1, 1, 1, \dots)$

   The response exactly equals the input; $\omega = 0$ is in the *passband*.

III.   Alternating $x(n)$ = $(\dots, 1, -1, 1, -1, 1, \dots)$

   Averaged Output $y(n)$ = $(\dots, 0, 0, 0, 0, 0, \dots)$

   The response is zero; the frequency $\omega = \pi$ is in the *stopband*.

We conclude that the averaging filter is *lowpass*. Low frequencies are mostly passed, high frequencies are mostly stopped. To understand the specifics of that word "mostly," we have to choose input frequencies between $\omega = 0$ and $\omega = \pi$. So the input signal will now be $x(n) = e^{i\omega n}$, at the pure frequency ω. The crucial point is that *the output signal $y(n)$ is also at frequency ω*:

$$x(n) = e^{in\omega} \text{ produces } y(n) = \frac{1}{2}e^{in\omega} + \frac{1}{2}e^{i(n-1)\omega} = \left[\frac{1}{2} + \frac{1}{2}e^{-i\omega}\right]e^{in\omega} . \quad (14)$$

The output frequency is the same ω, but the amplitude and phase are changed. The filter multiplies each frequency component of the input by the *frequency response* function

$$H(e^{i\omega}) = \frac{1}{2} + \frac{1}{2}e^{-i\omega} . \quad (15)$$

At $\omega = 0$, this frequency response is $H = 1$. Therefore the constant signal passes unchanged through the filter (as we know). At $\omega = \pi$, the frequency response is $H = 0$. Therefore the alternating signal is completely blacked out by the averaging filter. We now separate the function $H(e^{i\omega})$ into its amplitude and phase, to see what happens to each individual frequency:

$$H(e^{i\omega}) = \frac{1}{2} + \frac{1}{2}e^{-i\omega} = e^{-i\omega/2}\left(\frac{1}{2}e^{i\omega/2} + \frac{1}{2}e^{-i\omega/2}\right) = e^{-i\omega/2}\left(\cos\frac{\omega}{2}\right) . \quad (16)$$

The amplitude is $|H| = \cos\frac{\omega}{2}$. It drops from one at $\omega = 0$ to zero at $\omega = \pi$. The graph of $|H|$ from $-\pi$ to π is one arch of a cosine. It shows a passband and a stopband, where the multiplying factor $H(e^{i\omega})$ is near one and near zero. For this very short filter, the *transition band* in between is very wide. This is a somewhat crude lowpass filter, but extremely simple and inexpensive.

The phase of $H(e^{i\omega})$ in equation (16) is $\phi = -\omega/2$. For this filter, the phase depends linearly on ω. This property of *linear phase* follows directly from the symmetry of the filter coefficients $h(0) = \frac{1}{2}$ and $h(1) = \frac{1}{2}$. Reversing the order produces no change. More precisely, the coefficients are symmetric around their middle point (at $\frac{1}{2}$):

Symmetry $h(n) = h(1 - n)$ around $\frac{1}{2}$ produces the linear phase $-\frac{1}{2}\omega$.

This is one example of a general rule: *symmetry produces linear phase*. That is a highly important property in image processing, because the eye catches any failure of symmetry after an image is compressed.

Moving Difference (Highpass Filter)

A second quick example will reinforce these points, by setting up a contrast. Instead of an averaging filter (coefficients $\frac{1}{2}$ and $\frac{1}{2}$) we take differences:

$$y(n) = \frac{1}{2}x(n) - \frac{1}{2}x(n-1). \tag{17}$$

This is still FIR and causal. Suppose it acts on the three special input signals. The response to a unit impulse is the filter coefficients $\frac{1}{2}$ and $-\frac{1}{2}$. The all-ones input has zero differences, so the lowest frequency $\omega = 0$ is stopped. The highest frequency $\omega = \pi$ is passed without any change. This is a *highpass filter*:

I. Impulse	$x(n) = \delta(n)$	Impulse response	$y(0) = \frac{1}{2}$, $y(1) = -\frac{1}{2}$
II. Constant	$x(n) = 1$	Zero output	$y(n) = 0$
III. Alternating	$x(n) = (-1)^n$	Alternating output	$y(n) = (-1)^n$.

As for every linear time-invariant system, the response to a pure frequency signal $x(n) = e^{in\omega}$ is a multiple $y(n) = H(e^{i\omega})e^{in\omega}$ of that signal:

$$x(n) = e^{in\omega} \text{ produces } y(n) = \frac{1}{2}e^{in\omega} - \frac{1}{2}e^{i(n-1)\omega} = \left[\frac{1}{2} - \frac{1}{2}e^{-i\omega}\right]e^{in\omega}. \tag{18}$$

The multiplying factor is the frequency response function $H(e^{i\omega})$. Again we separate the phase factor from the amplitude $|H|$:

$$\frac{1}{2} - \frac{1}{2}e^{-i\omega} = e^{-i\omega/2}\left(\frac{1}{2}e^{i\omega/2} - \frac{1}{2}e^{-i\omega/2}\right) = e^{-i\omega/2}\, i\sin\frac{\omega}{2}. \tag{19}$$

The amplitude is $|H| = \sin\frac{\omega}{2}$. This is zero at $\omega = 0$ and one at $\omega = \pi$. The filter is highpass (but again not very sharp). The amplitude response is now a sine instead of a cosine.

The phase factor is $e^{-i\omega/2}i$. The extra factor $i = \sqrt{-1}$ appears because this filter is *antisymmetric*. We still call this "linear phase," although strictly speaking the linearity ends at $\omega = \pi$ where $\sin\frac{\omega}{2}$ changes sign. (At that point the magnitude changes to $-\sin\frac{\omega}{2}$. Therefore the phase must jump by π to produce a sign change $e^{i\pi} = -1$. So the true phase function $\phi(\omega)$ is linear with jumps.) In short: *linear phase filters are symmetric or antisymmetric around their centers.*

Let us emphasize the difference between lowpass and highpass. A lowpass filter preserves the *smooth* part of the signal. A highpass filter preserves the *rough and noisy* part. In wavelet language, lowpass gives averages and highpass gives details. In some applications those details are important to keep (like edges in an image). In other applications the high frequencies are mostly noise (from measurements). A good lowpass filter has many uses, so we look now at better filters with more coefficients.

4 Better FIR Filters

A causal FIR filter of order N has coefficients $h(0), h(1), \ldots, h(N)$. Notice that there are $N + 1$ coefficients; this is the *length* of the filter. These coefficients stay fixed for all time so the filter is time-invariant.

At each time step, the $N + 1$ coefficients multiply $N + 1$ samples from the input signal — the current sample $x(n)$, the previous sample $x(n - 1)$, continuing back to the sample $x(n - N)$. This weighted combination of input values produces the output $y(n)$:

$$y(n) = h(0)x(n) + h(1)x(n - 1) + \cdots + h(N)x(n - N). \tag{20}$$

This is the action of the filter in the time domain. We may write it compactly as a sum from $k = 0$ to $k = N$:

$$y(n) = \sum_{k=0}^{N} h(k)x(n - k). \tag{21}$$

A filter is a discrete convolution! It is the fundamental operation for discrete time-invariant systems. To implement this convolution in hardware, we only need three building blocks: *unit delay, multiplier,* and *adder.* To account for the typical term $h(1)x(n - 1)$ in equation (20), the three operations are represented in Figure 1:

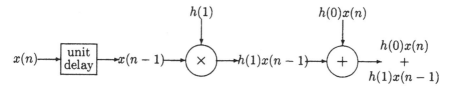

Figure 1: The three blocks that build every FIR filter: **delay, multiply, add.**

In a hardware implementation, the sample value $x(n - 1)$ is stored in memory for one clock cycle and released. A double delay to $x(n - 2)$ is a cascade of two unit delays. The N-unit delay in the convolution uses N memory cells and a circular buffer. Basically we have a shift register.

Modern DSP microprocessors often combine "multiply-add" into one special unit whose speed is critical. We multiply and accumulate, exactly like a dot product of vectors, $h \cdot x = h_1 x_1 + h_2 x_2 + \cdots + h_n x_n$. In numerical linear algebra this is executed in double precision to minimize the damage from cancellation of large numbers.

We can implement a filter in MATLAB using the convolution command **conv**:

$$x = 0:4+\sin(\text{pi}*(0:4)/2);$$
$$h = [0.25\ 0.5\ 0.25];$$
$$y = \text{conv}(h,x)$$

The 5-point input vector is linear plus sinusoidal: $x = [0\ 1\ 2\ 3\ 4] + [0\ 1\ 0 -1\ 0]$. The linear part is smooth. The sinusoid has discrete frequency $\omega = \frac{\pi}{2}$ because it takes two time intervals (not just one) to oscillate from its maximum to its minimum (from 1 to -1).

The filter $h = \left[\frac{1}{4}, \frac{1}{2}, \frac{1}{4}\right]$ is *lowpass* and *causal* and *linear phase* (it is symmetric around its center). The high frequency $\omega = \pi$ is eliminated because the alternating sum is $\frac{1}{4} - \frac{1}{2} + \frac{1}{4} = 0$. The all-ones signal is preserved because the filter coefficients sum to 1. Actually the linear signal $[0\ 1\ 2\ 3\ 4]$ is also preserved (except at the ends of the signal), but you will see how *the causal filter introduces a delay.* Let us display that part $y_{\text{linear}} = \text{conv}(0:4, h)$ as an ordinary multiplication of the polynomials $(0 + x + 2x^2 + 3x^3 + 4x^4)$ and $(\frac{1}{4} + \frac{1}{2}x + \frac{1}{4}x^2)$:

$x(n)$	0	1	2	3	4		
$h(n)$	$\frac{1}{4}$	$\frac{1}{2}$	$\frac{1}{4}$				
$h(0)x(n)$	0	$\frac{1}{4}$	$\frac{2}{4}$	$\frac{3}{4}$	$\frac{4}{4}$		
$h(1)x(n-1)$		0	$\frac{1}{2}$	$\frac{2}{2}$	$\frac{3}{2}$	$\frac{4}{2}$	
$h(2)x(n-2)$			0	$\frac{1}{4}$	$\frac{2}{4}$	$\frac{3}{4}$	$\frac{4}{4}$
$\text{conv}(h,x_{\text{linear}})$	0	$\frac{1}{4}$	1	2	3	$2\frac{3}{4}$	1
	end effect		delay to $x(n-1)$			end effect	

Notice that there are *seven* outputs. An input signal of length L convolved with an order N filter produces an output of length $L + N$. A fourth degree polynomial times a second degree polynomial yields a sixth degree polynomial. Five terms convolved with three terms produce seven terms.

At the center of the output, the linear inputs $1, 2, 3$ are preserved (with a delay). The reason is that not only $\sum h(k) = 1$ (which preserves constant signals) but also $\sum kh(k) = 1$ (which preserves linear signals).

The end effects have length $N = 2$, the order of the filter. The effect appears at the left end because we don't have the samples $x(-2)$ and $x(-1)$ that should contribute to $y(0)$ and $y(1)$. (If those missing samples are both zero than the left end is correct as it stands.) The effect appears at the right end because we don't have the samples $x(5)$ and $x(6)$. (We have run out of samples to match with the filter window.) One of the possible techniques to produce an output of length $L = 5$ instead of $L + N = 7$ is *wrap-around* or *cycling* or *periodicity*. The right end $(2\frac{3}{4}, 1)$ is added to $(0, \frac{1}{4})$ at the left end. This is *circular convolution* and for periodic signals it is natural.

Now we input the sinusoid $[0\ 1\ 0\ -1\ 0]$. The output from a filter should be a sinusoid of the same frequency, but with a different (reduced) amplitude. Again we have to overlook the end effects:

$x(n)$	0	1	0	-1	0			
$h(n)$	$\frac{1}{4}$	$\frac{1}{2}$	$\frac{1}{4}$					
$h(0)x(n)$	0	$\frac{1}{4}$	0	$-\frac{1}{4}$	0			
$h(1)x(n-1)$		0	$\frac{1}{2}$	0	$-\frac{1}{2}$	0		
$h(2)x(n-2)$			0	$\frac{1}{4}$	0	$-\frac{1}{4}$	0	
$\mathbf{conv}(h, x_{\sin})$	0	$-\frac{1}{4}$	$\frac{1}{2}$	0	$-\frac{1}{2}$	$-\frac{1}{4}$	0	
	left end		sinusoid			right end		

The output sinusoid has amplitude $\frac{1}{2}$. There is also a unit delay between input and output, because the filter is symmetric around $h(1)$. The center of the filter is at $n = 1$. The only causal filter centered at $n = 0$ is the identity filter $h(n) = \delta(n)$.

How should we have known that the amplitude is reduced by $\frac{1}{2}$ at the discrete frequency $\omega = \frac{\pi}{2}$? The answer for all sinusoids is contained in the frequency response function $H(\omega) = H(e^{i\omega})$. For the filter $h = (\frac{1}{4}, \frac{1}{2}, \frac{1}{4})$ the response is just a polynomial with these coefficients $h(0)$, $h(1)$, and $h(2)$:

$$H(\omega) = \frac{1}{4} + \frac{1}{2}e^{-i\omega} + \frac{1}{4}e^{-2i\omega}. \tag{22}$$

At $\omega = \frac{\pi}{2}$ this is $H = -\frac{1}{2}i$. *The magnitude at that frequency is* $|H| = \frac{1}{2}$, agreeing with the observed 50% reduction in amplitude. The factor $-i = e^{-i\pi/2}$ is responsible for the phase shift (which is the delay).

When a linear filter is applied to a *sum of inputs* (linear plus sinusoid), we get the *sum of outputs*. So when we know what the filter does to sinusoidal signals, we know everything. Those special signals take us into the frequency domain.

5 Filters in the Frequency Domain

In the time domain, a filter combines a signal with delays of that same signal. The signal $x(n - k)$ with k delays is multiplied by the filter coefficient $h(k)$. The combination from $k = 0$ to $k = N$ is the filtered output $y(n)$:

$$y(n) = \sum_{k=0}^{N} h(k)x(n - k). \tag{23}$$

That is a discrete convolution $y = h * x$ of two vectors.

How does a filter look in the frequency domain? The basic rule is that *a convolution becomes a multiplication*. We have seen that already for pure sinusoids; they are multiplied by $H(\omega)$. So the same must happen, by linearity, for any combination of sinusoids.

Algebraically, we multiply the Fourier series $H(\omega)$ with coefficients $h(k)$ and the Fourier series $X(\omega)$ with coefficients $x(m)$. The result is the series $Y(\omega)$ with coefficients $y(n)$. In the frequency domain, the output Y is the input X multiplied by the response function H:

$$\sum y(n)e^{-in\omega} = \left(\sum h(k)e^{-ik\omega}\right)\left(\sum x(m)e^{-im\omega}\right)$$
$$Y(\omega) = H(\omega)X(\omega) \tag{24}$$

To get $e^{-in\omega}$ in the product of $e^{-ik\omega}$ with $e^{-im\omega}$, we must have $k + m = n$. That is exactly what we see in the convolution (23). The indices k and $n - k$ add to n. The products $e^{-ik\omega}$ and $e^{-i(n-k)\omega}$ multiply to give $e^{-in\omega}$. Equations (23) and (24) match exactly.

This "convolution rule" is more than just algebra because equation (24) has a valuable scientific meaning:

Each component $X(\omega)$ of the input is amplified by the filter response $H(\omega)$ (or $H(e^{i\omega})$) at that frequency ω.

Now consider a combination of frequencies, integrating over ω. We want to see how such a combination can produce every signal $x(n)$. From its formula, $X(\omega) = \sum x(n)e^{-in\omega}$ is the (complex!) dot product of $\{x(n)\}$ with the pure exponential signal $\{e^{in\omega}\}$. The part of the signal in the direction of that harmonic is $X(\omega)e^{in\omega}$. When these harmonic parts are combined, by integrating from $\omega = -\pi$ to $\omega = \pi$ and dividing by 2π, we recover the original vector $x(n)$:

$$x(n) = \frac{1}{2\pi}\int_{-\pi}^{\pi} X(\omega)e^{in\omega}\, d\omega. \tag{25}$$

So $X(\omega)$ is the input component, and $H(\omega)X(\omega)$ is the output component, and we are truly in the frequency domain.

Example 1. *Suppose the input x is the unit impulse $\delta = (\ldots, 0, 0, 1, 0, 0, \ldots)$. This special signal combines all frequencies in equal amounts. Its transform $X(\omega)$ is a constant $\sum \delta(n)e^{-in\omega} = 1$, because the series has only one term $n = 0$. When we reassemble all parts — multiply the pure exponential vectors $\{e^{in\omega}\}$ by 1 and combine all frequencies by integration as in (25), we recover the impulse as expected:*

$$\frac{1}{2\pi}\int_{-\pi}^{\pi} 1 \cdot e^{in\omega}\, d\omega = \delta(n) = \left\{\begin{array}{ll} 1 & \text{if } n = 0 \\ 0 & \text{if } n \neq 0 \end{array}\right\}. \tag{26}$$

Now filter this input signal $\delta(n)$, the impulse. In the time domain, the convolution has only one term $k = n$:

$$y(n) = \sum h(k)\delta(n - k) = h(n).\qquad(27)$$

In the frequency domain, the rule $Y(\omega) = H(\omega)X(\omega)$ reduces to $Y(\omega) = H(\omega)$. As predicted, this matches $y(n) = h(n)$ in the time domain. **The transform of the impulse response $\{h(n)\}$ is the frequency response $H(\omega)$.** *This is the function that describes the filter in the frequency domain:*

$$H(\omega) = \sum_{k=0}^{N} h(k)e^{-ik\omega} = h(0) + h(1)e^{-i\omega} + \cdots + h(N)e^{-iN\omega}.\qquad(28)$$

Let us understand this function graphically. One particular frequency response $H(\omega)$ is shown by Figure 2. The filter is lowpass because low frequencies have $H(\omega) \approx 1$. They pass through the filter. The output $Y(\omega) \approx X(\omega)$ keeps those frequencies. High frequencies have $H(\omega) \approx 0$, so $H(\omega)X(\omega)$ is very small at these frequencies. We say that the output is a smoothed *version of the input. High frequencies (fast oscillations) are associated with noise, and the filter removes them. If we want to keep them then we use a highpass filter.*

You can see this smoothing in the time domain, but maybe not so well. The filter coefficients are the time domain outputs when the input is an impulse. The vector $h(k)$ of filter coefficients is certainly smoother than $\delta(n)$; the spike at the center is much less prominent. But the details of smoothing are clearer from $H(\omega)$ than $h(k)$. The symmetry of the filter is shown by the fact that the amplitude $|H(\omega)|$ is an even function:

$$h(k) = h(N - k) \quad means\ that \quad H(\omega) = (even\ function)e^{-i\omega N/2}.$$

6 Equiripple Lowpass Filters

Figure 2 is the graph of a good lowpass filter of order $N = 20$. Some designers might say it is the best. *The error in the passband $|\omega| \leq \omega_{pass}$ and in the stopband $|\pi - \omega| \leq \pi - \omega_{stop}$ is as small as possible.* That is *maximum error*, not mean-square error. We gave the inputs $N = 20$ and $\omega_{pass} = 0.44\pi$ and $\omega_{stop} = 0.56\pi$ to the MATLAB function remez.m and it designed the filter.

The output from remez.m is the set of filter coefficients $h(0), \ldots, h(20)$. We can multiply $H(e^{i\omega})$ by $e^{-i10\omega}$ to center it, with no change in amplitude. Then the cosine polynomial $|H(e^{i\omega})|$ has *degree ten*; it stops at $\cos 10\omega$.

The special feature of this minimax filter is its *equiripple* property. All "ripples" in the stopband plus passband have the same magnitude. This property tells us immediately that we cannot reduce all errors (the error is the distance from the ideal 1–0 filter in the graph). No correction could

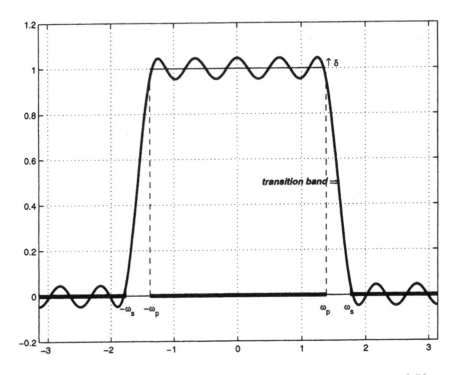

Figure 2: Frequency response for $N = 20$ and $\omega_{\text{pass}} = 0.44\pi$, $\omega_{\text{stop}} = 0.56\pi$.

have the required plus or minus sign at all twelve frequencies, because those signs are alternating and the correction would cross zero too often (11 times between 0 and π) for a polynomial of degree 10. This little bit of mathematics, where the degree of the polynomial limits how often it can alternate sign, is the foundation of minimax approximation theory.

The design method used by MATLAB is called the Remez algorithm or the Parks-McClellan algorithm. The Remez idea of iteratively reducing the largest error until the graph is equiripple was adapted by Parks and McClellan for filter design. Users often assign a heavier weight $W > 1$ to the errors in the stopband, and then the stopband ripples have smaller heights δ/W.

In a practical design problem, *we need to know what length of filter to ask for.* Therefore the relation of the error δ to the order N and the cutoff frequencies (ω_{pass} and ω_{stop}) is highly important in practice. Certainly δ decreases exponentially with N. The key exponent in $e^{-\beta N}$ was estimated by Jim Kaiser from numerical experiments, and his empirical formula has been adjusted by many authors. A recent paper [12] by Shen and the author gave the asymptotically correct formula

$$N \simeq \frac{20 \log_{10} (\pi\delta)^{-1} - 10 \log_{10} \log_{10} \delta^{-1}}{(10 \log_{10} e) \ln \cot \frac{\pi - \Delta\omega}{4}}. \tag{29}$$

Here $\Delta\omega = \omega_{\text{stop}} - \omega_{\text{pass}}$ is the *transition bandwidth*. (We use the awkward form $20 \log_{10}$ because this gives the quantity in *decibels*.) In the transition band between ω_{pass} and ω_{stop}, the dropoff from $H(\omega) \approx 1$ to $H(\omega) \approx 0$ is described by an error function. Shen's formula

$$H(\omega) \simeq \text{erf} \left(\sqrt{\frac{N\beta}{4}} \, \frac{\omega_m - \omega - S_N(W)}{\omega_{\text{stop}} - \omega_m} \right), \tag{30}$$

with

$$S_N(W) = \frac{\ln W + \frac{1}{2} \ln \ln W}{2N}.$$

gives a good fit in this "don't care region" around a critical frequency near $\omega_m = \frac{1}{2}(\omega_{\text{pass}} + \omega_{\text{stop}})$. As $N \to \infty$ the equiripple filters approach the ideal one-zero filter with cutoff at ω_{critical}.

Of course the equiripple filter is not the only candidate! If we accept larger ripples near the endpoints ω_{pass} and ω_{stop}, we can have smaller ripples inside the passband and stopband. A fairly extreme case (not usually recommended) is a *least squares filter design*. The filter minimizes the *energy* in the error, possibly weighted by W, instead of the maximum error:

$$\text{Choose } H(\omega) \text{ to minimize} \quad \int_{\text{passband}} |H(\omega) - 1|^2 \, d\omega + W \int_{\text{stopband}} |H(\omega)|^2 \, d\omega. \tag{31}$$

We know the result when the passband meets the stopband ($\omega_{\text{pass}} = \omega_{\text{stop}}$). Then $H(\omega)$ is approximating an ideal 1-0 square wave or brick wall filter. *There is a Gibbs oscillation at the discontinuity.* The maximum error E quickly approaches the Gibbs constant .09xxx. For large order N, the worst ripples have fixed height even though the integral of (ripples)^2 is as small as possible. This Gibbs phenomenon is one of the great obstacles to accurate computations near a shock front

7 Wavelet Transforms

A lowpass filter greatly reduces the high frequency components, which often represent noise in the signal. For some purposes that is exactly right. But suppose we want to *reconstruct* the signal. We may be storing it or transmitting it or operating on it, but we don't want to lose it. In this case we can use *two filters*, highpass as well as lowpass. That generates a "**filter bank**," which sends the signal through two or more filters.

The filter bank structure leads to a *Discrete Wavelet Transform*. This has become a guiding idea for so many problems in signal analysis and synthesis. In itself the transform is lossless! Its inverse (the synthesis step) is another transform of the same type — two filters that are fast to compute. Between the DWT and the inverse DWT we may compress and transmit the signal. This sequence of steps, *transform then compress then reconstruct*, is the key to more and more applications.

The word *wavelet* is properly associated with a *multiresolution into different scales*. The simplest change of scale comes from downsampling a signal — keeping only its even-numbered components $y(2n)$. This sampling operation is denoted by the symbol $\downarrow 2$:

$$(\downarrow 2) \begin{bmatrix} y(0) \\ y(1) \\ y(2) \\ y(3) \end{bmatrix} = \begin{bmatrix} y(0) \\ y(2) \end{bmatrix}.$$

Information is lost. But you will see how *doubling* the length by using two filters, then *halving* each output by $(\downarrow 2)$, can give a lossless transform. The input is at one time scale and the two half-length outputs are at another scale (an octave lower).

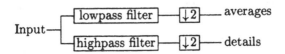

Figure 3: The discrete wavelet transform: averages and details.

Note that an input of even length L produces two outputs of length $L/2$, after downsampling. The lowpass filter H_0 and the highpass filter H_1 originally maintain length L, when we deal suitably with the samples at the ends (possibly by extending the signal to produce the extra components that the filter needs). Figure 3 shows this combination of two filters, with each output downsampled. The redundancy from $2L$ outputs is removed by $(\downarrow 2)$. Then the overall filter bank is L by L.

To simplify the theory we often pretend that $L = \infty$. This avoids any difficulty with the samples at the ends. But in reality signals have finite length.

The wavelet idea is to repeat the filter bank. The lowpass output in Figure 3 becomes the input to a second filter bank. The computation is cut in half because this input is half length. Typical applications of the wavelet transform go to four or five levels. We could interpret this multiscale transform as (quite long!) filters acting on the very first inputs. But that would miss the valuable information stored in the outputs (averages and details) along the way.

8   The Haar Transform

We now choose one specific example of a filter bank. At first there is no iteration (two scales only). Then we iterate to multiple scales. The example is associated with the name of Alfred Haar. It uses the *averaging filter* (moving average) and the *differencing filter* (moving difference). Those were our earliest and simplest examples of a lowpass filter and a highpass filter. So they combine naturally into the most basic example of a filter bank. It will be convenient (but not necessary) to reverse the sign of H_1.

The two filters are denoted by H_0 (lowpass) and H_1 (highpass):

$y_0 = H_0 x$ is the averaging filter $y_0(n) = \frac{1}{2}(x(n-1) + x(n))$

$y_1 = H_1 x$ is the differencing filter $y_1(n) = \frac{1}{2}(x(n-1) - x(n))$.

Suppose the input signal is zero except for four samples $x(1) = 6$, $x(2) = 4$, $x(3) = 5$, $x(4) = 1$. This input vector is $x = (6, 4, 5, 1)$. *We are looking for its coefficients in the Haar wavelet basis.* Those will be four numbers, $y_0(2)$ and $y_0(4)$ from subsampling the lowpass output together with $y_1(2)$ and $y_1(4)$ from highpass.

In reality we would not compute the odd-numbered components $y(1)$ and $y(3)$ since they are immediately destroyed by $(\downarrow 2)$. But we do it here to see the complete picture. Take averages and differences of $x = (6, 4, 5, 1)$:

	$y_0(1)$	$=$	3		

$$
\begin{array}{llcl@{\qquad}llcl}
& y_0(1) & = & 3 & & y_1(1) & = & -3 \\
& y_0(2) & = & 5 & & y_1(2) & = & 1 \\
\text{Averages} & y_0(3) & = & 4.5 & \tfrac{1}{2}\,(\text{Differences}) & y_1(3) & = & -0.5 \\
& y_0(4) & = & 3 & & y_1(4) & = & 2 \\
& y_0(5) & = & 0.5 & & y_1(5) & = & 0.5
\end{array}
$$

You might notice that the sum of the y_0 vector and the y_1 vector is the input $x = (6, 4, 5, 1)$ with a unit delay (to $x(n-1)$). This comes from a simple relation (average + difference) that is special to Haar:

$$\frac{1}{2}(x(n-1) + x(n)) + \frac{1}{2}(x(n-1) - x(n)) = x(n-1) .$$

It is more important to notice that the *differences tend to be smaller than the averages.* For a smooth input this would be even more true. So in a compression step, when we often lose information in the highpass coefficients, the loss is small using the wavelet basis. Here is a first look at the whole compression algorithm:

signal x $\xrightarrow[\text{[lossless]}]{}$ *wavelet coefficients* $\xrightarrow[\text{[lossy]}]{}$ *compressed coefficients* $\xrightarrow[\text{[lossless]}]{}$ *compressed signal* $\widehat{x}$

At this point we have eight or even ten coefficients in y_0 and y_1. They are redundant! They came from only four samples in x. *Subsample by* $\downarrow 2$ to keep only the even-numbered components:

$$y_0(2) = 5 \qquad y_1(2) = 1$$
$$y_0(4) = 3 \qquad y_1(4) = 2.$$

Those are the four "first-level wavelet coefficients" of the signal. The inverse transform (which is coming in the next section) will use those coefficients to reconstruct x. That will be the synthesis step. Computing the coefficients was the analysis step:

> *Analysis:* Find the wavelet coefficients
> (separate the signal into wavelets)
> *Synthesis:* Add up wavelets times coefficients
> (to reconstruct the signal).

It is like computing Fourier coefficients, and then summing the Fourier series. For wavelets, the *analysis filter bank* (H_0 and H_1 followed by $\downarrow 2$) computes the coefficients. The *synthesis filter bank* (this inverse wavelet transform will involve upsampling by $\uparrow 2$ and two filters) sums the wavelet series.

Now go from the lowpass $y_0(2) = 5$ and $y_0(4) = 3$ to the next scale by computing *averages of averages and differences of averages*:

$$z_0(2) = \frac{5+3}{2} = 4 \qquad z_1(2) = \frac{5-3}{2} = 1.$$

This completes the iteration of the Haar analysis bank. We can see the three scales (fine, medium, and coarse) in a block diagram that shows the tree of filters with subsampling:

Effectively, z comes from downsampling by 4. The vector $z_0(2n)$ contains the *low-low* coefficients, averages of averages. The high-low vector $z_1(2n)$ is also $\frac{1}{4}$ the length of the original signal. The highpass vector $y_1(2n)$ is half the original length, and $\frac{1}{4} + \frac{1}{4} + \frac{1}{2} = 1$ (this is critical length sampling).

You will ask, why not take averages and differences also of this first high-pass output $y_1(2n)$? That is certainly possible. A "wavelet packet" might choose to do it. But the basic wavelet tree assumes that the highpass coefficients are small and not worth the additional effort. They are candidates for compression. For a typical long smooth signal, iteration to four or five scale levels will further decorrelate the samples. Iterations beyond that are generally not worthwhile.

Summary: The input is $x = (6, 4, 5, 1)$. The wavelet coefficients are $(4, 1, 1, 2)$:

$$\text{(low-low } z_0 \text{, high-low } z_1 \text{, high } y_1) = (4, 1, 1, 2).$$

The special point of the wavelet basis is that you can pick off the highpass details (1 and 2 in y_1), before the coarse details in z_1 and the overall average in z_0. A picture will explain this *multiscale pyramid*:

Split $x = (6, 4, 5, 1)$ into averages and waves at small scale and then large scale:

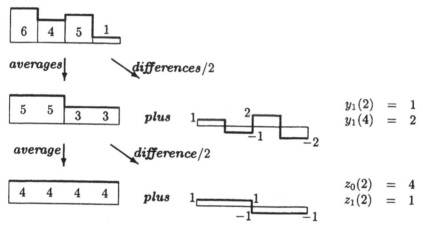

$$y_1(2) = 1$$
$$y_1(4) = 2$$

$$z_0(2) = 4$$
$$z_1(2) = 1$$

9 Reconstruction by the Synthesis Bank

The averaging-differencing filter (named for Haar) produces two outputs from two successive inputs:

$$\begin{aligned}
y_0(2n) &= \frac{1}{2}(x(2n-1) + x(2n)) \\
y_1(2n) &= \frac{1}{2}(x(2n-1) - x(2n))
\end{aligned} \tag{32}$$

It is easy to recover $x(2n-1)$ and $x(2n)$. In fact, the inputs are just sums and differences of these outputs:

$$\begin{aligned}
x(2n-1) &= y_0(2n) + y_1(2n) \\
x(2n) &= y_0(2n) - y_1(2n)
\end{aligned} \tag{33}$$

Thus the reconstruction step uses the same operations as the analysis step.

Other analysis banks don't have this simple "two at a time" block structure, but the basic principle still applies. We will show how filtering followed by downsampling is inverted by **upsampling followed by filtering**. This sequence of inverse operations (and the notation) is displayed in the following diagram. *This is the synthesis bank:*

$$\text{Lowpass channel: } y_0(2n) \longrightarrow \boxed{\uparrow 2} \!-\! \boxed{F_0}$$
$$\widehat{x}(n) = x(n - \ell).$$
$$\text{Highpass channel: } y_1(2n) \longrightarrow \boxed{\uparrow 2} \!-\! \boxed{F_1}$$

The goal is to recover the input exactly, when no compression has been applied to the wavelet transform $z(n)$. The synthesis bank does recover x but with ℓ delays (ℓ depends on the filter bank). Long delays are not desirable, but causal filters cannot avoid some delay. The filter only looks backward.

First, we show how the combination of upsampling and filtering recovers the input to the Haar filter (with one delay!).

$$(\uparrow 2)\, y_0(2n) = \begin{bmatrix} y_0(0) \\ 0 \\ y_0(2) \\ 0 \end{bmatrix} \longrightarrow \boxed{\begin{array}{c}\text{sum}\\\text{filter } F_0\end{array}} \longrightarrow \begin{bmatrix} y_0(0) \\ y_0(0) \\ y_0(2) \\ y_0(2) \end{bmatrix} \quad \begin{bmatrix} x(-1) \\ x(0) \\ x(1) \\ x(2) \end{bmatrix}.$$

$$(\uparrow 2)\, y_1(2n) = \begin{bmatrix} y_1(0) \\ 0 \\ y_1(1) \\ 0 \end{bmatrix} \longrightarrow \boxed{\begin{array}{c}\text{difference}\\\text{filter } F_1\end{array}} \longrightarrow \begin{bmatrix} y_1(0) \\ -y_1(0) \\ y_1(2) \\ -y_1(2) \end{bmatrix} \tag{34}$$

At the end we used equations (33) for the sum and the difference, to produce x.

The choice of synthesis filters F_0 and F_1 is directly tied to the analysis filters H_0 and H_1 (since the two filter banks are inverses). We will write down the rule for F_0 and F_1, and show that it succeeds (**perfect reconstruction**). F_0 comes from *alternating the signs* in H_1. This takes the highpass H_1 into a lowpass filter F_0. Similarly, the highpass F_1 comes from alternating the signs of the coefficients in H_0. An example will make the point more clearly:

$$h_0 = \frac{1}{8}(-1, 2, 6, 2, -1) \qquad f_0 = \frac{1}{2}(1, 2, 1)$$

$$h_1 = \frac{1}{2}(1, -2, 1) \qquad f_1 = \frac{1}{4}(1, 2, -6, 2, 1).$$

The coefficients of h_1 and f_1 add to zero; a zero-frequency signal (a constant DC signal) is killed by these highpass filters. The coefficients of h_0 add to 1 and the coefficients of f_0 add to 2. The filters $h_0 = (\frac{1}{2}, \frac{1}{2})$ and $f_0 = (1, 1)$ in the Haar example followed these rules.

We can verify that this analysis-synthesis filter bank gives perfect reconstruction with $\ell = 3$ delays: The output is $x(n - 3)$. Suppose the input is a sinusoid $x(n) = e^{in\omega}$ at frequency ω. Then H_0 and H_1 become multiplications by their response functions $H_0(e^{i\omega})$ and $H_1(e^{i\omega})$. The combination of $(\downarrow 2)$ followed by $(\uparrow 2)$ introduces zeros in the odd-numbered components, which means that the **alias frequency** $\omega + \pi$ appears together with ω:

$$(\uparrow 2)(\downarrow 2)\{e^{in\omega}\} = \begin{bmatrix} e^{in\omega} \\ 0(\text{odd } n) \\ e^{i(n+2)\omega} \\ 0(\text{odd } n) \end{bmatrix} = \frac{1}{2}\left[\{e^{in\omega}\} + \{e^{in(\omega+\pi)}\}\right]. \qquad (35)$$

The factor $e^{in\pi}$ is $+1$ for even n and -1 for odd n. This aliased frequency $\omega + \pi$ is appearing in both channels of the filter bank. It has to be cancelled at the last step by F_0 and F_1. This was the reason behind the alternating signs (between h_0 and f_1 and between h_1 and f_0). Those alternations introduce powers of $e^{i\pi}$ in the frequency response functions:

$$\begin{aligned} F_0(e^{i\omega}) &= \frac{1}{2}(1 + 2e^{-i\omega} + e^{-2i\omega}) \\ &= \frac{1}{2}(1 - 2e^{-i(\omega+\pi)} + e^{-2i(\omega+\pi)}) \\ &= 2H_1(e^{i(\omega+\pi)}) = 2H_1(e^{-i\omega}). \qquad (36) \end{aligned}$$

Similarly,

$$F_1(e^{i\omega}) = -2H_0(e^{-i(\omega+\pi)}) = -2H_0(e^{-i\omega}) \qquad (37)$$

Now follow the pure exponential signal through analysis and synthesis, substituting these expressions (36)–(37) for F_0 and F_1 when they multiply the signal (the last step in the filter bank):

$$H_0(e^{i\omega})e^{in\omega} \longrightarrow \boxed{\downarrow 2}\!-\!\boxed{\uparrow 2} \longrightarrow \frac{1}{2}\left[H_0(e^{i\omega})e^{in\omega} + H_0(-e^{i\omega})e^{in(\omega+\pi)}\right]$$

$$\longrightarrow \boxed{F_0} \longrightarrow H_1(-e^{i\omega})H_0(e^{i\omega})e^{in\omega} + H_1(-e^{i\omega})H_0(-e^{i\omega})e^{in(\omega+\pi)}$$

$$H_1(e^{i\omega})e^{in\omega} \longrightarrow \boxed{\downarrow 2}\!-\!\boxed{\uparrow 2} \longrightarrow \frac{1}{2}\left[H_1(e^{i\omega})e^{in\omega} + H_1(-e^{i\omega})e^{in(\omega+\pi)}\right]$$

$$\longrightarrow \boxed{F_1} \longrightarrow H_0(-e^{i\omega})H_1(e^{i\omega})e^{in\omega} - H_0(-e^{i\omega})H_1(-e^{i\omega})e^{in(\omega+\pi)}.$$

These are the outputs from the low and high channels. When we add, the alias terms with frequency $\omega + \pi$ cancel out. The choice of F_0 and F_1 gave "alias cancellation."

The condition for perfect reconstruction comes from the terms involving $e^{in\omega}$. Those don't cancel! The sum of these terms should be $e^{i(n-\ell)\omega}$, which produces the

PR Condition: $\qquad H_1(-e^{i\omega})H_0(e^{i\omega}) - H_0(-e^{i\omega})H_0(e^{i\omega}) = e^{-i\ell\omega}. \qquad (38)$

This must hold for all frequencies ω. If we replace ω by $\omega + \pi$, the left side of the equation changes sign. Therefore, the right side must change sign, and *the system delay ℓ is always odd.*

Actually, the left side of (38) is exactly **the odd part** of the product

$$P(e^{i\omega}) = H_1(-e^{i\omega})H_0(e^{i\omega}) = \frac{1}{2}F_0(e^{i\omega})H_0(e^{i\omega}). \tag{39}$$

The PR condition states that *only one odd-numbered coefficient of this low-pass filter is nonzero*. That nonzero coefficient is the ℓth. We now verify that the example satisfies this PR condition:

$$
\begin{aligned}
P(e^{i\omega}) &= \frac{1}{4}(1 + 2e^{-i\omega} + e^{-2i\omega})\frac{1}{8}(-1 + 2e^{-i\omega} + 6e^{-2i\omega} + 2e^{-3i\omega} - e^{-4i\omega}) \\
&= \frac{1}{32}(-1 + 9e^{-2i\omega} + 16e^{-3i\omega} + 9e^{-4i\omega} - e^{-6i\omega}).
\end{aligned}
$$

The coefficients of the product filter P are $-1, 0, 9, 16, 9, 0, -1$ divided by 32. The middle coefficient is $\frac{16}{32} = \frac{1}{2}$. *The zeros in the other odd-numbered coefficients give perfect reconstruction.* And it is the particular coefficients in $-1, 0, 9, 16, 9, 0, -1$ that make P a powerful lowpass filter, because it has a *fourth-order zero* at the top frequency $\omega = \pi$. Factoring this polynomial produces

$$P(e^{i\omega}) = \left(\frac{1 + e^{-i\omega}}{2}\right)^4 (-1 + 4e^{-i\omega} - e^{-2i\omega}). \tag{40}$$

Roughly speaking, the power $(1 + e^{-i\omega})^4$ makes P a good lowpass filter. The final factor is needed to produce zeros in the first and fifth coefficients of P.

Remark 1. *The PR condition applies to the product P and not the separate filters H_0 and H_1. So any other factorization of $P(e^{i\omega})$ into $H_1(-e^{i\omega})H_0(e^{i\omega})$ gives perfect reconstruction. Here are two other analysis-synthesis pairs that share the same product $P(e^{i\omega})$:*

Biorthogonal 6/2

$$h_0 = \frac{1}{16}(-1, 1, 8, 8, 1, -1) \quad f_0 = (1, 1)$$

$$h_1 = \frac{1}{2}(1, -1) \qquad\qquad f_1 = \frac{1}{8}(-1, 1, -8, 8, -1, 1)$$

Orthogonal 4/4

$$h_0 = \frac{1}{8}\left(1 + \sqrt{3}, 3 + \sqrt{3}, 3 - \sqrt{3}, 1 - \sqrt{3}\right)$$

$$f_0 = \frac{1}{4}\left(1 - \sqrt{3}, 3 - \sqrt{3}, 3 + \sqrt{3}, 1 + \sqrt{3}\right)$$

$$h_1 = \frac{1}{8}\left(1 - \sqrt{3}, -3 + \sqrt{3}, 3 + \sqrt{3}, -1 - \sqrt{3}\right)$$

$$f_1 = \frac{1}{4}\left(-1 - \sqrt{3}, 3 + \sqrt{3}, -3 + \sqrt{3}, 1 - \sqrt{3}\right)$$

The second one is orthogonal because the synthesis filters are just transposes (time-reversed flips) of the analysis filters. When the inverse is the same as *the transpose, a matrix or a transform or a filter bank is called orthogonal. If we wrote out the infinite matrix H_{bank} for this analysis pair, we would discover that F_{bank} is essentially the transpose. (There will be a factor of 2 and also a shift by 3 diagonals to make the matrix lower triangular and causal again. These shifts are responsible for the $\ell = 3$ system delays.)*

The algebra of PR filter banks was developed by Smith-Barnwell and Mintzer in the 1970s. The iteration of those filter banks led to wavelets in the 1980s. The great paper of Ingrid Daubechies [2] gave the theory of orthogonal wavelets, and the first wavelets she constructed came from the coefficients $(1 + \sqrt{3}, 3 + \sqrt{3}, 3 - \sqrt{3}, 1 - \sqrt{3})$ given above.

The other filter banks, not orthogonal, are called *biorthogonal*. Synthesis is biorthogonal to analysis. The rows of H_{bank} are orthogonal to the columns of F_{bank}—except that $(row\ k) \cdot (column\ k) = 1$. This is always true for a matrix and its inverse:

$$HF = I \quad \text{means} \quad (row\ j\ of\ H) \cdot (column\ k\ of\ F) = \delta(j - k) \, .$$

The rows of H and columns of H^{-1} are biorthogonal. The analysis and synthesis wavelets that come below from infinite iteration will also be biorthogonal. For these basis functions we often use the shorter word *dual*.

10   Scaling Functions and Refinement Equation

May I write down immediately the most important equation in wavelet theory? It is an equation in continuous time, and its solution $\phi(t)$ is the "**scaling function.**" This section will describe basic examples, which happen to be spline functions. In later sections the connection to a filter bank and to multiresolution will be developed. More precisely, the connection is to the lowpass operator $(\downarrow 2)H_0$ of filtering and downsampling.

This fundamentally new and fascinating equation is the **refinement equation** or **dilation equation**:

$$\phi(t) = 2 \sum_{k=0}^{N} h(k)\phi(2t - k) \, . \tag{41}$$

The factor 2 gives the right side the same integral as the left side, because each $\phi(2t - k)$ has half the area (from compression to $2t$) and $\sum h(k) = 1$. It is the inner factor 2, the one that rescales t to $2t$, that makes this equation so remarkable.

Equation (41) is linear. If $\phi(t)$ is a solution, so is any multiple of $\phi(t)$. The usual normalization is $\int \phi(t)\,dt = 1$. That integral extends over all time, $-\infty < t < \infty$, but we actually find that the solution $\phi(t)$ is zero outside the interval $0 \le t < N$. **The scaling function has compact support.** This localization of the function is one of its most useful properties.

Example 1. *Suppose $h(0) = h(1) = \frac{1}{2}$ as in the averaging filter. Then the refinement equation for the scaling function $\phi(t)$ is*

$$\phi(t) = \phi(2t) + \phi(2t - 1).\tag{42}$$

The solution is the unit box function, which equals 1 on the interval $0 \leq t < 1$ (and elsewhere $\phi(t) = 0$). The function $\phi(2t)$ on the right side of the equation is a half-box, reaching only to $t = \frac{1}{2}$. The function $\phi(2t - 1)$ is a shifted half-box, reaching from $t = \frac{1}{2}$ to $t = 1$. Together, the two half-boxes make up the whole box, and the refinement equation (42) is satisfied.

Notice in this example the two key operations of classical wavelet theory:

Dilation of the time scale by 2 or 2^j:
   The graph of $\phi(2t)$ is compressed by 2.

Translation of the time scale by 1 or k:
   The graph of $\phi(t - k)$ is shifted by k.

The combination of those operations, from t to $2^j t$ to $2^j t - k$, will later produce a family of wavelets $w(2^j t - k)$ from a single wavelet $w(t)$.

Example 2. *Suppose $h(0) = \frac{1}{4}$, $h(1) = \frac{1}{2}$, $h(2) = \frac{1}{4}$. The averaging filter has been repeated. Whenever we multiply filters (Toeplitz matrices), we are convolving their coefficients. Squaring a filter means convolving the vector of coefficients with itself, and this is verified by $(\frac{1}{2}, \frac{1}{2}) * (\frac{1}{2}, \frac{1}{2}) = (\frac{1}{4}, \frac{1}{2}, \frac{1}{4})$. This convolution is just a multiplication of a polynomial times itself:*

$$\left(\frac{1}{2} + \frac{1}{2}z^{-1}\right)^2 = \frac{1}{4} + \frac{1}{2}z^{-1} + \frac{1}{4}z^{-2}.$$

The solution $\phi(t)$, which is the scaling function for $h = (\frac{1}{4}, \frac{1}{2}, \frac{1}{4})$, is the old solution convolved with itself:

(box function) $*$ (box function) $=$ **hat function**.

The box is piecewise constant. The hat is piecewise linear, equal to t and $2 - t$ on the intervals $[0, 1]$ and $[1, 2]$. Figure 4 shows how three compressed and shifted hat functions combine into the full hat function. The picture verifies the refinement equation $\phi(t) = \frac{1}{2}\phi(2t) + \phi(2t - 1) + \frac{1}{2}\phi(2t - 2)$.

Example 3. *(Splines of degree $p - 1$). Every time an extra averaging filter is introduced, we are convolving $\phi(t)$ with the box function. This makes the filter coefficients and the scaling function smoother; the function has one additional derivative. From a product of p averaging filters, the frequency response is $(\frac{1+e^{-i\omega}}{2})^p$. Its kth coefficient will be the binomial number $\binom{p}{k} =$ "p choose k" divided by 2^p.*

*The scaling function for this filter is a **B-spline** on the interval $[0, p]$. This spline $\phi(t)$ is formed from polynomials of degree $p - 1$ that are joined*

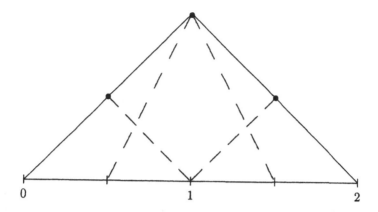

Figure 4: Three half-length hats combine into a full-length hat.

together smoothly (p − 2 continuous derivatives) at the nodes $t = 0, 1, \ldots, p$. This is the convolution of p box functions:

Filter:
$$\left(\frac{1}{2}, \frac{1}{2}\right)\left(\frac{1}{2}, \frac{1}{2}\right)\cdots\left(\frac{1}{2}, \frac{1}{2}\right) = \frac{1}{2^p}\left(1, p, \frac{p(p-1)}{2}, \ldots, p, 1\right)$$

Function: (box)(box) ··· (box) = B-spline of degree p − 1.

These are essentially all the examples in which the scaling function $\phi(t)$ has a simple form. If we take $h(0) = \frac{2}{3}$ and $h(1) = \frac{1}{3}$, we get a very singular and unbounded solution, with infinite energy. It blows up at $t = 0$, where (41) becomes $\phi(0) = \frac{4}{3}\phi(0)$, but it is not a standard delta function. After convolving with $(\frac{1}{2}, \frac{1}{2})$, the coefficients $h = (\frac{1}{3}, \frac{1}{2}, \frac{1}{6})$ are much better. The scaling function is smoother and in L^2. But still we have no elementary expression for $\phi(t)$ when the refinement equation has these coefficients. All we can do is to compute $\phi(t)$ at more and more binary points $t = m/2^j$, or find its Fourier transform (see below). We can graph the function as accurately as desired, but we don't know a simple formula.

Now we begin the connection with a filter bank. The output from a lowpass filter (convolution with h), followed by downsampling (n changes to $2n$), is but it is *not* a standard delta function. $\sum h(k)x(2n-k)$. Suppose this output goes back in as input. (**This is iteration.**) The process could be studied in discrete time, but continuous time is more interesting and revealing. So the original input is a box function, and we get $\phi^{(i+1)}(t)$ from $\phi^{(i)}(t)$:

Filter and rescale: $$\phi^{(i+1)}(t) = 2\sum_{k=0}^{N} h(k)\phi^{(i)}(2t - k).$$ (43)

This is the **cascade algorithm**. The lowpass filter is "cascaded." If this iteration converges, so that $\phi^{(i)}(t)$ approaches a limit $\phi(t)$ as $i \to \infty$, then that limit function satisfies the refinement equation (41).

In this sense the scaling function is a fixed point (or eigenfunction with $\lambda = 1$) of the filtering-downsampling operation $2(\downarrow 2)H_0$. The normalization by 2 is necessary because time is compressed by 2 at each iteration. Starting from the box function $\phi^{(0)}(t)$ on $[0, 1]$, the function $\phi^{(i)}(t)$ will be piecewise constant on intervals of length 2^{-i}. Thus continuous time gives a natural way to account for the downsampling ($n \to 2n$ and $t \to 2t$). This time compression is clearer for a function than for a vector.

When we apply the iteration (43) for the averaging filter $h = (\frac{1}{2}, \frac{1}{2})$, we find $\phi^{(1)} \equiv \phi^{(0)}$. The convergence is immediate because the box $\phi^{(0)}$ is the correct scaling function. When we use $h = (\frac{1}{4}, \frac{1}{2}, \frac{1}{4})$, every iteration produces a "staircase up and down" with steps of 2^{-i}. The function $\phi^{(i)}(t)$ is a piecewise constant hat, supported on the interval $[0, 2 - 2^{-i}]$. The limit as $i \to \infty$ is the correct piecewise linear hat function $\phi(t)$.

Similarly the B-spline comes from the cascade algorithm. But for other coefficients $h(k)$, always adding to 1, there may or may not be a limit function. The functions $\phi^{(i)}(t)$ can converge to a limit $\phi(t)$, or they can explode. If you try the bad filter $h = (\frac{2}{3}, \frac{1}{3})$, you find that $\phi^{(i)}(0) = (\frac{4}{3})^i$. And the iterations explode at infinitely many other points too (please try this example). A natural question is to find the requirement for the cascade algorithm to converge:

1. For which filters $h(k)$ does $\phi^{(i)}(t)$ approach a limit $\phi(t)$ in L^2?

A minimal condition is that $\sum(-1)^k h(k) = 0$. In the frequency domain, this is the response $H(e^{i\omega}) = \sum h(k)e^{-ik\omega}$ at the highest frequency $\omega = \pi$. The filter must have a "**zero at π**" for convergence to have a chance (and this eliminates many equiripple lowpass filters). The precise Condition E is on the eigenvalues of a matrix (as we expect in all iterations!):

Condition E: The matrix $T = 2(\downarrow 2)HH^T$ must have all eigenvalues $|\lambda| < 1$, except for a simple eigenvalue $\lambda = 1$.

This is a familiar condition for convergence of the powers T^i of a matrix to a rank one matrix (as in the "power method" to compute eigenvalues).

You might like to see three rows of T for the hat filter $h = (\frac{1}{4}, \frac{1}{2}, \frac{1}{4})$:

$$2(\downarrow 2)HH^T = T = \frac{1}{8} \begin{bmatrix} 1 & 4 & 6 & \begin{pmatrix} 4 & 1 & \\ 4 & 6 & 4 \\ & 1 & 4 \end{pmatrix} & 1 \\ & & 1 & & 6 & 4 & 1 \end{bmatrix}.$$

Note the double shift between rows which comes from ($\downarrow 2$). The coefficients $1, 4, 6, 4, 1$ come from HH^T. In polynomial terms, we are multiplying $H(z) = \frac{1}{4}(1 + 2z^{-1} + z^{-2})$ by its time-reversed flip $H(z^{-1}) = \frac{1}{4}(1 + 2t + z^2)$. The matrix product HH^T is an *autocorrelation* of the filter.

We have highlighted the crucial 3 by 3 matrix inside T. Its three eigenvalues are $\lambda = 1, \frac{1}{2}, \frac{1}{4}$. Because of the double shift, the rest of T will contribute $\lambda = \frac{1}{8}$ (twice) and $\lambda = 0$ (infinitely often). No eigenvalue exceeds $\lambda = 1$ and Condition E is clearly satisfied.

A very short MATLAB program will compute T from the coefficients $h(k)$ and find the eigenvalues. There will be $\lambda = 1, \frac{1}{2}, \frac{1}{4}, \dots, \frac{1}{2^{2p-1}}$, when the filter response $H(e^{i\omega})$ has a pth order zero at $\omega = \pi$. It is the *largest other eigenvalue* $|\lambda_{\max}|$ of T that decides the properties of $\phi(t)$. That eigenvalue $|\lambda_{\max}|$ must be smaller than 1. This is Condition E above for convergence of the cascade algorithm.

We mention one more important fact about $|\lambda_{\max}|$, without giving the proof here. This number determines the **smoothness** of the scaling function (and later also of the wavelet). A smaller number means a smoother function. We measure smoothness by the **number of derivatives** of $\phi(t)$ that have finite energy (belong to the space L^2). This number s need not be an integer. Thus the box function is in L^2, and its first derivative (delta functions $\delta(t) - \delta(t-1)$) is certainly not in L^2. But at a jump, all derivatives of order $s < \frac{1}{2}$ do have finite energy:

$$\text{Energy} = \int_{-\infty}^{\infty} \left| \frac{d^s \phi}{dt^s} \right|^2 dt = \int_{-\infty}^{\infty} \left| (i\omega)^s \widehat{\phi}(\omega) \right|^2 d\omega .$$

The Fourier transform of the box function has $|\widehat{\phi}(\omega)| = |\operatorname{sinc} \omega| \le const / |\omega|$. Then for every $s < \frac{1}{2}$,

$$\text{Energy in } s^{\text{th}} \text{ derivative} < const + const \int_1^{\infty} |\omega|^{2s-2} d\omega < \infty .$$

The hat function is one order smoother than the box function (one extra Haar factor). So it has derivatives with finite energy up to order $s = \frac{3}{2}$. The general formula for the smoothness of $\phi(t)$ is remarkably neat: All derivatives have finite energy, up to the order

$$s_{\max} = -\log_4 |\lambda|_{\max} . \tag{44}$$

For the hat function and its filter $h = (\frac{1}{4}, \frac{1}{2}, \frac{1}{4})$, the double eigenvalue $\lambda = \frac{1}{8}$ was mentioned for the matrix T above. This is $\lambda_{\max}$ and the formula (44) correctly finds the smoothness $s = -\log_4(\frac{1}{8}) = \frac{3}{2}$.

We can explicitly solve the refinement equation in the frequency domain, for each separate ω. The solution $\widehat{\phi}(\omega)$ is an infinite product. The equation connects the scales t and $2t$, so its Fourier transform connects the frequencies ω and $\omega/2$:

The transform of $\phi(2t - k)$ is $\int \phi(2t - k)e^{-i\omega t} dt = \frac{1}{2}e^{-i\omega k/2}\widehat{\phi}(\frac{\omega}{2})$.

Then the refinement equation $\phi(t) = 2 \sum h(k)\phi(2t - k)$ transforms to

$$\widehat{\phi}(\omega) = \left(\sum h(k)e^{-i\omega k/2} \right) \widehat{\phi}\left(\frac{\omega}{2}\right) = H\left(\frac{\omega}{2}\right) \widehat{\phi}\left(\frac{\omega}{2}\right) . \tag{45}$$

By iteration (which is the cascade algorithm!), the next step gives $\widehat{\phi}(\omega) = H(\frac{\omega}{2})H(\frac{\omega}{4})\widehat{\phi}(\frac{\omega}{4})$. Infinite iteration gives the *infinite product formula*

$$\widehat{\phi}(\omega) = \prod_{j=1}^{\infty} H\left(\frac{\omega}{2^j}\right) \qquad (\text{since } \widehat{\phi}(0) = 1). \qquad (46)$$

This product gives a finite answer for each frequency ω, so we always have a formula for the Fourier transform of $\phi(t)$. But if $|\lambda_{\max}| > 1$ and Condition E fails to hold, the inverse transform from $\widehat{\phi}(\omega)$ to $\phi(t)$ will produce a wild function with smoothness $s < 0$ and infinite energy.

In the case of box functions and splines, this infinite product simplifies to sinc functions and their powers:

$$\widehat{\phi}_{\text{box}}(\omega) = \left(\frac{1 - e^{-i\omega}}{i\omega}\right) \quad \text{and} \quad \widehat{\phi}_{\text{spline}}(\omega) = \left(\frac{1 - e^{-i\omega}}{i\omega}\right)^p. \qquad (47)$$

In an interesting recent paper, Blu and Unser [15] use this formula to define splines for *noninteger p*. These *fractional splines* are not polynomials, and their support is not compact, but they interpolate in a natural way between the piecewise polynomial splines (where p is an integer).

The Daubechies 4-tap filter produces a scaling function with $s_{\max} = 1$. The "lazy filter" with only one coefficient $h(0) = 1$ produces $\phi(t) = $ Dirac delta function. You can verify directly that $\delta(t) = 2\delta(2t)$, so the refinement equation is satisfied. In the frequency domain, $H(\omega) \equiv 1$ so that $\widehat{\phi}(\omega) \equiv 1$. The smoothness of the delta function is one order lower ($s = -\frac{1}{2}$) than the step function: $H = I$ and $T = 2I$ and $|\lambda_{\max}| = 2$ and $s = -\log_4 2 = -\frac{1}{2}$.

11   Multiresolution and the Spaces A

Our first explanation of the refinement equation $\phi(t) = 2\sum h(k)\phi(2t - k)$ was by analogy. This is a continuous time version of filtering (by h) and downsampling (t to $2t$). When we iterate the process, we are executing the cascade algorithm. When the initial $\phi^{(0)}(t)$ is the box function, all iterates $\phi^{(i)}(t)$ are piecewise constant and we are really copying a discrete filter bank. We are iterating forever! In this description, the refinement equation gives the limit of an infinite filter tree.

Now we give a different and deeper explanation of the same equation. We forget about filters and start with functions. Historically, this is how the refinement equation appeared. (Later it was realized that there was a close connection to the filter banks that had been developed in signal processing, and the two histories were joined. Wavelets and lowpass filters come from highpass and lowpass filters. An M-channel filter bank would lead to one scaling function, with M's instead of 2's in the refinement equation, and $M - 1$ wavelets.) But the first appearance of $\phi(t)$ was in the idea of **multiresolution**.

Multiresolution looks at different time scales, t and $2t$ and every $2^j t$. These correspond to octaves in frequency. When a problem splits naturally into several time scales or frequency scales, this indicates that the wavelet approach (multiresolution) may be useful. Wavelets are another tool in time-frequency analysis comparable to the windowed "short-time" Fourier analysis that transforms $f(t)$ to a function of the window position t and the frequency ω. For wavelets, it is a *time-scale analysis*, with j for the scale level and k for the position. The wavelet coefficients produce a "scalogram" instead of a spectrogram.

Start with Mallat's basic rules for a multiresolution [10]. There is a space V_0, containing all combinations of the basis functions $\phi(t - k)$. This shift-invariant basis produces a shift-invariant subspace of functions:

Property 1: $f(t)$ is in V_0 if and only if $f(t - 1)$ is in V_0.

Now compress each basis function, rescaling the time from t to $2t$. The functions $\phi(2t - k)$ are a new basis for a new space V_1. This space is again shift-invariant (with shifts of $\frac{1}{2}$). And between V_0 and V_1 there is an automatic scale-invariance:

Property 2: $f(t)$ is in V_0 if and only if $f(2t)$ is in V_1.

Now we come to the crucial third requirement. **We insist that the subspace V_1 contains V_0.** This requires in particular that the subspace V_1 contains the all-important function $\phi(t)$ from V_0. In other words, $\phi(t)$ must be a combination of the functions $\phi(2t - k)$ that form a basis for V_1:

Property 3. (*Refinement Equation*) $\phi(t) = \sum c_k \phi(2t - k)$.

When this holds, all other basis functions $\phi(t - k)$ at the coarse scale will be combinations of the basis functions $\phi(2t - k)$ at the fine scale. In other words $V_0 \subset V_1$.

I hope that every reader recognizes the connection to filters. *Those coefficients c_k in Property 3 will be exactly the filter coefficients $2h(k)$.* Previously we constructed $\phi(t)$ from the refinement equation. Now we are starting from V_0 and V_1 and their bases, and the requirement that V_1 contains V_0 gives the refinement equation.

The extension to a whole "scale of subspaces" is immediate. The space V_j consists of all combinations of $\phi(2^j t)$ and its shifts $\phi(2^j t - k)$. When we replace t by $2^j t$ in the refinement equation (Property 3), we discover that V_j is contained in V_{j+1}. Thus the sequence of subspaces is "nested" and $V_0 \subset V_1 \subset V_2 \subset \cdots$.

There is one optional property, which was often included in the early papers but is not included now. That is the *orthogonality* of the basis functions $\phi(t - k)$. The Daubechies functions possess this property, but the spline functions do not. (The only exception is the box function = spline of degree zero = Haar scaling function = first Daubechies function.) In the orthogonal case, the coefficients c_k in the refinement equation come from an *orthogonal filter bank*. Certainly orthogonality is a desirable property and these bases are very frequently used — but then $\phi(t)$ cannot have the symmetry that is desirable in image processing.

To complete the definition of a multiresolution, we should add a requirement on the basis functions. We don't insist on orthogonality, but we do require a stable basis (also known as a Riesz basis). We think of V_0 as a subspace of L^2 (the functions with finite energy $\|f\|^2 = \int |f(t)|^2 \, dt$). The basis functions $\phi(t - k)$ must be "uniformly" independent:

Property 4. (*Stable Basis*) $\|\sum_{-\infty}^{\infty} a_k \phi(t - k)\|^2 \geq c \sum a_k^2$ with $c > 0$.

This turns out, beautifully and happily, to be equivalent to Condition E on the convergence of the cascade algorithm. An earlier equivalent condition on $\phi(t)$ itself was discovered by Cohen. We can also establish that the spaces V_j approximate every finite energy function $f(t)$: the projection onto V_j converges to $f(t)$ as $j \to \infty$. And one more equivalent condition makes this a very satisfactory theory: the scaling functions $\phi(t - k)$ together with the normalized wavelets $2^{j/2} w(2^j t - k)$ at all scales $j = 0, 1, 2, \dots$ also form a stable basis.

We now have to describe these wavelets. For simplicity we do this in the orthogonal case. First a remark about "multiwavelets" and a word about the order of approximation to $f(t)$ by the subspaces V_j.

Remark 2. *Instead of the translates $\phi(t - k)$ of one function, the basis for V_0 might consist of the translates $\phi_1(t - k), \dots, \phi_r(t - k)$ of r different functions. All of the properties 1–4 are still desired. The refinement equation in Property 3 now takes a vector form, with r by r matrix coefficients c_k: V_0 is a subspace of V_1 when all the functions $\phi_1(t), \dots, \phi_r(t)$ are combinations of the basis functions $\phi_1(2t - k), \dots, \phi_r(2t - k)$:*

$$\begin{bmatrix} \phi_1(t) \\ \vdots \\ \phi_r(t) \end{bmatrix} = \sum c_k \begin{bmatrix} \phi_1(2t - k) \\ \vdots \\ \phi_r(2t - k) \end{bmatrix}.$$

The corresponding lowpass filter becomes a "multifilter," with vectors of r samples as inputs and r by r matrices in place of the numbers $h(0), \dots, h(N)$. The processing is more delicate but multifilters have been successful in denoising.

The new feature of multiwavelets is that the basis functions can be both symmetric and orthogonal (with short support). A subclass of "balanced" multifilters needing no preprocessing of inputs has been identified by Vetterli-LeBrun [9] and by Selesnick [11].

12 Polynomial Reproduction and Accuracy of Approximation

The requirements for a good basis are easy to state, and important:

1. *Fast computation* (wavelets are quite fast)

2. *Accurate approximation* (by relatively few basis functions).

We now discuss this question of accuracy. When $f(t)$ is a smooth function, we ask how well it is approximated by combining the translates of $\phi(t)$. Why not give the answer first:

$$\left\| f(t) - \sum_{k=-\infty}^{\infty} a_{jk}\phi(2^j t - k) \right\| \leq C 2^{-jp} \left\| f^{(p)}(t) \right\|. \tag{48}$$

This says that with time steps of length $h = 2^{-j}$, the degree of approximation is h^p. This number p is still the *multiplicity of the zero* of $H(e^{i\omega})$ at $\omega = \pi$.

This error estimate (48) is completely typical of numerical analysis. The error is *zero* when $f(t)$ is a polynomial of degree $p - 1$. (We test Simpson's Rule and every quadrature rule on polynomials.) The Taylor series for $f(t)$ starts with such a polynomial (perfect approximation). Then the remainder term involves the pth derivative $f^{(p)}(t)$ and the power h^p. This leads to the error estimate (48). So the degree of the first polynomial t^p that is not reproduced by the space V_0 determines the crucial exponent p.

The box function has $p = 1$. The error is $O(h)$ in piecewise constant approximation. The hat function has $p = 2$ because $H(e^{i\omega}) = (\frac{1+e^{-i\omega}}{2})^2$. The error is $O(h^2)$ in piecewise linear approximation. The 4-tap Daubechies filter also has $p = 2$, and gives an error that is $O(h^2)$. Daubechies created a whole family with $p = 2, 3, 4, \ldots$ of orthogonal filters and scaling functions and wavelets. They have $2p$ filter coefficients and approximation errors $O(h^p)$. They are zero outside the interval from $t = 0$ to $t = 2p - 1$.

We can briefly explain how the order p of the zero at $\omega = \pi$ turns out to be decisive. **That multiple zero guarantees that all the polynomials $1, t, t^2, \ldots, t^{p-1}$ can be exactly produced as combinations of $\phi(t - k)$.** In the frequency domain, the requirement on $\widehat{\phi}(\omega)$ for this polynomial reproduction is known as the Strang-Fix condition:

$$\frac{d^j \widehat{\phi}}{d\omega^j}(2\pi n) = 0 \qquad \text{for } n \neq 0 \text{ and } j < p. \tag{49}$$

Thus $\widehat{\phi}(\omega)$ at the particular frequencies $\omega = 2\pi n$ is the key.

Formula (46) for $\widehat{\phi}(\omega)$ is an infinite product of $H(\omega/2^j)$. At frequency $\omega = 2\pi n$, we write $n = 2^\ell q$ with q odd. Then the factor in the infinite product with $j = \ell + 1$ is $H(2\pi n/2^{\ell+1}) = H(q\pi)$. By periodicity this is

$H(\pi)$. A pth order zero of $\widehat{\phi}$ lies at $\omega = 2\pi n$. Thus the Strang-Fix condition on $\widehat{\phi}$ is equivalent to Condition A_p on the filter:

   Condition A_p: $H(e^{i\omega})$ has a pth order zero at $\omega = \pi$.

For the filter coefficients, a zero at π means that $\sum(-1)^n h(n) = 0$. A zero of order p translates into $\sum(-1)^n n^k h(n) = 0$ for $k = 0, \ldots, p-1$.

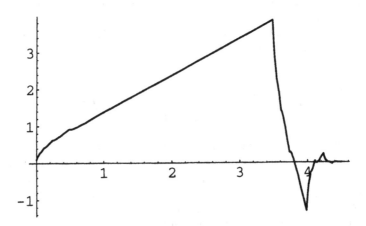

Figure 5: With $p = 2$, the D4 scaling functions $\sum k\phi(t-k)$ produces a straight line.

It is remarkable to see in Figure 5 how combinations of $\phi(t-k)$ produce a straight line, when $\phi(t)$ is the highly irregular Daubechies scaling function coming from $h = \frac{1}{8}(1+\sqrt{3}, 3+\sqrt{3}, 3-\sqrt{3}, 1-\sqrt{3})$. The response $H(e^{i\omega})$ has a *double zero* at $\omega = \pi$. The eigenvalues of $T = 2(\downarrow 2)HH^T$ are $\lambda = 1, \frac{1}{2}, \frac{1}{4}, \frac{1}{4}, \frac{1}{8}$, which includes the $1, \frac{1}{2}, \frac{1}{4}, \frac{1}{8}$ that always come with $p = 2$. The smoothness $s = 1$ of the D4 scaling function is governed by that extra $\frac{1}{4} = \lambda_{\text{max}}$.

13   Multiresolution and Wavelets

When j increases by 1, the mesh size $h = 2^{-j}$ is multiplied by $\frac{1}{2}$. The error of approximation by scaling functions is reduced by $(\frac{1}{2})^p$. There are twice as many functions available when approximation at scale j uses the translates of a compressed $\phi(2t)$:

   The approximation space V_j contains all combinations of the shifted and compressed scaling functions $\phi(2^j t - k)$, $-\infty < k < \infty$.

The special feature of these spaces is that V_1 contains V_0 (and every V_{j+1} contains V_j). This comes directly from $\phi(t) = 2\sum h(k)\phi(2t - k)$. *The right* *side of the refinement equation is a function in V_1, at scale $j = 1$.* Therefore the left side $\phi(t)$ lies in V_1. So do all its translates, and all their linear combinations. Thus the whole of V_0 lies in V_1.

Now we look at the functions that are in V_1 but *not in V_0*. This will lead us to the *wavelet space W_0*. The sum of the subspaces $V_0 + W_0$ will equal V_1.

The increasing sequence of spaces $V_0 \subset V_1 \subset V_2 \subset \cdots$ leads us to a **multiresolution** of any finite-energy function $f(t)$. Suppose first that the functions $\phi(t - k)$ are orthogonal. Let $f_j(t)$ be the perpendicular projection of $f(t)$ onto the space V_j. Thus $f_j(t)$ is the nearest function to $f(t)$ in V_j. We will write $f(t)$ as a telescoping sum

$$f(t) = f_0(t) + [f_1(t) - f_0(t)] + [f_2(t) - f_1(t)] + \cdots . \tag{50}$$

This is the sum of a lowest-level "coarse function" $f_0(t)$ and an infinite series of "fine details" $d_j(t) = f_{j+1}(t) - f_j(t)$. Those details will lie in the wavelet subspaces W_j. So the splitting of one function $f(t)$ indicates the splitting of the whole space of all L^2 functions:

$$L^2 = V_0 + W_0 + W_1 + \cdots \tag{51}$$

Notice that each detail $d_j(t)$ also lies in the next space V_{j+1} (because f_{j+1} and f_j are both in V_{j+1}). Thus W_j is in V_{j+1}. Furthermore $d_j(t)$ is orthogonal to V_j, by simple reasoning:

$f(t) - f_j(t)$ is orthogonal to V_j because f_j is the projection

$f(t) - f_{j+1}(t)$ is similarly orthogonal to V_{j+1} and therefore to V_j.

Subtraction shows that the detail $d_j(t) = f_{j+1}(t) - f_j(t)$ is orthogonal to V_j.

We assign the name W_j to the *wavelet subspace* containing all those "detail functions" in V_{j+1} that are orthogonal to V_j. Then the finer space V_{j+1} is a sum of coarse functions $f_j(t)$ in V_j and details $d_j(t)$ in W_j:

For functions: $f_j(t) + d_j(t) = f_{j+1}(t)$
For subspaces: $V_j + W_j = V_{j+1}$

The function $f(t)$ is a combination of the compressed translates $\phi(2^j t - k)$, which are a basis for V_j. We want the detail $d_j(t)$ to be a combination of compressed translates $w(2^j t - k)$, which are a basis for W_j. **The fundamental detail function $w(t)$ will be the wavelet.**

The wavelet has a neat construction using the highpass coefficients $h_1(k)$:

Wavelet Equation: $\quad w(t) = 2 \sum h_1(k) \phi(2t - k) . \tag{52}$

The right side is certainly in V_1. To prove that $w(t)$ is orthogonal to $\phi(t)$ and its translates, we use the relation $h_1(k) = (-1)^k h_0(N - k)$ between the lowpass and highpass coefficients. We omit this verification here, and turn instead to the most basic example.

The *Haar wavelet* is $w(t) = \phi(2t) - \phi(2t - 1) =$ up-and-down box function. It is a *difference* of narrow boxes. Its integral is zero. (Always $\int w(t)\,dt = 0$ because $\sum h_1(k) = 0$.) Furthermore the Haar wavelets $w(2^j t - k)$, compressed and shifted versions of the square wave $w(t)$, are *all* orthogonal to each other and to the box function $\phi(t)$. This comes from the orthogonality of the Haar filter bank.

A biorthogonal filter bank will have four filters H_0, F_0, H_1, F_1. Then H_0 and F_0 yield two different scaling functions $\tilde{\phi}(t)$ and $\phi(t)$, in analysis and synthesis. Similarly H_1 and F_1 yield analysis wavelets $\tilde{w}(t)$ and synthesis wavelets $w(t)$. The orthogonality property becomes *biorthogonality* between analysis and synthesis:

$$\int \phi(t)\tilde{w}(2^j t - k)\,dt = \int \tilde{\phi}(t)w(2^j t - k)\,dt = 0.$$

14 Good Basis Functions

This section returns to one of the fundamental ideas of linear algebra — a basis. It often happens that one basis is more suitable than another for a specific application like compression. The whole idea of a *transform* (this paper concentrates on the Fourier transform and wavelet transform) is exactly a change of basis. Each term in a Fourier series or a wavelet series is a basis vector, a sinusoid or a wavelet, times its coefficient. A change of basis gives a *new representation of the same function*.

Remember what it means for the vectors $w_1, w_2, \ldots, w_n$ to be a basis for an n-dimensional space $\mathbf{R}^n$:

1. The w's are linearly independent.

2. The $n \times n$ matrix W with these column vectors is invertible.

3. Every vector x in $\mathbf{R}^n$ can be written in exactly one way as a combination of the w's:

$$\mathbf{x} = c_1 w_1 + c_2 w_2 + \cdots + c_n w_n.$$

Here is the key point: those coefficients $c_1, \cdots, c_n$ completely describe the vector. In the original basis (standard basis), the coefficients are just the samples $x(1), \ldots, x(n)$. In the new basis of w's, the same x is described by $c_1, \ldots, c_n$. It takes n numbers to describe each vector and it *also* takes a choice of basis:

$$x = \begin{bmatrix} x(1) \\ x(2) \\ x(3) \\ x(4) \end{bmatrix}_{\text{standard basis}} \quad \text{and} \quad x = \begin{bmatrix} z_0(2) \\ z_1(2) \\ y_1(2) \\ y_1(4) \end{bmatrix}_{\text{Haar basis}}.$$

A basis is a set of axes for $\mathbf{R}^n$. The coordinates $c_1, \ldots, c_n$ tell how far to go along each axis. The axes are at right angles when the basis vectors $w_1, \ldots, w_n$ are orthogonal.

The Haar basis is orthogonal. This is a valuable property, shared by the Fourier basis. For the vector $x = (6, 4, 5, 1)$ with four samples, we need four Haar basis vectors. Notice their orthogonality (inner products are zero).

$$
w_1 = \begin{bmatrix} 1 \\ 1 \\ 1 \\ 1 \end{bmatrix} \qquad
w_2 = \begin{bmatrix} 1 \\ 1 \\ -1 \\ -1 \end{bmatrix} \qquad
w_3 = \begin{bmatrix} 1 \\ -1 \\ 0 \\ 0 \end{bmatrix} \qquad
w_4 = \begin{bmatrix} 0 \\ 0 \\ 1 \\ -1 \end{bmatrix}.
$$

That first basis vector is not actually a wavelet, it is the very useful flat vector of all ones. It represents a constant signal (the DC component). Its coefficient c_1 is the overall average of the samples this low-low coefficient c_1 was $\frac{1}{4}(6 + 4 + 5 + 1) = 4$ for our example vector.

The Haar wavelet representation of the signal is simply $x = Wc$. The input x is given in the standard basis of samples. The basis vectors $w_1, \ldots, w_4$ go into the columns of W. They are multiplied by the coefficients $c_1, \ldots, c_4$ (this is how matrix multiplication works!). The matrix-vector product Wc is exactly $c_1 w_1 + c_2 w_2 + c_3 w_3 + c_4 w_4$. Here are the numbers in $x = Wc$:

$$
\begin{bmatrix} 6 \\ 4 \\ 5 \\ 1 \end{bmatrix} =
\begin{bmatrix} 1 & 1 & 1 & 0 \\ 1 & 1 & -1 & 0 \\ 1 & -1 & 0 & 1 \\ 1 & -1 & 0 & -1 \end{bmatrix}
\begin{bmatrix} 4 \\ 1 \\ 1 \\ 2 \end{bmatrix}. \tag{53}
$$

Those coefficients $c = (4, 1, 1, 2)$ *are* $W^{-1}x$. This is what the analysis filter bank must do; a change of basis involves the inverse of the basis matrix. The analysis step computes coefficients $c = W^{-1}x$, and the synthesis step multiplies coefficients times basis vectors to reconstruct $x = Wc$.

The point of wavelets is that both steps, analysis and synthesis, are executed quickly by filtering and subsampling. Let me extend this point to repeat a more general comment on good bases. Orthogonality gives a simple relation between the transform and the inverse transform (they are "transposes" of each other). But I regard two other properties of a basis as more important in practice than orthogonality:

Speed: The coefficients $c_1, \ldots, c_n$ are fast to compute.

Accuracy: A small number of basis vectors and their coefficients can represent the signal very accurately.

For the Fourier and wavelet bases, the speed comes from the FFT and FWT: Fast Fourier Transform and Fast Wavelet Transform. The FWT is exactly the filter bank tree that we just illustrated for the Haar wavelets.

The accuracy depends on the signal! We want to choose a basis appropriate for the class of signals we expect. (It is usually too expensive to choose a basis adapted to each individual signal.) Here is a rough guideline:

For *smooth* signals with no jumps, the *Fourier basis* is hard to beat.
For *piecewise smooth* signals with jumps, a *wavelet basis* can be better.

The clearest example is a step function. The Fourier basis suffers from the *Gibbs phenomenon*: oscillations (ringing) near the jump, and very slow $\frac{1}{n}$ decay of the Fourier coefficients c_n. The wavelet basis is much more localized, and only the wavelets crossing the jump will have large coefficients. Of course the Haar basis could be perfect for a step function, but it has slow decay for a simple linear ramp. Other wavelets of higher order can successfully capture a jump discontinuity in the function or its derivatives.

To emphasize the importance of a good basis (an efficient representation of the data), we will list some important choices. Each example is associated with a major algorithm in applied mathematics. We will mention bases of functions (continuous variable x or t) although the algorithm involve vectors (discrete variables n and k).

1. Fourier series (sinusoidal basis)

2. Finite Element Method (piecewise polynomial basis)

3. Spline Interpolation (smooth piecewise polynomials)

4. Radial Basis Functions (functions of the distance $\|x - x_i\|$ to interpolation points x_i in d dimensions)

5. Legendre polynomials, Bessel functions, Hermite functions, ... (these are orthogonal solutions of differential equations)

6. Wavelet series (functions $w(2^j x - k)$ from dilation and translation of a wavelet).

There is a similar list of discrete bases. Those are vectors, transformed by matrices. It is time to see the matrix that represents a lowpass filter (convolution matrix), and also the matrix that represents a filter bank (wavelet transform matrix).

15 Filters and Filter Banks by Matrices

We begin with the matrix for the averaging filter $y(n) = \frac{1}{2}(x(n-1) + x(n))$. The matrix is lower triangular, because this filter is causal. The output at time n does not involve inputs at times later than n. The averaging matrix has the entries $\frac{1}{2}$ on its main diagonal (to produce $\frac{1}{2}x(n)$), and it has $\frac{1}{2}$ on its first subdiagonal (to produce $\frac{1}{2}x(n-1)$). We regard the vectors x and y

as doubly infinite, so the index n goes from $-\infty$ to ∞. Then the matrix that executes this filter is $H = H_{\text{averaging}}$:

$$
y = Hx = \begin{bmatrix} \cdot & & & \\ \frac{1}{2} & \frac{1}{2} & & \\ & \frac{1}{2} & \frac{1}{2} & \\ & & \frac{1}{2} & \frac{1}{2} \\ & & & \cdot \end{bmatrix} \begin{bmatrix} \cdot \\ x(n-1) \\ x(n) \\ x(n+1) \\ \cdot \end{bmatrix}
$$

Multiplying along the middle row of Hx gives the desired output $y(n) = \frac{1}{2}(x(n-1) + x(n))$. Multiplying the pure sinusoid with components $x(n) = e^{i\omega n}$ gives the desired output $y(n) = (\frac{1}{2} + \frac{1}{2}e^{-i\omega})e^{i\omega n}$. Thus the frequency response function $H(e^{i\omega}) = \frac{1}{2} + \frac{1}{2}e^{-i\omega}$ is the crucial multiplying factor. In linear algebra terms, the sinusoid $x = e^{i\omega n}$ is an eigenvector of every filter and the frequency response $H(e^{i\omega})$ is the eigenvalue λ.

We might ask if this averaging matrix is invertible. The answer is no, because the response $\frac{1}{2} + \frac{1}{2}e^{-i\omega}$ is *zero* at $\omega = \pi$. The alternating sinusoid $x(n) = e^{i\pi n} = (-1)^n$ averages to zero (each sample is the opposite of the previous sample). Therefore $Hx = 0$ for this oscillating input.

When the input is the impulse $x = (\ldots, 0, 1, 0, \ldots)$, the response Hx is exactly *a column of the matrix*. This is the impulse response $h = (\frac{1}{2}, \frac{1}{2}, 0, \ldots)$.

Now look at the matrix for any causal FIR filter. It is still lower triangular (causal) and it has a finite number of nonzero diagonals (FIR). Most important, the matrix is "shift-invariant." Each row is a shift of the previous row. If $(h(0), h(1), h(2), h(3))$ are the coefficients (a, b, c, d) of a four-tap filter, then those coefficients go down the four nonzero diagonals of the matrix. The coefficient $h(k)$ is the entry along the kth diagonal of H:

$$
H = \begin{bmatrix} \cdot & \cdot & \cdot & & & \\ d & c & b & a & & \\ & d & c & b & a & \\ & & d & c & b & a \\ & & & \cdot & \cdot & \cdot & \cdot \end{bmatrix} . \tag{54}
$$

This is the *Toeplitz matrix* or *convolution matrix*.

Otto Toeplitz actually studied finite constant-diagonal matrices. Those are more difficult because the matrix is cut off at the top row and bottom row. The pure sinusoids $x = e^{i\omega n}$ are no longer eigenvectors; Fourier analysis always has trouble at boundaries. The problem for Toeplitz was to understand the properties of his matrix as $N \to \infty$. The conclusion (see Szegö and Grenander [7]) was that the limit is a *singly* infinite matrix, with one boundary, rather than our doubly infinite matrix (no boundaries).

The multiplication $y = Hx$ executes a discrete convolution of x with h:

$$
y(n) = \sum_{k=0}^{n} h(k)x(n-k) . \tag{55}
$$

This impulse response $h = (a, b, c, d)$ is in each column of H.

The eigenvectors are still the special exponentials $x(n) = e^{i\omega n}$. The eigenvalue is the frequency response $H(e^{i\omega})$. Just multiply x by a typical row of H, and factor out $e^{i\omega n}$:

$$
\begin{aligned}
y(n) &= de^{i\omega(n-3)} + ce^{i\omega(n-2)} + be^{i\omega(n-1)} + ae^{i\omega n} \\
&= \left(de^{-i3\omega} + ce^{-i2\omega} + be^{-i\omega} + a \right) e^{i\omega n} \\
&= H(e^{i\omega})e^{i\omega n} .
\end{aligned}
$$

To say this in linear algebra language, the Fourier basis (pure sinusoids) will diagonalize any constant-diagonal matrix. *The convolution $Hx = h * x$ becomes a multiplication by $H(e^{i\omega})$.* This is the *convolution rule* on which filter design is based.

A single filter is not intended to be inverted (and is probably not invertible).

To move toward filter banks we introduce *downsampling*. This is the length-reducing operation $(\downarrow 2)y(n) = y(2n)$. Since it is linear, it can also be represented by a matrix. But that matrix does not have constant diagonals! It is not a Toeplitz matrix. Downsampling is not shift-invariant, because a unit delay of the signal produces a completely different set of even-numbered samples (they previously had odd numbers). In some way the matrix for $\downarrow 2$ is short and wide:

$$
(\downarrow 2)y = \begin{bmatrix} \cdot & 1 & & & & \\ \cdot & 0 & 0 & 1 & & \\ \cdot & 0 & 0 & 0 & 0 & 1 \end{bmatrix} \begin{bmatrix} \cdot \\ y(0) \\ y(1) \\ y(2) \\ y(3) \\ y(4) \end{bmatrix} = \begin{bmatrix} y(0) \\ y(2) \\ y(4) \end{bmatrix} . \tag{56}
$$

One interesting point about this matrix is its transpose (tall and thin). Multiplying by $(\downarrow 2)^T$ will put zeros back into odd-numbered samples. This is exactly the process of *upsampling* (denoted by $\uparrow 2$):

$$
(\uparrow 2)z = (\downarrow 2)^T z = \begin{bmatrix} \cdot & \cdot & \cdot \\ 1 & 0 & 0 \\ 0 & 0 & \\ 1 & 0 & \\ 0 & & \\ & & 1 \end{bmatrix} \begin{bmatrix} z(0) \\ z(1) \\ z(2) \end{bmatrix} = \begin{bmatrix} \cdot \\ z(0) \\ 0 \\ z(1) \\ 0 \\ z(2) \end{bmatrix} . \tag{57}
$$

The product $(\uparrow 2)(\downarrow 2)$, with downsampling first, is different from $(\downarrow 2)(\uparrow 2)$:

$$
(\uparrow 2)(\downarrow 2)y = \begin{bmatrix} y(0) \\ 0 \\ y(2) \\ 0 \\ y(4) \end{bmatrix} \qquad \text{but} \qquad (\downarrow 2)(\uparrow 2)z = z .
$$

The lossy equation $(\downarrow 2)$ has a right inverse $(\uparrow 2)$ but not a left inverse.

Now combine filtering with downsampling. The "2" enters into the convolution:

$$(\downarrow 2)Hx(n) = y(2n) = \sum_{k=0}^{N} h(k)x(2n - k). \tag{58}$$

This is the crucial combination for filter banks and wavelets. It is linear but not shift-invariant. To execute $(\downarrow 2)H$ by a matrix, *we remove the odd-numbered rows of* H. This leaves a short wide matrix whose rows have a *double shift*:

$$(\downarrow 2)H = \begin{bmatrix} \cdot & \cdot & & & & \\ d & c & b & a & & \\ & d & c & b & a & \\ & & d & c & b & a \\ & & & & \cdot & \cdot \end{bmatrix}. \tag{59}$$

The matrix multiplication $(\downarrow 2)H_0 x$ produces the subsampled lowpass output $y_0(2n)$, which is half of the wavelet transform of x.

The other half of the transform is from the highpass channel. It begins with the second filter H_1. The output $y_1 = H_1 x$ is subsampled to produce $y_1(2n) = (\downarrow 2)H_1 x$. The two half-length vectors z_0 and z_1 give the "averages" and "details" in the input signal x. There is overlap between them at all frequencies because the filters are not ideal (they don't have brick wall perfect cutoffs). With properly chosen filters we can still reconstruct x perfectly from $y_0(2n)$ and $y_1(2n)$.

The analysis filter bank consists of $(\downarrow 2)H_0$ and $(\downarrow 2)H_1$. To express the wavelet transform in one square matrix, we combine those two matrices (each containing a double shift):

$$H_{\text{bank}} = \begin{bmatrix} (\downarrow 2)H_0 \\ (\downarrow 2)H_1 \end{bmatrix} = \begin{bmatrix} \cdot & \cdot & & & \\ h_0(3) & h_0(2) & h_0(1) & h_0(0) & \\ & h_0(3) & h_0(2) & h_0(1) & h_0(0) \\ \cdot & \cdot & \cdot & \cdot & \cdot \\ h_1(3) & h_1(2) & h_1(1) & h_1(0) & \\ & h_1(3) & h_1(2) & h_1(1) & h_1(0) \end{bmatrix}. \tag{60}$$

The diagonal of H_{bank} is not constant and the filter bank is not shift-invariant. But it is *double-shift invariant!* If the input x is delayed by two time steps, then the output is also delayed by two time steps — and otherwise unchanged. We have a *block* FIR time-invariant system.

Notice that the columns containing even-numbered coefficients are separated from the columns containing odd-numbered coefficients. The filter bank is dealing with two phases, x_{even} and x_{odd}, which are producing two outputs, $y_0(2n)$ and $y_1(2n)$.

For a single filter, we selected the special input $x(n) = e^{i\omega n}$. The filter multiplied it by a scalar $H(e^{i\omega})$. The pure frequency ω was preserved. For a filter bank, the corresponding idea is to select an even phase $x_{\text{even}}(n) = x(2n) = \alpha e^{i\omega n}$ and an odd phase $x_{\text{odd}}(n) = x(2n - 1) = \beta e^{i\omega n}$. Then the output in each channel again has this frequency ω:

$$
H_{\text{bank}}
\begin{bmatrix}
\vdots \\
\alpha e^{i\omega(n-1)} \\
\beta e^{i\omega n} \\
\alpha e^{i\omega n} \\
\beta e^{i\omega(n+1)} \\
\vdots
\end{bmatrix}
=
\begin{bmatrix}
H_{0,\text{even}} & H_{0,\text{odd}} \\
H_{1,\text{even}} & H_{1,\text{odd}}
\end{bmatrix}
\begin{bmatrix}
X_{\text{even}} \\
X_{\text{odd}}
\end{bmatrix}. \tag{61}
$$

The crucial point is to identify that 2 by 2 *polyphase matrix*, which is the block analog of the frequency response function $H(e^{i\omega})$ for a single filter. The polyphase matrix is

$$
H_{\text{poly}}(e^{i\omega}) =
\begin{bmatrix}
H_{0,\text{even}} & H_{0,\text{odd}} \\
H_{1,\text{even}} & H_{1,\text{odd}}
\end{bmatrix}
=
\begin{bmatrix}
\sum h_0(2k)e^{ik\omega} & \sum h_0(2k+1)e^{ik\omega} \\
\sum h_1(2k)e^{ik\omega} & \sum h_1(2k+1)e^{ik\omega}
\end{bmatrix}. \tag{62}
$$

In the analysis of filter banks , the polyphase matrix plays the leading part. Let us mention the most important point:

The polyphase matrix for the synthesis filter bank is the inverse of H_{poly}:

$$
H_{\text{bank}}^{-1} \qquad \text{corresponds to} \qquad H_{\text{poly}}^{-1}(e^{i\omega})
$$

Therefore the synthesis filter bank is FIR only if H_{poly}^{-1} is a finite polynomial (not an infinite series in $e^{i\omega}$). Since the inverse of a matrix involves division by the *determinant*, the condition for an FIR analysis bank (wavelet transform) to have an FIR synthesis bank (inverse wavelet transform) is this:

The determinant of $H_{\text{poly}}(e^{i\omega})$ must be a monomial $Ce^{iL\omega}$ (one term only).

Then we can divide by the determinant, and the inverse matrix is a finite polynomial. This is the requirement for a perfect reconstruction FIR filter bank. From $H^{-1}(e^{i\omega})$ we construct the inverse transform, which is the synthesis bank.

Remark 1. *You can see the blocks more clearly by interlacing the rows of* $(\downarrow 2)H_0$ *and* $(\downarrow 2)H_1$. *This will interlace the output vectors* $y_0(2n)$ *and* $y_1(2n)$. *We are doing this not in the actual implementation, but in order to recognize the* block Toeplitz matrix *that appears:*

$$
H_{\text{block}} =
\begin{bmatrix}
\ddots & & \ddots & & & \\
h_0(3) & h_0(2) & h_0(1) & h_0(0) & & \\
h_1(3) & h_1(2) & h_1(1) & h_1(0) & & \\
& & h_0(3) & h_0(2) & h_0(1) & h_0(0) \\
& & h_1(3) & h_1(2) & h_1(1) & h_1(0) \\
& & & & \ddots & \ddots
\end{bmatrix}. \tag{63}
$$

The same 2 by 2 matrix is repeated down each block diagonal.

Remark 2. *Matrix multiplication is optional for time-invariant filters and doubly infinite signals. Many non-specialists in signal processing are comfortable with matrices as a way to understand the action of the transform (and block matrices correspond perfectly to block transforms). Specialists may prefer the description in the frequency domain, where a filter multiplies by the simple scalar response function $H(e^{i\omega})$. In popular textbooks like Oppenheim and Schafer's* Discrete-Time Signal Processing, *there is scarcely a single matrix. I meet those students in my wavelet class, and they thoroughly understand filters.*

For the finite-length signals that enter in image processing, and for time-varying filter banks of all kinds, I believe that matrix notation is valuable. This approach is especially helpful for non-specialists (who think first of the signal in time or space, instead of in the frequency domain). So we want to comment briefly on the matrices that describe *filtering for a signal of finite length L.*

The "interior" of the matrix will have constant diagonals as before. The entry in the kth diagonal is the filter coefficient $h(k)$. At the top rows and/or bottom rows (the ends of the signal) some change in this pattern is necessary. We must introduce boundary conditions to tell us what to do when the convolution $y(n) = \sum h(k)x(n - k)$ is asking for samples of x that are outside the image boundary (and therefore not defined). We may change $h(k)$ near the boundary or we may extend $x(n)$ beyond the boundary — but we have to do something! And it is often clearest to see $y = H_L x$ as an L by L matrix multiplication.

A crude approach is *zero-padding*, which assumes that those unknown samples are all zero:

$$\text{Define } x(n) = 0 \quad \text{for} \quad n < 0 \quad \text{and} \quad n \geq L.$$

In this case the infinite Toeplitz matrix is chopped off to a finite Toeplitz matrix. We have an L by L section of the original matrix. The result is a discontinuity in the signal and a visible artifact (sometimes a dark band at the boundary) in the reconstruction. Most software does better than this.

Cyclic convolution, which assumes a *periodic signal* with $x(L) = x(0)$, is more popular. It preserves the Fourier structure, the DFT of length L. There is perfect agreement between *circulant matrices* (Toeplitz with wrap-around of the diagonals) and diagonalization by the DFT. The eigenvectors of every circulant matrix are the DFT basis vectors $v_k = (1, w^k, w^{2k}, \ldots)$ with $w = \exp(2\pi i/L)$. Note that $w^L = 1$, which assures the periodic wrap-around. The corresponding eigenvalues of the circulant matrix are the numbers $H(w^k) = H(e^{2\pi i k/L})$ for $k = 0, \ldots, L - 1$. We are seeing L discrete frequencies with equal spacing $2\pi/L$, instead of the whole continuum of frequencies $|\omega| \leq \pi$.

An example will put this idea into matrix notation. The cyclic averaging filter for length $L = 4$ is represented by a 4 by 4 circulant matrix:

$$H_{\text{cyclic}} = \frac{1}{2} \begin{bmatrix} 1 & 0 & 0 & 1 \\ 1 & 1 & 0 & 0 \\ 0 & 1 & 1 & 0 \\ 0 & 0 & 1 & 1 \end{bmatrix}.$$

The entry $\frac{1}{2}$ is on the main diagonal and the first subdiagonal — which cycles around into the top right corner of the matrix. If we apply H_{cyclic} to an input x, the output y is the moving average (with the cyclic assumption that $x(-1) = x(3)$):

$$y = H_{\text{cyclic}} x = \frac{1}{2} \begin{bmatrix} x(0) & + & x(3) \\ x(1) & + & x(0) \\ x(2) & + & x(1) \\ x(3) & + & x(2) \end{bmatrix}.$$

You can see that the zero-frequency input $x = (1, 1, 1, 1)$ is an eigenvector with eigenvalue $H(e^{i0}) = 1$. The next eigenvector is $x = (1, i, -1, -i)$ containing powers of $e^{2\pi i/L} = i$. The eigenvalue is $\frac{1}{2}(1 - i)$, which agrees with the response $H(e^{i\omega}) = \frac{1}{2} + \frac{1}{2}e^{-i\omega}$ at the frequency $\omega = \pi/2$. Thus circulant matrices are a perfect match with cyclic convolution and the DFT.

A third approach is preferred for symmetric (or antisymmetric) filters. *We extend the signal symmetrically.* At each boundary, there are two possible *symmetric reflections*:

"Whole-sample"= Reflection with no repeat: $x(2), x(1), x(0), x(1), x(2)$

"Half-sample"= Reflection with a repeat: $x(2), x(1), x(0), x(0), x(1), x(2)$

We choose the reflection that agrees with the filter. Symmetric even-length filters have repeats as in a, b, b, a; the symmetric odd-length filter a, b, c, b, a has no repeat. This extension process is easy to implement because the filter can act normally on the extended signal, and then the output y is reduced back to length L (and subsampled, when the filter bank includes ↓ 2). Chapter 8 of our joint textbook *Wavelets and Filter Banks* (Wellesley-Cambridge Press 1996) gives the details of symmetric extension [14].

The top and bottom rows of the filter matrix H are modified by this "fold-across" coming from symmetric extension of the input signal. The matrix remains banded but the diagonal entries change in a systematic way near the boundaries. In an FIR filter bank with FIR inverse, the key point is that both transforms (analysis and synthesis, deconstruction and reconstruction) remain banded. This bandedness means a fast transform with a fast inverse. So wavelets have the two essential properties, speed of execution and accuracy of approximation, that make them a good transform.

16 Filters and Wavelets for 2D Images

It is easy to describe the most popular and convenient 2-dimensional filters. They come directly from 1-dimensional filters. Those are applied to each separate row of the image. Then the output is reorganized into columns, and the 1-dimensional filters are applied to each column.

This produces four outputs instead of the usual two. The low-low output (lowpass filter on rows and then columns) normally contains most of the energy and the information. This quarter-size image is the input to the next iteration. Effectively, one step of the analysis bank has downsampled the signal by four.

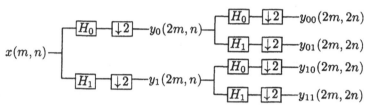

It is useful to follow through the Haar example. The low-low output will be an average over four samples — two neighbors from each row and then averaged over two neighboring rows:

$y_{00}(2m, 2n)$

$$= \frac{1}{4} \left[x(2m, 2n) + x(2m - 1, 2n) + x(2m, 2n - 1) + x(2m - 1, 2n - 1) \right]$$

The other three outputs y_{10} and y_{01} and y_{11} involve the same four input samples. There are minus signs from taking differences instead of averages. The four sign patterns give the same four outputs as a 2-dimensional Discrete Fourier Transform on 2×2 points:

$$
\begin{array}{cccccccc}
+ \ + & \quad & + \ - & \quad & + \ + & \quad & + \ - \\
+ \ + & \quad & + \ - & \quad & - \ - & \quad & - \ +
\end{array}
$$

The middle two outputs pick up vertical and horizontal edges. The last output includes (but not so well) any diagonal edges. The 2D version of longer filters with more zeros at $\omega = \pi$ will be much better, but they still have this bias to the vertical and horizontal directions. These are "separable" filters or "tensor product" filters — they separate into 1D times 1D.

When we reconstruct from these four pieces of the wavelet transform, we get four output images. With perfect reconstruction and no compression, they add up to the input image. Each piece uses a quarter of the transform coefficients. It is usual to display the four quarter-size reconstructions in one original-size image (and the low-low part in one corner is always recognizable as a blurred version of the input).

low-low (close to original)	high-low (vertical edges)
low-high (horizontal edges)	high-high (little energy)

The next iteration (which effectively has downsampling by 16) splits the low-low quarter of this display into four smaller pieces.

It is also possible to develop genuinely two-dimensional filters. We think of the samples as positioned on a lattice in the two-dimensional plane. We could use all gridpoints (m, n), as above, or we could use the "quincunx lattice" that keeps only half of those points ($m + n$ is even) on a staggered mesh. The nonseparable filters can be more isotropic (less dependent on xy orientation of the image). But these filters quickly become complicated if we want higher accuracy. The separable filters are much simpler to implement, and more popular.

In continuous time, for functions instead of vectors, we have **one scaling function and three wavelets**. The scaling function comes from a two-dimensional refinement equation. The coefficients are $h_0(k, \ell)$ for a nonseparable 2D lowpass filter, and they are $h_0(k)h_0(\ell)$ when the filter is separable. In that case the refinement equation is

$$\phi(x, y) = 4 \sum \sum h_0(k)h_0(\ell)\phi(2x - k, 2y - \ell) \tag{64}$$

The solution will be the product $\phi(x)\phi(y)$ of two 1-dimensional scaling functions.

Similarly the three wavelet equations will have coefficients $h_0(k)h_1(\ell)$ and $h_1(k)h_0(\ell)$ and $h_1(k)h_1(\ell)$. The resulting wavelets are $\phi(x)w(y)$ and $w(x)\phi(y)$ and $w(x)w(y)$. Their translates form a basis for the wavelet subspace W_0.

17 Applications: Compression and Denoising

We have concentrated so far on the wavelet transform and its inverse, which are the linear (and lossless) steps in the action diagram. This final section will discuss the nonlinear (and lossy) step, which replaces the wavelet coefficients $y(2n)$ by approximations $\hat{y}(2n)$. Those are the coefficients that we use in reconstruction. So the output $\hat{x}$ is different from the input x.

We hope that the difference $x - \hat{x}$ is too small to be heard (in audio) and too small to be seen (in video). We begin by recalling the big picture of transform plus processing plus inverse transform:

$$signal\ x \xrightarrow[\text{[lossless]}]{} \begin{array}{c} wavelet \\ coefficients \end{array} \xrightarrow[\text{[lossy]}]{} \begin{array}{c} compressed \\ coefficients \end{array} \xrightarrow[\text{[lossless]}]{} \begin{array}{c} compressed \\ signal \end{array} \hat{x}$$

A basic processing step is **thresholding**. We simply set to zero all coefficients that are smaller than a specified threshold α:

$$\text{Hard threshholding: } \widehat{y}(2n) = 0 \text{ if } |y(2n)| \leq \alpha. \tag{65}$$

This normally removes all, or almost all, the high frequency components of the signal. Those components often contain "noise" from random errors in the inputs. This process of **denoising** is highly important in statistics and Donoho [5] has proposed a suitable threshold $\alpha = \sigma\sqrt{2N}$, with standard deviation σ and N components.

Notice that thresholding is not a linear operation. We are keeping the largest components and not the first N components. This is simple to execute, but the mathematical analysis by DeVore and others requires the right function spaces (they are Besor spaces). We refer to [4] for an excellent summary of this theory.

A more subtle processing step is **quantization**. The inputs $y_0(2n)$ are real numbers. The outputs $\widehat{y}_0(2n)$ are strings of zeros and ones, ready for transmission or storage. Roughly speaking, the numbers $y_0(2n)$ are sorted into a finite set of bins. The compression is greater when more numbers go into the "zero bin." Then we transmit only the bin numbers of the coefficients, and use a value $\widehat{y}$ at the center of the bin for the reconstruction step. A vector quantization has M dimensional bins for packets of M coefficients at a time.

This transform coding is of critical importance to the whole compression process, It is a highly developed form of roundoff, and we mention two basic references [6, 8]. I believe that quantization should be studied and applied in a wide range of numerical analysis and scientific computing.

The combination of linear transform and nonlinear compression is fundamental. The transform is a change to a better basis — a more efficient representation of the signal. The compression step takes advantage of that improved representation, to reduce the work. New transforms and new bases will be needed in processing new types of signals. The needs of modern technology impose the fundamental message that has been the starting point of this paper: **Signal processing is for everyone**.

References

[1] A. Cohen and R.D. Ryan, *Wavelets and Multiscale Signal Processing*, Chapman and Hall, 1995.

[2] I. Daubechies, Orthonormal bases of compactly supported wavelets, *Comm. Pure Appl. Math.* **41** (1988) 909–996.

[3] I. Daubechies, *Ten Lectures on Wavelets*, SIAM, 1993.

[4] R. DeVore and B. Lucier, Wavelets, *Acta Numerica* **1** (1991) 1–56.

[5] D. Donoho, De-noising by soft thresholding, *IEEE Trans. Inf. Theory* **41** (1995) 613–627.

[6] R.M. Gray, *Source Coding Theory*, Kluwer, 1990.

[7] U. Grenander and G. Szegö, *Toeplitz Forms and Their Applications*, University of California Press, Berkeley, 1958.

[8] N.J. Jayant and P. Noll, *Digital Coding of Waveforms*, Prentice-Hall, 1984.

[9] J. LeBrun and Vetterli, Balanced wavelets: Theory and design, *IEEE Trans. Sig. Proc.* **46** (1998) 1119–1124; Higher order balanced multi-wavelets, *ICASSP Proceedings* (1998).

[10] S. Mallat, Multiresolution approximation and wavelet orthonormal bases of L^2 (R), *Trans. Amer. Math. Soc.* **315** (1989) 69–87.

[11] I. Selesnick, Multiwavelets with extra approximation properties, *IEEE Trans. Sig. Proc.* **46** (1998) 2898–2909; Balanced multiwavelet bases based on symmetric FIR filters, *IEEE Trans. Sig. Proc.*, to appear.

[12] J. Shen and G. Strang, The asymptotics of optimal (equiripple) filters, *IEEE Trans. Sig. Proc.* **47** (1999) 1087–1098.

[13] G. Strang, The Discrete Cosine Transform, *SIAM Review* **41** (1999) 135–147.

[14] G. Strang and T. Nguyen, *Wavelets and Filter Banks*, Wellesley-Cambridge Press, 1996.

[15] M. Unser and T. Blu, Fractional splines and wavelets, *SIAM Review* **41** (1999), to appear.

LIST OF PARTICIPANTS

Aruliah Dhavide   dhavide@cs.ubc.ca
Ballestra Luca Vincenzo   Ballestr@mat.unimi.i
Bandeira Luis   lbandeira@mail.telepac.pt
Bernat Josep   bernatj@aia.ptv.es
Bicho Luis   luisbicho@excite.com
Bothelho Machado Ulla   ullam@maths.lth.se
Bui Tien   bui@cs.concordia.ca
Burkard Rainer E.   burkard@opt.math.tu-graz.ac.at
Capasso Vincenzo   vincenzo.capasso@mat.unimi.it
Carfora Maria Francesca   carfora@iamna.iam.na.cnr.it
Carita Graca   gcarita@dmat.uevora.pt
Catinas Daniela Dorina   daniela.catinas@math.utcluj.ro
Catinas Emil Adrian   ecatinas@ictp.math.ubbcluj.ro
Costa Joao   JoaoCosta77@yahoo.com
De Cesare Luigi   irmald01@area.ba.cnr.it
Deuflhard Peter   deuflhard@zib.de
Diehl Moritz   Moritz.Diehl@iwr.uni-heidelberg.de
Di Gennaro Davide   digennaro@logikos.it
Di Liddo Andrea   irmaad02@area.ba.cnr.it
Engl Heinz W.   engl@indmath.uni-linz.ac.at
Estatico Claudio   estatico@dima.unige.it
Felici Thomas   felici@indmath.uni-linz.ac.at
Gaito Sabrina   gaito@mat.unimi.it
Gradinaru Vasile   gradinar@na.uni-tuebingen.de
Halldin Roger   rogerh@maths.lth.se
Hohage Thorsten   hohage@indmath.uni-linz.ac.at
Jameson Antony   jameson@baboon.stanford.edu
Kindelan Manuel   kinde@ing.uc3m.es
Kuegler Philipp   kuegler@indmath.uni-linz.ac.at
Leaci Antonio   leaci@ingle01.unile.it
Lindquist Daniel   danlin@math.chalmers.se
Lorenzi Alfredo   alfredo.lorenzi@mat.unimi.it
Lorenzi Luca   luca@vmimat.mat.unimi.it
Malmberg Anders   andersm@maths.lth.se
Mancini Alberto   mancini@math.unifi.it
Mattheij Robert M.   mattheij@win.tue.nl
Mazzei Luca   lmazzei@leland.stanford.edu
Mellado Jose Damian   damian@ing.uc3m.es
Micheletti Alessandra   alessandra.micheletti@mat.unimi.it
Mininni Rosamaria   mininni@pascal.dm.uniba.it
Nordstroem Fredrik   fn@maths.lth.se
Notarnicola   irmafn06@area.ba.cnr.it
Ornelas Antonio   ornelas@dmat.uevora.pt
Periaux Jacques   periaux@rascasse.inria.fr
Plexousakis Michael   plex@nada.kth.se
Revuelta Bayod   arevuelt@math.uc3m.es
Salani Claudia   salani@mat.unimi.it

Sangalli Giancarlo   Sangalli@dimat.unipv.it
Santos telma   telmasantos@yahoo.com
Strang Gilbert   gs@math.mit.edu
Tchakoutio Paul   paul@mathematik.uni-freiburg.de
Tzafriri Rami   ramitz@cs.huji.ac.il
Valdman Jan   jva@numerik.uni-kiel.de
Veeser Andreas   andy@mathematik.uni-freiburg.de
Vu Tuan Giao   giao@math.chalmers.se
Zucca Cristina   zucca@dm.unito.dm

LIST OF C.I.M.E. SEMINARS

4. Lecture Notes are printed by photo-offset from the master-copy delivered in camera-ready form by the authors. Springer-Verlag provides technical instructions for the preparation of manuscripts. Macro packages in T_EX, L^AT_EX2e, L^AT_EX2.09 are available from Springer's web-pages at

http://www.springer.de/math/authors/b-tex.html.

Careful preparation of the manuscripts will help keep production time short and ensure satisfactory appearance of the finished book.

The actual production of a Lecture Notes volume takes approximately 12 weeks.

5. Authors receive a total of 50 free copies of their volume, but no royalties. They are entitled to a discount of 33.3 % on the price of Springer books purchase for their personal use, if ordering directly from Springer-Verlag.

Commitment to publish is made by letter of intent rather than by signing a formal contract. Springer-Verlag secures the copyright for each volume. Authors are free to reuse material contained in their LNM volumes in later publications: A brief written (or e-mail) request for formal permission is sufficient.

Addresses:

Professor F. Takens, Mathematisch Instituut,
Rijksuniversiteit Groningen, Postbus 800,
9700 AV Groningen, The Netherlands
E-mail: F.Takens@math.rug.nl

Professor B. Teissier
Université Paris 7
UFR de Mathématiques
Equipe Géométrie et Dynamique
Case 7012
2 place Jussieu
75251 Paris Cedex 05
E-mail: Teissier@math.jussieu.fr

Springer-Verlag, Mathematics Editorial, Tiergartenstr. 17,
D-69121 Heidelberg, Germany,
Tel.: *49 (6221) 487-701
Fax: *49 (6221) 487-355
E-mail: lnm@Springer.de